Digital control systems

Digital control systems

P. N. PARASKEVOPOULOS

PRENTICE HALL

London   New York   Toronto   Sydney   Tokyo   Singapore
Madrid   Mexico City   Munich

First published 1996 by
Prentice Hall Europe
Campus 400, Maylands Avenue
Hemel Hempstead
Hertfordshire, HP2 7EZ
A division of
Simon & Schuster International Group

Typeset in 10/12 pt Times
by Mathematical Composition Setters Ltd
Salisbury, Wiltshire

Printed and bound in Great Britain by
Redwood Books, Trowbridge, Wiltshire

Library of Congress Cataloging-in-Publication Data

Paraskevopoulos, P. N.
    Digital control systems / P.N. Paraskevopoulos.
        p. cm.
    Includes bibliographical references and index.
    ISBN 0-13-341876-6 (pbk. : alk. paper)
    1. Digital control systems. I. Title.
TJ223.M53P36 1996
629.8–dc20   95–42734
                                                  CIP

British Library Cataloguing in Publication Data

A catalogue record for this book is available from
the British Library

ISBN 0-13-341876-6 (pbk)

---

1 2 3 4 5   00 99 98 97 96

To my wife Mary

Contents

Preface

Automatic control is one of today's most significant areas of science and technology. This is attributed to the fact that automation is linked with the development of almost every form of technology.

Automatic control has developed rapidly over the last 60 years. An impressive boost to this development was given by the Second World War, as well as by space exploration. In the last 20 years, automatic control has undergone a significant and rapid development mainly due to digital computers. Indeed, recent developments in digital computers, and particularly their low cost, facilitate their use in controlling complex systems and processes.

The area of digital control systems, or computer-controlled systems or sampled-data control systems, is the technological area which aims to develop controller design methods based on digital computers. The results reported thus far are significant from both the theoretical and the practical point of view. From the theoretical point of view, these results are presented in great depth, covering a wide variety of modern digital control problems such as optimal and stochastic control, adaptive control, state observers, Kalman filters, systems identification and others. From the practical point of view these results have been successfully implemented in numerous practical systems and processes. As an indication, we mention the following applications: control of position, velocity, voltage, temperature, pressure, fluid level, electrical energy plants, industrial plants producing paper, cement, steel, sugar, plastics, clothes and food, nuclear and chemical reactors, ground, sea and air transportation systems, robots, space applications, farming, biotechnology, medicine, etc.

This text is based on the introductory course on digital control systems which I teach to senior undergraduate students in the Department of Electrical and Computer Engineering of the National Technical University of Athens. The text starts with a description and analysis of discrete-time systems. Next, classical (Bode and Nyquist diagrams, root locus, etc.) as well as modern controller design techniques (pole placement, deadbeat control, state observers, optimal control, etc.) are presented. There follows an introduction to system identification and adaptive control and hardware and software methods for controller realization. Finally, a brief introduction to fuzzy control is presented. This text, therefore, is appropriate for both undergraduate as well as first-year graduate courses in digital control systems and is also useful for practising engineers.

The book has 11 chapters, organized as follows:

Chapter 1 is an introduction to digital control systems.

Chapter 2 refers to the Z-transform which is necessary mathematical background for the study of discrete-time systems.

Chapter 3 refers to the description and analysis of discrete-time systems and sampled data using the following mathematical models: difference equations, transfer function, impulse response and state-space equations.

Chapter 4 refers to the important notions of stability, controllability and observability of discrete-time systems. The Jury and Lyapunov stability criteria are presented. Furthermore, the controllability and observability criteria are included, as is the Kalman decomposition.

Chapter 5 covers classical discrete-time controller design techniques. In particular, those techniques using the indirect approach, the root-locus method and Bode and Nyquist diagrams are presented. Emphasis is placed on the PID controller. The chapter ends with a treatment of the steady state errors.

Chapters 6 and 7 cover modern discrete-time controller design techniques carried out in state space. Chapter 6 covers the design problems of pole assignment, deadbeat control and state observers. The closed-loop system design using state observers is also studied.

Chapter 7 covers the problem of optimal control of discrete-time systems.

Chapter 8 refers to discrete-time system identification. The basic algorithms for OFF-LINE and ON-LINE parametric identification are given.

Chapter 9 covers discrete-time system adaptive control. The following four adaptive schemes are presented: the gradient method (MIT rule), adaptive control using the Lyapunov design, model reference adaptive control, and self-tuning regulators.

Chapter 10 presents the hardware and software methods for the realization of discrete-time controllers. Special attention is given to quantization errors.

Chapter 11 presents a brief introduction to fuzzy controllers. This material aims to familiarize the reader with this new approach to control which is very useful in practice. particularly when the complexity of the system under control defies the conventional methods presented in the previous chapters.

Two appendices are included: Appendix A covers the essentials of matrix theory and Appendix B gives Z-transform tables.

At the end of each chapter, appropriate references are cited (both books and papers). These references are numbered to facilitate cross-referencing in various points of the text.

In this book, lower case bold letters indicate vectors (e.g. $\mathbf{x}$, $\mathbf{y}$, $\mathbf{u}$, $\dot{\mathbf{x}}$, etc.), while upper case bold letters indicate matrices (e.g. $\mathbf{A}$, $\mathbf{B}$, $\mathbf{C}$, $\mathbf{D}$, $\mathbf{R}$, $\mathbf{S}$, etc.).

Acknowledgements

I would like to thank very much my former undergraduate students Dr G. Georgiou, S.C. Magoulas and A. Grafakou, my current graduate students G. Panagiotakis, A. Tsirikos, L. Toubanaki and H. Frantzikinakis, and my former graduate students Dr F.N. Koumboulis and Dr K. Arvanitis for their help in preparing this book. Special thanks are also due to Dr Georgiou for his assistance in formulating the material of Chapter 9 and of Sections 10.3 and 10.4 and to my colleague Professor R.E. King for his numerous suggestions. Last, but not least, I would like to thank Professors T. Koussiouris and A. Alexandridis and Drs K. Diamandaras, A. Soldatos and A. Bounas for their helpful suggestions.

P.N. Paraskevopoulos
National Technical University of Athens
October 1995

1

Introduction to digital control systems

1.1 Introduction

An automatic control system is one whose various components, which include the sensors, process, controllers and actuators, are appropriately designed and subsequently suitably interconnected so that the overall system behaves in a desired manner. Initially analog, with the availability and reduced cost of digital electronics, the control of such systems evolved into the more flexible digital form which is now used almost exclusively. The advent of digital computers and micro-computers in particular led to their application for industrial production, consumer appliances, communications and transportation amongst others. Computers are of particular importance to modern society and production, which depend on them critically for their internal control. Today it is doubtful whether there is a single area of industrial control in which computers have not had a significant impact.

Though modern computer control systems may appear to differ widely, they have common underlying similarities and analytical techniques which unify them. The control systems of a nuclear reactor, of a power distribution system and of a large communications system may appear to have little in common with those of a space vehicle or a robot assembling a motor vehicle. Yet all these systems share such functions as data acquisition, decision making and execution of prespecified control algorithms, communication to and enforcement of these decisions on the physical process with a view to ensuring that the overall system behaves in the desired manner. The overall closed loop system thus has the familiar feedback structure which characterizes all control systems.

Instances of computer-controlled systems abound. Some examples are: numerically controlled machine tools (CNC) which produce high precision parts; flexible manufacturing systems (FMS) in which the mutual interactions and transport of components from one machine to another are controlled by computers; large-scale industrial plants producing everything from cement to pharmaceuticals, from beverages to canned fruits, whose supervisory control and data acquisition (SCADA) are dependent entirely on computers; electric power management systems

1

(EPMS) which control every phase of the energy system from the generators to the distribution network and which would be inconceivable without modern high speed computers and digital control; aircraft and space vehicles which depend critically on computers and digital control for their flight systems; modern motor vehicles which depend on computers for fuel control and ABS (anti-skid breaking system); home appliances such as air-conditioners, refrigerators and vacuum cleaners which use computers to achieve economy and performance hitherto unattainable.

The ultimate objectives of any computer-controlled system are to improve product quality and productivity, reduce energy demand, and improve production flexibility and overall system and operator safety, factors which are significant in reducing production costs in the competitive world in which we live. Automation and automatic control are essential to both developed as well as developing countries which must save scarce energy resources and raw materials as they strive to improve the efficiency and productivity of environmentally dangerous and polluting industrial processes.

The principal scope of this text is to lay the foundations of digital (sometimes termed 'discrete time') control theory, which is essential for controlling systems using computers. Emphasis is given to modern methods of analysis and synthesis of digital control and to a lesser extent to the hardware and software aspects necessary for the implementation of the control algorithms.

1.2   A brief historical review of control systems

Control systems have been in existence since ancient times. A well-known ancient automatic control system is that of Heron of Alexandria in Egypt. This system was constructed to open a temple gate automatically when a fire was lit at the altar and to close it when the fire went out. Most probably, this device was developed more to impress believers than to be functional, as it was hidden underground.

In the eighteenth century, the use of automatic control started to advance. In 1769, James Watt invented the first centrifugal speed regulator (or flyball governor) which was widely used in practice, particularly for the automatic control of locomotives. Maxwell, in 1868 [84], [85], and Vyshnegradskii in 1877 [56], developed the first mathematical background for control which they applied to Watt's centrifugal regulator. Moreover, Routh's results on stability in 1877 were especially significant [51].

Automatic control theory and its applications have developed rapidly in the last 60 years or so [1–50], [52–55], [57–83], [86–94]. The period 1930 to 1940 was important in the history of control since remarkable theoretical and practical results, such as those of Nyquist [86], [87] and Black [62], [63] were reported.

During the following years and until about 1957, further significant research and development was reported, due mainly to Nichols [94], Bode [14], Wiener [57] and Evans [19], [66]. All the results of the last century, and up to about 1957, constitute what has been termed **classical control**. Progress from 1957 to date has been especially impressive from both the theoretical and the practical point of view. This last period has been characterized as that of **modern control**, the most significant results of which have been due to Astrom [2–4], Athans [5], [59–61], Bellman [7], [8], Brockett [15], [64], Doyle [65], [68], Francis [65], 68], Jury [26], [27], Kailath [28], [69], Kalman [29], [30], [70–81], Luenberger [37], [82], [83], MacFarlane [38], Rosenbrock [49], [50], Saridis [52], Wonham [58], [91], [92], Zames [93] and many others. For more on the historical development of control the reader can refer to [39], [42], [67] and [88–90].

A significant boost to the development of classical control methods was given by the Second World War, whereas for modern control techniques the launch of Sputnik in 1957 by the former Soviet Union and the American Apollo project, which put men on the moon in 1969, played important roles. In recent years, an impressive development in control systems has taken place with the ready availability of digital computers. Their power and flexibility have made it possible to control complex systems efficiently, using techniques which were hitherto unknown.

This book is devoted to this new and promising area of digital control, or discrete-time control, or computer-controlled systems. For related references in the area the reader can also refer to [1], [3], [6], [9–13], [16], [17], [20–27], [31–36], [40], [41], [44–48], [54] and [55]. Of these references, we strongly suggest the following for further reading: [1], [3], [21], [23], [24], [31], [34], [41], [44] and [45].

1.3 The basic structure of digital control systems

Figure 1.1 shows the basic structure of a simple digital control system. The system (plant) under control is a continuous-time system (e.g. a motor, electrical power plant, robot, etc.). The 'heart' of the controller is the digital computer. The A/D converter converts a continuous-time signal into a discrete-time signal at times specified by a clock. The D/A converter, in contrast, converts the discrete-time signal output of the computer to a continuous-time signal to be fed to the plant. The D/A converter normally includes a hold circuit (for more on A/D and D/A converters see Sections 3.3.1 and 3.3.2). The quantizer Q converts a discrete-time signal to binary digits.

The controller may be designed to satisfy simple, as well as complex, specifications. For this reason, it may operate as a simple logic device as in programmable logic controllers (PLCs), or make dynamic and complicated processing operations on the error $e(kT)$ in order to produce a suitable input $u(t)$ to control the plant. This control input $u(t)$ to the plant must be such that the behaviour of the closed-loop system (i.e. the output $y(t)$) satisfies desired specifications.

The problem of realizing a digital controller is mainly one of developing a computer program. Digital controllers present significant advantages over classical analog controllers. Some of these advantages are as follows:

1. Greater flexibility in modifying the controller's features. Indeed, the controller's features may be readily programmed. For classical analog controllers, any change in the characteristics of the controller is usually laborious and expensive, since it requires changes in the structure and/or the elements of the controller.

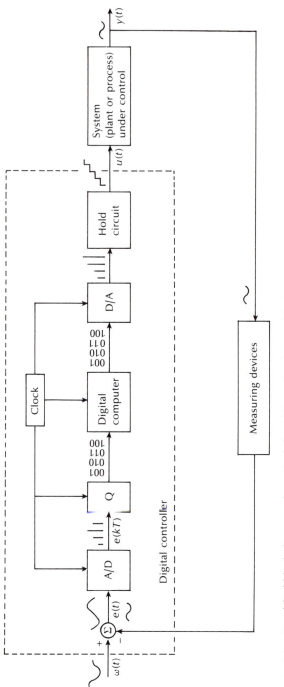

Figure 1.1 Simplified block diagram of a typical closed-loop digital control system.

2. Processing of data is simple. Complex computations may be performed in a fast and convenient way. Analogue controllers do not have this characteristic.
3. They are superior to analog controllers with regard to the following characteristics:
   (a) sensitivity
   (b) drift effects
   (c) internal noise
   (d) reliability
4. They are cheaper than analog controllers.
5. They are considerably smaller in size than analog controllers.

Nevertheless, digital controllers have certain disadvantages compared with analog controllers. The most significant disadvantage is due to the error introduced during sampling of the analog signals, as well as during quantization of the discrete-time signals.

As we have already noted, the system under control may be any type of system combining discrete and continuous elements. Indeed, digital control techniques can be applied to a great variety of systems and processes, such as the following:

1. Servomechanisms (position, velocity, etc.).
2. Voltage control, temperature control, pressure control, fluid level control, fluid flow velocity control, etc.
3. Electrical apparatus and plants (motors, generators, electrical power plants, etc.).
4. Industrial control (paper, cement, sugar, steel, cars, clothing, food, etc.).
5. Reactors (nuclear and chemical).
6. Ground transportation systems (cars, trains, etc.).
7. Sea transportation systems (ships, submarines, etc.).
8. Air transportation systems (helicopters, aircraft, etc.).
9. Weapon systems (missiles, fire control systems, etc.).
10. Robots (for assembly, welding, etc.).
11. Space applications (space telescopes, spacecraft, space stations, etc.).
12. Digital computers (printers, disk drives, magnetic tape, etc.).
13. Farming (greenhouses, irrigation, etc.).
14. Biotechnology, medicine, biology, etc.
15. In many other applications, such as traffic control, office automation, computer-integrated manufacturing, energy management of buildings, etc.

In this book, digital control techniques will be developed and implemented for several of the above practical systems.

1.4 References

Books

[1] J. Ackermann, *Sampled Data Control Systems*, Springer Verlag, New York, 1985.
[2] K.J. Åström, *Introduction to Stochastic Control Theory*, Academic Press, New York, 1970.

[3] K.J. Åström and B. Wittenmark, *Computer Controlled Systems*, Prentice Hall, Englewood Cliffs, New Jersey, 1984.

[4] K.J. Åström and B. Wittenmark, *Adaptive Control*, Addison-Wesley, New York, 1989.

[5] M. Athans and P.L. Falb, *Optimal Control*, McGraw-Hill, New York, 1966.

[6] D.M. Auslander and P. Sagues, *Microprocessors for Measurement and Control*, Osborne/McGraw-Hill, Berkeley, California, 1989.

[7] R. Bellman, *Dynamic Programming*, Princeton University Press, Princeton, New Jersey, 1957.

[8] R. Bellman, *Adaptive Control Processes, A Guided Tour*, Princeton University Press, Princeton, New Jersey, 1961.

[9] S. Bennet, *Real-Time Computer Control, An Introduction*, Prentice Hall, New York, 1988.

[10] S. Bennet and D. Linkens, *Computer Control of Industrial Processes*, Peter Peregrinus, Stevenage, 1982.

[11] J. Billingsley, *Controlling with Computers*, McGraw-Hill, New York, 1989.

[12] A.B. Bishop, *Introduction to Linear Discrete Control*, Academic Press, New York, 1975.

[13] W.S. Blaschke and J. McGill, *The Control of Industrial Processes by Digital Techniques*, Elsevier, New York, 1976.

[14] H.W. Bode, *Network Analysis and Feedback Amplifier Design*, Van Nostrand, New York, 1945.

[15] R.W. Brockett, *Finite Dimensional Linear Systems*, John Wiley, New York, 1970.

[16] J.A. Cadzow, *Discrete-Time Systems, An Introduction with Interdisciplinary Applications*, Prentice Hall, Englewood Cliffs, New Jersey, 1973.

[17] J.A. Cadzow and H.R. Martens, *Discrete-Time Systems and Computer Control Systems*, Prentice Hall, Englewood Cliffs, New Jersey, 1970.

[18] R.C. Dorf, *Modern Control Analysis*, Addison-Wesley, London, 1993.

[19] W.R. Evans, *Control Systems Dynamics*, McGraw-Hill, New York, 1954.

[20] D.G. Fisher and D.E. Seborg, *Multivariable Computer Control: A Case Study*, North-Holland, Amsterdam, 1976.

[21] G.F. Franklin, J.D. Powell and M.L. Workman, *Digital Control of Dynamic Systems*, Addison-Wesley, London, 1990 (Second Edition).

[22] G.H. Hostetter, *Digital Control System Design*, Holt Rinehart & Winston, New York, 1988.

[23] C.H. Houpis and G.B. Lamont, *Digital Control Systems*, McGraw-Hill, New York, 1985.

[24] R. Iserman, *Digital Control Systems*, Volumes I and II, Springer Verlag, Berlin, 1989.

[25] R. Jacquot, *Modern Digital Control Systems*, Marcel Dekker, New York, 1981.

[26] E.I. Jury, *Theory and Application of the Z-Transform Method*, John Wiley, New York, 1964.

[27] E.I. Jury, *Sampled Data Control Systems*, John Wiley, New York, 1958, Robert E. Krieger, Huntington, New York, 1973 (Second Edition).

[28] T. Kailath, *Linear Systems*, Prentice Hall, Englewood Cliffs, New Jersey, 1980.

[29] R.E. Kalman, *The Theory of Optimal Control and the Calculus of Variations*, ed. R. Bellman, *Mathematical Optimization Techniques*, University of California Press, Berkeley, California, 1963.

[30] R.E. Kalman, P. L. Falb and M. A. Arbib, *Topics in Mathematical System Theory*, McGraw-Hill, New York, 1969.

[31] P. Katz, *Digital Control Using Microprocessors*, Prentice Hall, London, 1981.

[32] V. Kucera, *Discrete Linear Control, The Polynomial Equation Approach*, John Wiley, New York, 1979.

[33] B.C. Kuo, *Analysis and Synthesis of Sampled-Data Control Systems*, Prentice Hall, Englewood Cliffs, New Jersey, 1963.

[34] B.C. Kuo, *Digital Control Systems*, Holt-Saunders, Tokyo, 1980.

[35] D.P. Lindorf, *Theory of Sampled-Data Control Systems*, John Wiley, New York, 1965.

[36] J.R. Leigh, *Applied Digital Control. Theory, Design and Implementation*, Prentice Hall, Hemel Hempstead, 1985.

[37] D.G. Luenberger, *Optimization by Vector Space Methods*, John Wiley, New York, 1969.

[38] A.G.C. MacFarlane, *Dynamic System Models*, George H. Harrap, London, 1970.

[39] O. Mayr, *The Origins of Feedback Control*, MIT Press, Cambridge, Massachusetts, 1970.

[40] R.H. Middleton and G.C. Goodwin, *Digital Control and Estimation: A Unified Approach*, Prentice Hall, New York, 1990.

[41] K. Ogata, *Discrete-Time Control Systems*, Prentice Hall, Englewood Cliffs, New Jersey, 1987.

[42] M. Otto, *Origins of Feedback Control*, MIT Press, Cambridge, Massachusetts, 1971.

[43] P.N. Paraskevopoulos, *Introduction to Automatic Control Systems*, in preparation.

[44] G.H. Perdikaris, *Computer Controlled Systems, Theory and Applications*, Kluwer, London, 1991.

[45] C.L. Phillips and H.T. Nagle Jr, *Digital Control System Analysis and Design*, Prentice Hall, Englewood Cliffs, New Jersey, 1984.

[46] J.R. Ragazzini and G.F. Franklin, *Sampled Data Control Systems*, McGraw-Hill, New York, 1958.

[47] G.V. Rao, *Complex Digital Control Systems*, Van Nostrand Reinhold, New York, 1979.

[48] J.G. Reid, *Linear Systems Fundamentals, Continuous and Discrete, Classic and Modern*, McGraw-Hill, New York, 1983.

[49] H.H. Rosenbrock, *State Space and Multivariable Theory*, Nelson, London, 1970.

[50] H.H. Rosenbrock, *Computer Aided Control System Design*, Academic Press, New York, 1974.

[51] E.J. Routh, *A Treatise on the Stability of a Given State of Motion*, Macmillan, London, 1877.

[52] G.N. Saridis, *Self-organizing Control of Stochastic Systems*, Marcel Dekker, New York, 1977.

[53] V. Strejc, *State Space Theory of Discrete Linear Control*, Academia, Prague and John Wiley, New York, 1981.

[54] J.T. Tou, *Digital and Sampled-Data Control Systems*, McGraw-Hill, New York, 1959.

[55] J.T. Tou, *Optimum Design of Digital Control Systems*, Academic Press, 1963.

[56] I.A. Vyshnegradskii, *On Controllers of Direct Action*, Izv. SPB Teckhnolog. Inst., 1877.

[57] N. Wiener, *The Extrapolation, Interpolation and Smoothing of Stationary Time Series*, John Wiley, New York, 1949.

[58] W.M. Wonham, *Linear Multivariable Control: A Geometric Approach*, Springer Verlag, New York, 1979 (Second Edition).

Papers

[59] M. Athans, 'Status of Optimal Control Theory and Applications for Deterministic Systems', *IEEE Transactions on Automatic Control*, Vol. AC-11, pp. 580–596, 1966.

[60] M. Athans, 'The Matrix Minimum Principle', *Information and Control*, Vol. 11, pp. 592–606, 1968.

[61] M. Athans, 'The Role and Use of the Stochastic Linear Quadratic-Gaussian Problem in Control System Design', *IEEE Transactions on Automatic Control*, Vol. AC-16, pp. 529–551, 1971.

[62] H.S. Black, 'Stabilized Feedback Amplifiers', *Bell System Technical Journal*, 1934.

[63] H.S. Black, 'Inverting the Negative Feedback Amplifier', *IEEE Spectrum*, Vol. 14, pp. 54–60, 1977.

[64] R.W. Brockett, 'Poles, Zeros and Feedback: State Space Interpretation', *IEEE Transactions on Automatic Control*, Vol. AC-10, pp. 129–135, 1965.

[65] J.C. Doyle, K. Glover, P.P. Khargonekar and B.A. Francis, 'State Space Solutions to Standard H_2 and H_∞ Control Problems', *IEEE Transactions on Automatic Control*, Vol. 34, pp. 831–847, 1989.

[66] W.R. Evans, 'Control System Synthesis by Root Locus Method', *AIEE Transactions*, Part II, Vol. 69, pp. 66–69, 1950.

[67] A.T. Fuller, 'The Early Development of Control Theory', *Transactions of the ASME (Journal of Dynamic Systems, Measurement & Control)*, Vol. 96G, pp. 109–118, 1976.

[68] B.A. Francis and J.C. Doyle, 'Linear Control Theory with an H_∞ Optimally Criterion', *SIAM Journal of Control and Optimizaion*, Vol. 25, pp. 815–844, 1987.

[69] T. Kailath, 'A View of Three Decades of Linear Filtering Theory', *IEEE Transactions on Information Theory*, Vol. 20, pp. 146–181, 1974.

[70] R.E. Kalman, 'A New Approach to Linear Filtering and Prediction Problems', *Transactions of the ASME (Journal of Basic Engineering)*, Vol. 82D, pp. 35–45, 1960.

[71] R.E. Kalman, 'On the General Theory of Control Systems', *Proceedings First International Congress, IFAC, Moscow, USSR*, pp. 481–492, 1960.

[72] R.E. Kalman, 'Contributions to the Theory of Optimal Control', *Proceedings 1959 Mexico City Conference on Differential Equations*, Mexico City, pp. 102–119, 1960.

[73] R.E. Kalman, 'Canonical Structure of Linear Dynamical Systems', *Proceedings National Academy of Science*, Vol. 48, pp. 596–600, 1962.

[74] R.E. Kalman, 'Mathematical Description of Linear Dynamic Systems', *SIAM Journal on Control*, Vol. 1, pp. 152–192, 1963.

[75] R.E. Kalman, 'When is a Linear Control System Optimal?', *Transactions of the ASME (Journal of Basic Engineering)*, Vol. 86D, pp. 51–60, 1964.

[76] R.E. Kalman and J.E. Bertram, 'A Unified Approach to the Theory of Sampling Systems', *Journal of the Franklin Institute*, pp. 405–425, 1959.

[77] R.E. Kalman and J.E. Bertram, 'Control Systems Analysis and Design via Second Method of Liapunov: II Discrete-Time Systems', *Transactions of the ASME (Journal of Basic Engineering)*, Series D, pp. 371–400, 1960.

[78] R.E. Kalman and R.S. Bucy, 'New Results in Linear Filtering and Prediction Theory', *Transactions of the ASME (Journal of Basic Engineering)*, Vol. 83D, pp. 95–108, 1961.

[79] R.E. Kalman and R.W. Koepcke, 'Optimal Synthesis of Linear Sampling Control

Systems Using Generalized Performance Indices', *Transactions of the ASME*, Vol. 80, pp. 1820–1826, 1958.

[80] R.E. Kalman, Y.C. Ho and K.S. Narendra, 'Controllability of Linear Dynamic Systems', *Contributions to Differential Equations*, Vol. 1, pp. 189–213, 1963.

[81] R.E. Kalman *et al.*, 'Fundamental Study of Adaptive Control Systems', *Wright-Patterson Air Force Base Technical Report,* ASD-TR-61-27, April 1962.

[82] D.G. Luenberger, 'Observing the State of a Linear System', *IEEE Transactions on Military Electronics*, Vol. MIL-8, pp. 74–80, 1964.

[83] D.G. Luenberger, 'An Introduction to Observers', *IEEE Transactions on Automatic Control*, Vol. AC-16, pp. 596–602, 1971.

[84] J.C. Maxwell, 'On Governors', *Proceedings of the Royal Society of London*, Vol. 16, 1868. See also *Selected Papers on Mathematical Trends in Control Theory*, Dover, New York, 1964, pp. 270–283.

[85] J.C. Maxwell, 'On Governors', *Philosophical Magazine*, Vol. 35, pp. 385–398, 1868.

[86] H. Nyquist, 'Certain Topics in Telegraph Transmission Theory', *Transactions of the AIEE*, Vol. 47, pp. 617–644, 1928.

[87] H. Nyquist, 'Regeneration Theory', *Bell System Technical Journal,* Vol. 11, pp. 126–147, 1932.

[88] 'Special Issue on Linear-Quadratic-Gaussian Problem', *IEEE Transactions on Automatic Control*, Vol. AC-16, December 1971.

[89] 'Special Issue on Identification and System Parameter Estimation', *Automatica*, Vol. 17, 1981.

[90] 'Special Issue on Linear Multivariable Control Systems', *IEEE Transactions on Automatic Control*, Vol. AC-26, 1981.

[91] W.M. Wonham, 'On Pole Assignment in Multi-Input Controllable Systems', *IEEE Transactions on Automatic Control*, Vol. AC-12, pp. 660–665, 1967.

[92] W.M. Wonham, 'On the Separation Theorem of Stochastic Control', *SIAM Journal on Control*, Vol. 6, pp. 312–326, 1968.

[93] G. Zames, 'Feedback and Optimal Sensitivity: Model Reference Transformations, Multiplicative Seminorms and Approximate Inverses', *IEEE Transactions on Automatic Control*, Vol. 26, pp. 301–320, 1981.

[94] J.G. Ziegler and N.B. Nichols, 'Optimum Settings for Automatic Controllers', *Transactions of the ASME*, Vol. 64, pp. 759–768, 1942.

2

The Z-transform

2.1 Introduction

It is well known that the Laplace transform is a mathematical tool which facilitates the study and design of linear time-invariant continuous-time systems. The reason for this is that it transforms the **differential equation** which describes the system under control to an **algebraic equation** which is much easier to manipulate. The corresponding technique for discrete-time systems is the Z-transform. As in the case of the Laplace transform, the Z-transform facilitates the study and design of linear time-invariant discrete-time systems since it transforms the **difference equation** which describes the system under control to an **algebraic equation**. Since the study of an algebraic equation is much easier than that of a difference equation, the Z-transform has been used extensively as a basic analysis and design tool for discrete-time systems.

The present chapter is devoted to the Z-transform, covering the basic theory, together with several examples of applications. More specifically, we begin the chapter with definitions of certain basic discrete-time signals. Subsequently, we present the definition and some basic properties and theorems of the Z-transform. Then, the inverse Z-transform is defined and some illustrative examples are given. The chapter closes with certain interesting examples of applications from various branches of science and technology. For more on the Z-transform, see [2] and [3].

2.2 The basic discrete-time control signals

In this section we present definitions for the following basic discrete-time signals: the unit pulse sequence, the unit step sequence, the unit gate sequence, the ramp sequence, the exponential sequence, the alternating sequence and the sine sequence. These signals are of major importance for control applications.

10

1. The unit pulse sequence

The unit pulse sequence is designated by $\delta(k - k_0)$ and is defined as follows:

$$\delta(k - k_0) = \begin{cases} 1 & \text{for } k = k_0 \\ 0 & \text{for } k \ne k_0 \end{cases} \tag{2.1}$$

The graphical representation of $\delta(k - k_0)$ is shown in Figure 2.1.

2. The unit step sequence

The unit step sequence is designated by $\beta(k - k_0)$ and is defined as follows:

$$\beta(k - k_0) = \begin{cases} 1 & \text{for } k \geqslant k_0 \\ 0 & \text{for } k < k_0 \end{cases} \tag{2.2}$$

The graphical representation of $\beta(k - k_0)$ is shown in Figure 2.2.

3. The unit gate (window) sequence

The unit gate sequence is designated by $g_\pi(k) = \beta(k - k_1) - \beta(k - k_2)$ and is defined as follows:

$$g_\pi(k) = \begin{cases} 1 & \text{for } k_1 \leqslant k \leqslant k_2 \\ 0 & \text{for } k < k_1 \text{ and for } k > k_2 \end{cases} \tag{2.3}$$

Figure 2.3 shows the graphical representation of $g_\pi(k)$. The unit gate sequence is usually used to zero all values of another sequence outside a certain time interval. Consider, for example, the sequence $f(k)$. Then, the sequence $y(k) = f(k)g_\pi(k)$ becomes

$$y(k) = f(k)g_\pi(k) = \begin{cases} f(k) & \text{for } k_1 \leqslant k \leqslant k_2 \\ 0 & \text{for } k < k_1 \text{ and for } k > k_2 \end{cases}$$

Figure 2.1 The unit pulse sequence $\delta(k - k_0)$.

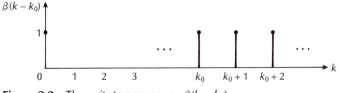

Figure 2.2 The unit step sequence $\beta(k - k_0)$.

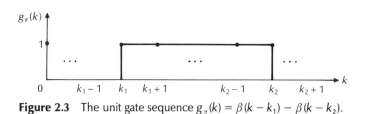

Figure 2.3   The unit gate sequence $g_\pi(k) = \beta(k - k_1) - \beta(k - k_2)$.

4. The ramp sequence

The ramp sequence is designated by $r(k - k_0)$ and is defined as follows:

$$r(k - k_0) = \begin{cases} k - k_0 & \text{for } k \geq k_0 \\ 0 & \text{for } k < k_0 \end{cases} \tag{2.4}$$

Figure 2.4 shows the graphical representation of $r(k - k_0)$.

5. The exponential sequence

The exponential sequence is designated by $g(k)$ and is defined as follows:

$$g(k) = \begin{cases} a^k & \text{for } k \geq 0 \\ 0 & \text{for } k < 0 \end{cases} \tag{2.5}$$

Figure 2.5 shows the graphical representation of $g(k) = a^k$. Clearly, when $a > 1$ the values of $g(k)$ increase as k increases, whereas for $a < 1$ the values of $g(k)$ decrease as k increases. For $a = 1$, $g(k)$ remains constant and equal to 1. In this last case, $g(k)$ becomes the unit step sequence $\beta(k)$.

6. The alternating sequence

The alternating sequence is designated by $\varepsilon(k)$ and is defined as follows:

$$\varepsilon(k) = \begin{cases} (-1)^k & \text{for } k \geq 0 \\ 0 & \text{for } k < 0 \end{cases} \tag{2.6}$$

Figure 2.6 shows the graphical representation of $\varepsilon(k)$.

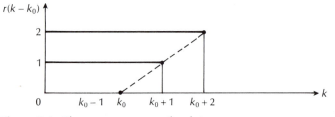

Figure 2.4   The ramp sequence $r(k - k_0)$.

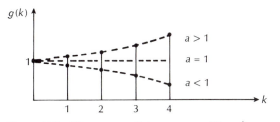

Figure 2.5   The exponential sequence $g(k) = a^k$.

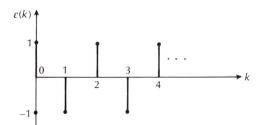

Figure 2.6   The alternating sequence $\varepsilon(k)$.

7. The sine sequence

The sine sequence is defined as follows:

$$f(k) = \begin{cases} A \sin \omega_0 k & \text{for } k \geq 0 \\ 0 & \text{for } k < 0 \end{cases} \qquad (2.7)$$

Figure 2.7 shows the graphical representation of $f(k) = A \sin \omega_0 k$, with $\omega_0 = 2\pi/12$.

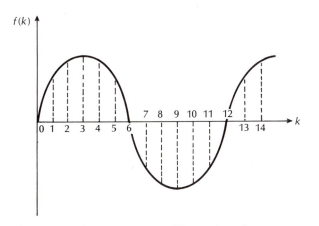

Figure 2.7   The sine sequence $f(k) = A \sin \omega_0 k$.

2.3 The Z-transform

2.3.1 *Introduction to the Z-transform*

The Z-transform of a discrete-time function $f(k)$ is designated by $F(z)$ and is defined as follows:

$$F(z) = \mathscr{Z}[f(k)] = \sum_{k=-\infty}^{\infty} f(k)z^{-k} \tag{2.8}$$

If the discrete-time function $f(k)$ is causal, i.e. $f(k) = 0$ for $k < 0$, then the definition (2.8) becomes

$$F(z) = \mathscr{Z}[f(k)] = \sum_{k=0}^{\infty} f(k)z^{-k} \tag{2.9}$$

In practice, the discrete-time sequence is often the result of periodic sampling of a continuous-time function $f(t)$. The conversion of $f(t)$ to $f(kT)$, where T represents the period between two consecutive points of $f(kT)$, is achieved through a sampler, as shown in Figure 2.8. The sampler is actually a switch which closes instantaneously and with frequency $f_s = 1/T$. The resulting output $f(kT)$ represents a discrete-time function with amplitude equal to the amplitude of $f(t)$ at the sampling instants kT, $k = 0, 1, 2, \ldots$.

The following theorem refers to the criterion for choosing the sampling frequency $f_s = 1/T$ (this issue was originally investigated by Nyquist but Shannon gave the complete proof of the theorem).

■ **THEOREM 2.3.1**   Let f_1 be the highest frequency in the frequency spectrum of $f(t)$. Then, for $f(t)$ to be recovered from $f(kT)$, it is necessary that $f_s \geqslant 2f_1$.

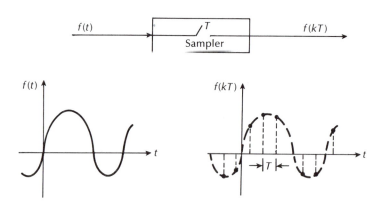

Figure 2.8   The operation of a sampler.

It is noted that the function $f(t)$ may be reconstructed from $f(kT)$ using a hold circuit (see Section 3.3.2) in series with a low-frequency filter which smooths out the form of the signal.

Let $\delta_T(t)$ denote the infinite sequence of unit pulse functions (or Dirac functions) shown in Figure 2.9. In addition, let $f^*(t)$ be the function

$$f^*(t) = f(t)\delta_T(t) = \sum_{k=-\infty}^{\infty} f(kT)\delta(t-kT) \quad \text{where } \delta_T(t) = \sum_{k=-\infty}^{\infty} \delta(t-kT) \tag{2.10}$$

When $f(t)$ is causal, equation (2.10) becomes

$$f^*(t) = \sum_{k=0}^{\infty} f(kT)\delta(t-kT) \tag{2.11}$$

The Laplace transform of (2.11) is

$$F^*(s) = \mathscr{L}[f^*(t)] = \sum_{k=0}^{\infty} f(kT) \int_0^{\infty} \delta(t-kT)\,e^{-st}\,dt = \sum_{k=0}^{\infty} f(kT)\,e^{-kTs} \tag{2.12}$$

The Z-transform of $f(kT)$ is

$$F(z) = \mathscr{Z}[f(kT)] = \sum_{k=0}^{\infty} f(kT)z^{-k} \tag{2.13}$$

If we use the mapping

$$z = e^{Ts} \quad \text{or} \quad s = T^{-1}\ln z \tag{2.14}$$

then

$$F(z) = F^*(s)\big|_{s=T^{-1}\ln z} \tag{2.15}$$

Equation (2.15) shows the relation between the sequence $f(kT)$ and the function $f^*(t)$ described in the z- and s-domains, respectively. A continuous-time system with input $f(t)$ and output $f^*(t)$ is called an ideal sampler (see, also, Section 3.3.2).

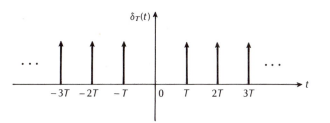

Figure 2.9 The Dirac functions $\delta_T(t)$.

The inverse Z-transform of a function $F(z)$ is denoted as $f(kT)$ and is defined as

$$f(kT) = \mathscr{L}^{-1}[F(z)] = \frac{1}{2\pi j} \oint F(z) z^{k-1} \, dz \tag{2.16}$$

Equations (2.13) and (2.16) constitute the Z-transform pair.

2.3.2 *Properties and theorems of the Z-transform*

1. Linearity

The Z-transform is linear, i.e. the following relation holds:

$$\mathscr{L}[c_1 f_1(kT) + c_2 f_2(kT)] = c_1 \mathscr{L}[f_1(kT)] + c_2 \mathscr{L}[f_2(kT)]$$
$$= c_1 F_1(z) + c_2 F_2(z) \tag{2.17}$$

where c_1 and c_2 are constants and

$$F_i(z) = \mathscr{L}[f_i(kT)] \text{ for } i = 1, 2 \tag{2.18}$$

■ **PROOF**   Apply the Z-transform definition (2.13) to relation (2.17) to yield

$$\mathscr{L}[c_1 f_1(kT) + c_2 f_2(kT)] = \sum_{k=0}^{\infty} [c_1 f_1(kT) z^{-k} + c_2 f_2(kT) z^{-k}]$$

$$= c_1 \sum_{k=0}^{\infty} f_1(kT) z^{-k} + c_2 \sum_{k=0}^{\infty} f_2(kT) z^{-k} = c_1 F_1(z) + c_2 F_2(z)$$

---

2. Shift in the time domain

The discrete-time functions $f(kT - \sigma T)$ and $f(kT + \sigma T)$ are actually the function $f(kT)$, shifted σT time units to the right and to the left, respectively. From definition (2.13), we have

(a) $\quad \mathscr{L}[f(kT - \sigma T)] = \sum_{k=0}^{\infty} f(kT - \sigma T) z^{-k} = \sum_{m=-\sigma}^{\infty} f(mT) z^{-m} z^{-\sigma}$

$$= z^{-\sigma} \left(\sum_{m=0}^{\infty} f(mT) z^{-m} + \sum_{m=-\sigma}^{-1} f(mT) z^{-m} \right)$$

$$= z^{-\sigma} F(z) + \sum_{m=-\sigma}^{-1} f(mT) z^{-(\sigma + m)} \tag{2.19a}$$

(b) $\quad \mathcal{Z}[f(kT - \sigma T)\beta(kT - \sigma T)] = \sum_{k=0}^{\infty} f[(k-\sigma)T]\beta[(k-\sigma)T]z^{-k}$

$$= z^{-\sigma}\left(\sum_{k=0}^{\infty} f[(k-\sigma)T]\beta[(k-\sigma)T]z^{-(k-\sigma)}\right)$$

$$= z^{-\sigma} \sum_{m=-\sigma}^{\infty} f(mT)\beta(mT)z^{-m} = z^{-\sigma} \sum_{m=0}^{\infty} f(mT)z^{-m}$$

$$= z^{-\sigma} F(z) \qquad\qquad (2.19b)$$

(c) $\quad \mathcal{Z}[f(kT + \sigma T)] = \sum_{k=0}^{\infty} f(kT + \sigma T)z^{-k} = z^{\sigma} \sum_{k=0}^{\infty} f[(k+\sigma)T]z^{-(k+\sigma)} = z^{\sigma} \sum_{m=\sigma}^{\infty} f(mT)z^{-m}$

$$= z^{\sigma}\left(\sum_{m=0}^{\infty} f(mT)z^{-m} - \sum_{m=0}^{\sigma-1} f(mT)z^{-m}\right) = z^{\sigma}\left(F(z) - \sum_{m=0}^{\sigma-1} f(mT)z^{-m}\right)$$

$$= z^{\sigma}\left(F(z) - \sum_{k=0}^{\sigma-1} f(kT)z^{-k}\right) \qquad\qquad (2.19c)$$

where, in the last step, we have set $m = k$.

From the foregoing equations the following special cases are obtained:

$$\mathcal{Z}[f(t+T)] = zF(z) - zf(0) \qquad\qquad (2.20a)$$

$$\mathcal{Z}[f(t+2T)] = z^2 F(z) - z^2 f(0) - zf(T) \qquad\qquad (2.20b)$$

$$\mathcal{Z}[f(t-T)] = z^{-1} F(z) + f(-T) \qquad\qquad (2.20c)$$

$$\mathcal{Z}[f(t-2T)] = z^{-2} F(z) + z^{-1} f(-T) + f(-2T) \qquad\qquad (2.20d)$$

3. Change in the z-scale

Consider the function $a^{\mp t} f(t)$. Then, according to definition (2.13),

$$\mathcal{Z}[a^{\mp t} f(t)] = \sum_{k=0}^{\infty} a^{\mp kT} f(kT)z^{-k} = \sum_{k=0}^{\infty} f(kT)(a^{\pm T}z)^{-k} = F(a^{\pm T}z) \qquad\qquad (2.21)$$

4. The Z-transform of a sum

Consider the finite sum

$$\sum_{i=0}^{k} f(iT)$$

This represents the summation of the first $k+1$ terms of the sequence $f(kT)$. Defining

$$g(kT) = \sum_{i=0}^{k} f(iT), \ g[(k-1)T] = \sum_{i=0}^{k-1} f(iT), \ldots$$

the discrete-time function $g(kT)$ may be described by the difference equation

$$g(kT+T) = g(kT) + f(kT+T)$$

Applying the Z-transform to this function yields

$$z[G(z) - g(0)] = G(z) + z[F(z) - f(0)]$$

where use was made of equation (2.20a). Since $g(0) = f(0)$, this relation becomes

$$G(z) = \mathscr{L}\left[\sum_{i=0}^{k} f(iT)\right] = \left(\frac{z}{z-1}\right)F(z) \tag{2.22}$$

5. Multiplication by k

Consider the discrete-time function $f(kT)$. Then the Z-transform of the function $kf(kT)$ is

$$\mathscr{L}[kf(kT)] = \sum_{k=0}^{\infty} kf(kT)z^{-k} = z\sum_{k=0}^{\infty} f(kT)(kz^{-k-1}) = -z\sum_{k=0}^{\infty} f(kT)\frac{dz^{-k}}{dz}$$

$$= -z\frac{d}{dz}\left(\sum_{k=0}^{\infty} f(kT)z^{-k}\right) = -z\frac{d}{dz}F(z) \tag{2.23}$$

6. Convolution of two discrete-time functions

Consider the causal discrete-time functions $f(kT)$ and $h(kT)$. The convolution of these two functions is designated by $y(kT) = f(kT) * h(kT)$ and is defined as

$$y(kT) = f(kT) * h(kT) = \sum_{i=0}^{\infty} f(iT)h(kT - iT) = \sum_{i=0}^{\infty} h(iT)f(kT - iT)$$

The Z-transform of the function $y(kT)$ is

$$Y(z) = \mathscr{L}[y(kT)] = \mathscr{L}[f(kT) * h(kT)] = \sum_{k=0}^{\infty}\left(\sum_{i=0}^{\infty} f(iT)h(kT - iT)\right)z^{-k}$$

$$= \sum_{k=0}^{\infty}\left(\sum_{i=0}^{\infty} h(iT)f(kT - iT)\right)z^{-k}$$

Reversing the summing order, we have

$$Y(z) = \sum_{i=0}^{\infty} h(iT) \sum_{k=0}^{\infty} f(kT - iT)z^{-k} = \sum_{i=0}^{\infty} h(iT)z^{-i} \sum_{k=0}^{\infty} f(kT - iT)z^{-(k-i)}$$

$$= \left(\sum_{i=0}^{\infty} h(iT)z^{-i} \right) \left(\sum_{m=-i}^{\infty} f(mT)z^{-m} \right)$$

Since $f(mT)$ is a causal function, i.e. $f(mT) = 0$ for $m < 0$, it follows that

$$Y(z) = \left(\sum_{i=0}^{\infty} h(iT)z^{-i} \right) \left(\sum_{m=0}^{\infty} f(mT)z^{-m} \right) = H(z)F(z) \tag{2.24}$$

7. Discrete-time periodic functions

A discrete-time function $f(kT)$ is called periodic with period p if the following relation holds:

$$f(kT) = f(kT + pT) \quad \text{for every } k = 0, 1, 2, \ldots$$

Let $F_1(z)$ be the Z-transform of the first period of $f(kT)$, i.e.

$$F_1(z) = \sum_{k=0}^{p-1} f(kT)z^{-k}$$

Then the Z-transform of the periodic function $f(kT)$ is

$$\mathcal{Z}[f(kT)] = F(z) = \mathcal{Z}[f(kT + pT)] = z^p \left(F(z) - \sum_{k=0}^{p-1} f(kT)z^{-k} \right) = z^p[F(z) - F_1(z)]$$

where relation (2.19c) was used. Hence

$$F(z) = \left(\frac{z^p}{z^p - 1} \right) F_1(z) \tag{2.25}$$

8. The initial value theorem

The following relation holds:

$$f(0) = \lim_{z \to \infty} F(z) \tag{2.26}$$

■ **PROOF** The Z-transform of $f(kT)$ may be written as

$$F(z) = \sum_{k=0}^{\infty} f(kT)z^{-k} = f(0) + f(kT)z^{-1} + f(kT)z^{-2} + \ldots$$

Taking the limits of both sides of this equation, as $z \to \infty$, we immediately arrive at relation (2.26).

---

9. Final value theorem

The following relation holds true:

$$\lim_{k \to \infty} f(kT) = \lim_{z \to 1} (1 - z^{-1})F(z) \qquad (2.27)$$

under the assumption that the function $(1 - z^{-1})F(z)$ does not have any poles outside or on the unit circle.

■ **PROOF**   Consider the Z-transform of $f(kT + T) - f(kT)$:

$$\mathscr{Z}[f(kT + T) - f(kT)] = \lim_{m \to \infty} \sum_{k=0}^{m} [f(kT + T) - f(kT)]z^{-k}$$

Using equations (2.13) and (2.20a) we obtain

$$zF(z) - zf(0) - F(z) = \lim_{m \to \infty} \sum_{k=0}^{m} [f(kT + T) - f(kT)]z^{-k}$$

or

$$(1 - z^{-1})F(z) - f(0) = \lim_{m \to \infty} \sum_{k=0}^{m} [f(kT + T) - f(kT)]z^{-k-1}$$

Taking the limits of both sides of the above equation, as $z \to 1$, we obtain

$$\lim_{z \to 1} (1 - z^{-1})F(z) - f(0) = \lim_{m \to \infty} \sum_{k=0}^{m} [f(kT + T) - f(kT)]$$

$$= \lim_{m \to \infty} \{[F(T) - f(0)] + [f(2T) - f(T)] + \ldots + [f(mT + T) - f(mT)]\}$$

$$= \lim_{m \to \infty} [-f(0) + f(mT + T)] = -f(0) + f(\infty)$$

Hence

$$\lim_{k \to \infty} f(kT) = \lim_{z \to 1} (1 - z^{-1})F(z)$$

All the properties and theorems given above are summarized in Appendix B.

■ **EXAMPLE 2.3.1**   Find the Z-transform of the impulse sequence $\delta(kT - \sigma T)$.

■ **SOLUTION**   Using definition (2.13), we have

$$\mathscr{Z}[\delta(kT - \sigma T)] = \sum_{k=0}^{\infty} \delta(kT - \sigma T)z^{-k} = z^{-\sigma}$$

■ **EXAMPLE 2.3.2**   Find the Z-transform of the step sequence $\beta(kT - \sigma T)$.

■ **SOLUTION**   Here

$$\mathscr{L}[\beta(kT - \sigma T)] = z^{-\sigma}\mathscr{L}[\beta(kT)] = z^{-\sigma}\sum_{k=0}^{\infty}\beta(kT)z^{-k} = z^{-\sigma}\left(\sum_{k=0}^{\infty}(z^{-1})^k\right) = z^{-\sigma}\left(\frac{1}{1-z^{-1}}\right)$$

$$= \frac{z^{-\sigma+1}}{z-1}$$

where use is made of property (2.19c) and of the relation

$$\sum_{k=0}^{\infty}z^i = \frac{1}{1-z} \quad \text{for} \quad |z| < 1$$

■ **EXAMPLE 2.3.3**   Find the Z-transform of the exponential sequence $g(kT) = a^{kT}$.

■ **SOLUTION**   Here

$$\mathscr{L}[g(kT)] = \sum_{k=0}^{\infty}a^{kT}z^{-k} = \sum_{k=0}^{\infty}(a^Tz^{-1})^k = \frac{1}{1-a^Tz^{-1}} = \frac{z}{z-a^T} \quad \text{for} \quad |az^{-1}| < 1$$

■ **EXAMPLE 2.3.4**   Find the Z-transform of the ramp sequence $r(kT - \sigma T)$.

■ **SOLUTION**   Here

$$\mathscr{L}[r(kT - \sigma T)] = z^{-\sigma}\mathscr{L}[r(kT)] = z^{-\sigma}\sum_{k=0}^{\infty}r(kT)z^{-k} = Tz^{-\sigma}\sum_{k=0}^{\infty}kz^{-k}$$

$$= Tz^{-\sigma}(-z)\frac{d}{dz}\mathscr{L}[\beta(kT)] = -Tz^{-\sigma+1}\frac{d}{dz}\left(\frac{z}{z-1}\right) = \frac{Tz^{-\sigma+1}}{(z-1)^2}$$

where use was made of property (2.23).

■ **EXAMPLE 2.3.5**   Find the Z-transform of the alternating sequence $\varepsilon(kT) = (-1)^{kT}$.

■ **SOLUTION**   Here

$$\mathscr{L}[\varepsilon(kT)] = \sum_{k=0}^{\infty}(-1)^{kT}z^{-k} = \sum_{k=0}^{\infty}[(-1)^Tz^{-1}]^k = \frac{1}{1-(-1)^Tz^{-1}} = \frac{z}{z-(-1)^T}$$

■ **EXAMPLE 2.3.6**   Find the Z-transform of the function $y(kT) = e^{bkT}$.

■ **SOLUTION**   Setting $a = e^b$ in Example 2.3.3, we obtain

$$\mathscr{L}[e^{bkT}] = \frac{z}{z-e^{bT}}$$

■ **EXAMPLE 2.3.7** Find the Z-transform of the functions $f(kT) = \sin \omega_0 kT$ and $f(kT) = \cos \omega_0 kT$.

■ **SOLUTION** Here

$$\mathscr{Z}[e^{j\omega_0 kT}] = \frac{z}{z - e^{j\omega_0 T}} = \frac{z}{z - \cos \omega_0 T - j \sin \omega_0 T}$$

$$= \frac{z(z - \cos \omega_0 T + j \sin \omega_0 T)}{(z - \cos \omega_0 T)^2 + \sin^2 \omega_0 T}$$

$$= \left(\frac{z(z - \cos \omega_0 T)}{z^2 - 2z \cos \omega_0 T + 1} \right) + j \left(\frac{z \sin \omega_0 T}{z^2 - 2z \cos \omega_0 T + 1} \right)$$

Since $e^{j\theta} = \cos \theta + j \sin \theta$, it follows that

$$\mathscr{Z}[\cos \omega_0 kT] = \frac{z(z - \cos \omega_0 T)}{z^2 - 2z \cos \omega_0 T + 1}$$

$$\mathscr{Z}[\sin \omega_0 kT] = \frac{z \sin \omega_0 T}{z^2 - 2z \cos \omega_0 T + 1}$$

■ **EXAMPLE 2.3.8** Find the Z-transform of the function $f(kT) = e^{-bkT} \sin \omega kT$.

■ **SOLUTION** Setting $a = e^b$ in equation (2.21), we have

$$\mathscr{Z}[e^{-bkT} f(kT)] = F(e^{bT} z)$$

Using the results of Example 2.3.7, we obtain

$$\mathscr{Z}[e^{-bkT} \sin \omega_0 kT] = \frac{ze^{bT} \sin \omega_0 T}{z^2 e^{2bT} - 2ze^{bT} \cos \omega_0 T + 1}$$

$$= \frac{ze^{-bT} \sin \omega_0 T}{z^2 - 2ze^{-bT} \cos \omega_0 T + e^{-2bT}}$$

2.4 The inverse Z-transform

The determination of the inverse Z-transform (as in the case for the inverse Laplace transform) is usually based upon the expansion of a rational function $F(z)$ into partial fractions whose inverse transform can be directly found in the tables of the Z-transform pairs given in Appendix B. It is noted that in cases where the numerator of $F(z)$ involves the term z, it is more convenient to expand the function $F(z)/z$, instead of $F(z)$, into partial fractions and subsequently to determine $F(z)$ from the relation $z[F(z)/z]$.

It is also noted that there are several other techniques for the determination of the inverse Z-transform: for example, the method of continuous fraction expansion, the direct implementation of the definition of the inverse Z-transform given by equation (2.16), and others. The method of partial fraction expansion appears to be computationally simpler compared with the other methods and for this reason it is almost always used for the determination of the inverse Z-transform.

■ **EXAMPLE 2.4.1**   Find the inverse Z-transform of the function

$$F(z) = \frac{-3z}{(z-1)(z-4)}$$

■ **SOLUTION**   Expanding $F(z)/z$ into partial fractions, we obtain

$$\frac{F(z)}{z} = \frac{-3}{(z-1)(z-4)} = \frac{1}{(z-1)} - \frac{1}{(z-4)}$$

and hence

$$F(z) = \frac{z}{(z-1)} - \frac{z}{(z-4)}$$

From the table of the Z-transform pairs (Appendix B) we find that

$$\mathscr{L}^{-1}\left[\frac{z}{z-1}\right] = \beta(kT) \quad \text{and} \quad \mathscr{L}^{-1}\left[\frac{z}{z-4}\right] = 4^k \quad \text{where} \quad T = 1$$

Hence

$$f(kT) = \mathscr{L}^{-1}[F(z)] = \beta(kT) - 4^k = 1 - 4^k$$

■ **EXAMPLE 2.4.2**   Find the inverse Z-transform of the function

$$F(z) = \frac{z(z-4)}{(z-2)^2(z-3)}$$

■ **SOLUTION**   Expanding $F(z)/z$ into partial fractions we obtain

$$\frac{F(z)}{z} = \frac{1}{z-2} + \frac{2}{(z-2)^2} - \frac{1}{z-3}$$

and hence

$$F(z) = \frac{z}{z-2} + \frac{2z}{(z-2)^2} - \frac{z}{z-3}$$

Since for the case $T = 1$

$$\mathscr{L}^{-1}\left[\frac{z}{z-2}\right] = 2^k, \quad \mathscr{L}^{-1}\left[\frac{2z}{(z-2)^2}\right] = k2^k \quad \text{and} \quad \mathscr{L}^{-1}\left[\frac{z}{z-3}\right] = 3^k$$

it follows that

$$f(kT) = \mathscr{L}^{-1}[F(z)] = 2^k + k2^k + 3^k = (k+1)2^k + 3^k$$

■ **EXAMPLE 2.4.3** Find the inverse Z-transform of the function

$$F(z) = \frac{2z^3 + z}{(z-2)^2(z-1)}$$

■ **SOLUTION** Expanding $F(z)/z$ into partial fractions we obtain

$$\frac{F(z)}{z} = \frac{9}{(z-2)^2} - \frac{1}{z-2} + \frac{3}{z-1}$$

and hence

$$F(z) = \frac{9z}{(z-2)^2} - \frac{z}{z-2} + \frac{3z}{z-1}$$

Since for $T = 1$

$$\mathscr{L}^{-1}\left[\frac{z}{z-2}\right] = 2^k, \quad \mathscr{L}^{-1}\left[\frac{z}{(z-2)^2}\right] = k2^{k-1} \quad \text{and} \quad \mathscr{L}^{-1}\left[\frac{z}{z-1}\right] = 1$$

it follows that

$$f(kT) = \mathscr{L}^{-1}[F(z)] = 9k2^{k-1} - 2^k + 3$$

■ **EXAMPLE 2.4.4** Find the inverse Z-transform of the function

$$F(z) = \frac{z^2}{z^2 - 2z + 2}$$

■ **SOLUTION** Examining the form of the denominator $z^2 - 2z + 2$ we observe that $F(z)$ may be the Z-transform of a function of the type e^{-akT} $(c_1 \sin \omega_0 kT + c_2 \cos \omega_0 kT)$, where c_1 and c_2 are constants. To verify this observation we proceed as follows. The constant term 2 is equal to the exponential e^{-2aT}, in which case $a = -(\ln 2)/2T$. The coefficient -2 of the z term must be equal to the function $-2e^{-aT} \cos \omega_0 T$, in which case $\cos \omega_0 T = 1/\sqrt{2}$ and $\omega_0 = \pi/4T$. Consequently, the denominator of $F(z)$ can be written as follows:

$$z^2 - 2z + 2 = z^2 - 2z \exp\left[-\left(\frac{-\ln 2}{2T}\right)T\right]\cos\left(\frac{\pi}{4T}\right)T + \exp\left[-2\left(\frac{-\ln 2}{2T}\right)T\right]$$

The numerator can be written as $z^2 = (z^2 - z) + z$. Since

$$\exp\left[-\left(\frac{-\ln 2}{2T}\right)T\right]\cos\left(\frac{\pi}{4T}\right)T = \exp\left[-\left(\frac{-\ln 2}{2T}\right)T\right]\sin\left(\frac{\pi}{4T}\right)T = 1$$

it follows that the numerator $(z^2 - z) + z$ may be written as

$$z^2 = \left\{z^2 - z\exp\left[-\left(\frac{-\ln 2}{2T}\right)T\right]\cos\left(\frac{\pi}{4T}\right)T\right\} + z\exp\left[-\left(\frac{-\ln 2}{2T}\right)T\right]\sin\left(\frac{\pi}{4T}\right)T$$

Hence, the function $F(z)$ may finally be written as

$$F(z) = \frac{z^2 - z}{z^2 - 2z + 2} + \frac{z}{z^2 - 2z + 2}$$

$$= \frac{z^2 - z\exp[-(-\ln 2/2T)T]\cos(\pi/4T)T}{z^2 - 2z\exp[-(-\ln 2/2T)T]\cos(\pi/4T)T + \exp[-2(-\ln 2/2T)T]}$$

$$+ \frac{z\exp[-(-\ln 2/2T)T]\sin(\pi/4T)T}{z^2 - 2z\exp[-(-\ln 2/2T)T]\cos(\pi/4T)T + \exp[-2(-\ln 2/2T)T]}$$

From the table of the Z-transform pairs it follows that

$$f(kT) = \mathscr{Z}^{-1}[F(z)] = \exp\left[-\left(\frac{-\ln 2}{2T}\right)kT\right]\left[\cos\left(\frac{\pi}{4T}\right)kT + \sin\left(\frac{\pi}{4T}\right)kT\right]$$

$$= \exp\left(\frac{\ln 2}{2}k\right)\left[\cos\left(\frac{\pi k}{4}\right) + \sin\left(\frac{\pi k}{4}\right)\right] \quad \text{for} \quad T = 1$$

2.5 Examples of certain interesting Z-transform applications

In this section, certain interesting problems will be presented which are not directly related to discrete-time control systems. The reason for presenting them is that they give a simple and rather attractive introduction to the important issue of describing a system via difference equations and solving them using the Z-transform.

1. The Fibonacci sequence

The Fibonacci sequence is a number sequence where each term in the sequence is the sum of the previous two terms. The first two terms of the sequence are given as 0 and 1. Let us determine the general expression for any arbitrary term of the sequence.

In order to solve the problem, let $y(k)$ represent an arbitrary term of the sequence. Then, the Fibonacci sequence may be described by the following difference equation:

$$y(k+2) = y(k+1) + y(k) \tag{2.28}$$

with initial conditions $y(0) = 0$ and $y(1) = 1$. Equation (2.28) may of course be solved either in the k-domain using methods equivalent to those used to solve differential equations in the t-domain or using the Z-transform. The Z-transform is much simpler over the k-domain methods since, as already noted, the Z-transform converts a difference equation into an algebraic one (similar results are derived in the case where the Laplace transform is applied to a differential equation). Applying the Z-transform to equation (2.28), we have

$$\mathscr{Z}[y(k+2)] = \mathscr{Z}[y(k+1)] + \mathscr{Z}[y(k)] \tag{2.29}$$

From equations (2.20a and b) we obtain

$$\mathscr{Z}[y(k+1)] = zY(z) - zy(0) = zY(z)$$

$$\mathscr{Z}[y(k+2)] = z^2 Y(z) - z^2 y(0) - zy(1) = z^2 Y(z) - z$$

Thus, equation (2.29) becomes

$$z^2 Y(z) - z = zY(z) + Y(z)$$

Hence

$$Y(z) = \frac{z}{z^2 - z - 1}$$

In order to determine the inverse Z-transform $y(k) = \mathscr{Z}^{-1}[Y(z)]$, $Y(z)$ is first expanded into partial fractions to yield

$$Y(z) = \frac{1}{\sqrt{5}} \left(\frac{z}{z - z_1} - \frac{z}{z - z_2} \right)$$

where

$$z_1 = \frac{1 + \sqrt{5}}{2} \quad \text{and} \quad z_2 = \frac{1 - \sqrt{5}}{2}$$

Hence

$$y(k) = \frac{1}{\sqrt{5}} \left[\left(\frac{1 + \sqrt{5}}{2} \right)^k - \left(\frac{1 - \sqrt{5}}{2} \right)^k \right] \quad \text{for} \quad k = 0, 1, \ldots \tag{2.30}$$

2. A mathematical model for rabbit populations

A simplified approach in determining a mathematical model to describe the increase in a rabbit population can be developed by making the following assumptions [1]:

1. A pair of rabbits gives birth to two young, a male and a female, at the end of every month.

2. A newly born couple of rabbits gives birth to its first two young in exactly 2 months' time.
3. Two rabbits pair, remain faithful to each other for ever(!) and give birth to young as in 1 and 2.

Next, assume that initially there exists only a single pair of rabbits. Let $y(k)$ represent the number of rabbit pairs produced from the initial rabbit pair, at the end of the kth month Then, we have

$$y(k) = y(k-1) + y(k-2) \tag{2.31}$$

where $y(k-1)$ represents the number of rabbit pairs whose age exceeds a single month at the end of the kth month and $y(k-2)$ represents the number of newly born pairs at the end of the kth month. The initial conditions are $y(-1) = 0$ and $y(0) = 1$. Thus, the difference equation (2.31) is a mathematical model describing the increase in the rabbit population. A closer examination of equation (2.31) shows that this difference equation is very similar to the Fibonacci sequence difference equation (2.28). To solve equation (2.31), we make use of the Z-transform. From equations (2.20c and d) we obtain

$$\mathcal{L}[y(k-1)] = z^{-1}Y(z) + y(-1)$$

$$\mathcal{L}[y(k-2)] = z^{-2}Y(z) + z^{-1}y(-1) + y(-2)$$

Thus, equation (2.31) becomes

$$Y(z) = z^{-1}Y(z) + y(-1) + z^{-2}Y(z) + z^{-1}y(-1) + y(-2)$$

or

$$Y(z) = \frac{z^{-1}y(-1) + y(-1) + y(-2)}{1 - z^{-1} - z^{-2}} = \frac{y(-2)}{1 - z^{-1} - z^{-2}} = \frac{y(-2)z^2}{z^2 - z - 1}$$

where the initial condition $y(-1) = 0$ was used. Applying the initial value theorem (2.26), we have

$$y(0) = \lim_{z \to \infty} Y(z) = y(-2)$$

Hence, choosing $y(-2) = y(0) = 1$, it follows that

$$Y(z) = \frac{z^2}{z^2 - z - 1}$$

Expanding $Y(z)$ into partial fractions yields

$$Y(z) = \frac{1}{\sqrt{5}} \left(\frac{z_1 z}{z - z_1} - \frac{z_2 z}{z - z_2} \right)$$

where

$$z_1 = \frac{1 + \sqrt{5}}{2} \quad \text{and} \quad z_2 = \frac{1 - \sqrt{5}}{2}$$

Hence

$$y(k) = \frac{1}{\sqrt{5}} \left[\left(\frac{1+\sqrt{5}}{2} \right)^{k+1} - \left(\frac{1-\sqrt{5}}{2} \right)^{k+1} \right] \quad \text{for} \quad k = -1, 0, 1, 2, \ldots \qquad (2.32)$$

3. Networks with recursive structure

Consider the network shown in Figure 2.10. All 41 resistors have a resistance of 1 ohm. Let us determine the voltage V_{R_o} across the output resistor R_o, where $R_o = 1$ ohm.

In order to solve this problem, let $v(0)$, $v(1)$, $v(2)$, ..., $v(k)$, ..., $v(21)$ be the node voltages of the network. Then, for an arbitrary node $k+1$, shown in Figure 2.11, application of Kirchhoff's current law results in the following:

$$i_a + i_b = i_c$$

or

$$[v(k) - v(k+1)] + [v(k+2) - v(k+1)] = v(k+1)$$

or

$$v(k) - 3v(k+1) + v(k+2) = 0 \qquad (2.33)$$

with initial conditions $v(0) = 10$ and $v(1)$ unknown. In order to determine V_{R_o}, it suffices to solve the difference equation (2.33). To this end, we apply the Z-transform to equation (2.33) to yield

$$V(z) - 3[zV(z) - zv(0)] + z^2 V(z) - z^2 v(0) - zv(1) = 0$$

or

$$V(z) = \frac{10z(z-3) + zv(1)}{z^2 - 3z + 1} \qquad (2.34)$$

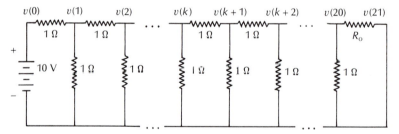

Figure 2.10   A network with recursive structure.

Figure 2.11   The currents at the node $k + 1$.

where use was made of the initial condition $v(0) = 10$. Equation (2.34) can be rewritten as

$$V(z) = \frac{10z(z - 1.5)}{z^2 - 3z + 1} - \frac{[15 - v(1)]z}{z^2 - 3z + 1}$$

Using the table of Z-transform pairs given in Appendix B we obtain

$$\mathcal{Z}^{-1}\left[\frac{z(z - 1.5)}{z^2 - 2z(1.5) + 1}\right] = \cosh \omega_0 kT \quad \text{where } \cosh \omega_0 T = 1.5 \text{ and } \omega_0 T = 0.9624$$

$$\mathcal{Z}^{-1}\left[\frac{z(\sqrt{5}/2)}{z^2 - 2z(1.5) + 1}\right] = \sinh \omega_0 kT \quad \text{where } \sinh \omega_0 T = \sqrt{5}/2$$

Hence

$$v(k) = 10 \cosh(0.9624k) - \frac{2[15 - v(1)]}{\sqrt{5}} \sinh(0.9624k) \tag{2.35}$$

We may now proceed to estimate the initial condition $v(1)$ as follows. From the network we observe that $v(21) = 0$, which, when substituted in equation (2.35), yields

$$v(21) = 0 = 10 \cosh[(0.9624)(21)] - \frac{2[15 - v(1)]}{\sqrt{5}} \sinh[(0.9624)(21)]$$

or

$$v(1) = 15 - 5\sqrt{5} \tanh^{-1} 20.2109$$

Substituting this value of $v(1)$ in equation (2.35), we obtain the general solution of the difference equation (2.33):

$$v(k) = 10[\cosh(0.9624k) - \tanh^{-1} 20.2109 \sinh(0.9624k)] \tag{2.36}$$

The voltage across the resistor R_0 is

$$V_{R_0} = v(20) - v(21) = v(20)$$

Therefore, the desired voltage can be estimated if we set $k = 20$ in equation (2.36), in which case

$$V_{R_0} = v(20) = 10(\cosh 19.24847 - \tanh^{-1} 20.21089 \sinh 19.24847)$$

■ **REMARK 2.3.1** The value of V_{R_0} is extremely small and of the order of 10^{-8} volts. If we had 11 resistors instead of 41, the general expression of $v(k)$ would be

$$v(k) = 10[\cosh(0.9624k) - \tanh^{-1} 5.7745 \sinh(0.9624k)]$$

and the voltage across R_o would be

$$V_{R_o} = v(5) - v(6) = v(5) = 10[\cosh(4.81211) - \tanh^{-1} 5.7745 \sinh(4.81211)]$$

$$= 10[61.500 - (1.000\,012\,929)(61.401\,87)] = 10[61.500 - 61.493\,055\,5]$$

$$= 0.069\,44 \text{ volts}$$

2.6 Problems

1. Find the Z-transform of the following functions:

(a) k^3 (b) $\sin^2(\omega k)$ (c) $\cos(\omega k + \phi)$
(d) $\sin(\omega_1 k)\cos(\omega_2 k)$ (e) $\cosh(\omega k + \phi)$ (f) $k \sin(\omega k)$

2. Find the inverse Z-transform of the functions:

(a) $\dfrac{z^{-2} - 1}{z^{-2} - 4}$ (b) $\dfrac{1 - z^{-N}}{1 - z^{-1}}$ (c) $\dfrac{1 - (1/2)z^{-1} - (3/4)z^{-2}}{1 - (1/4)z^{-1} - (1/8)z^{-2}}$

(d) $\dfrac{1 - az^{-1}}{a - z^{-1}}$ (e) $\dfrac{a - z^{-1}}{1 - az^{-1}}$ (f) $\dfrac{z^2}{(z + 2)(z + 3)}$

(g) $\dfrac{z^2 + 5z + 4}{z^2 + z + 1}$ (h) $\dfrac{z}{(z - e^{-aT})(z - e^{-bT})}$ (i) $\dfrac{z - 1}{z^2 - 2z + 1}$

3. Find A, B, C, D, E and the inverse Z-transform of each of the following:

(a) $\dfrac{z^3 - a}{(z - 1)^3} = \dfrac{A}{(z - 1)^3} + \dfrac{B}{(z - 1)^2} + \dfrac{C}{z - 1}$

(b) $\dfrac{z^4 - a}{(z - 1)^4} = \dfrac{A}{(z - 1)^4} + \dfrac{B}{(z - 1)^3} + \dfrac{C}{(z - 1)^2} + \dfrac{D}{z - 1}$

(c) $\dfrac{z^5 - a}{(z - 1)^5} = \dfrac{A}{(z - 1)^5} + \dfrac{B}{(z - 1)^4} + \dfrac{C}{(z - 1)^3} + \dfrac{D}{(z - 1)^2} + \dfrac{E}{z - 1}$

4. Solve the difference equations using the Z-transform method:

(a) $y(k + 2) + 5y(k + 1) + 6y(k) = u(k + 1)$
   with $y(0) = 0$ and $y(1) = 0$ and $u(k) = \beta(k) = $ unit step function.
(b) $y(k + 2) - y(k) = u(k)$
   with $y(0) = y(1) = 0$ and $u(k) = \beta(k) = $ unit step function.

5. Obtain the Z-transform of the sequences:

(a) $f(k) = \{1, 1, 1, 1/2, 1/2, 1/2, 1/4, 1/4, 1/4, 1/8, 1/8, 1/8, 1/16, 1/16, 1/16, \ldots\}$

(b) $f(k) = \{1, 1, 0, 0, 1/2, 1/2, 0, 0, 1/4, 1/4, 0, 0, 1/8, 1/8, 0, 0, 1/16, 1/16, \ldots\}$

6. Show that the transfer function of an all-pass filter

$$H(z) = \frac{(z^{-1} - \bar{\lambda}_1)(z^{-1} - \bar{\lambda}_2)\ldots(z^{-1} - \bar{\lambda}_n)}{(1 - \lambda_1 z^{-1})(1 - \lambda_2 z^{-1})\ldots(1 - \lambda_n z^{-1})}$$

where $|\lambda_i| < 1$, satisfies the conditions

$$|H(e^{j\theta})| = 1$$

for every θ.

7. The autocorrelation function of a signal $x(k)$, $k = 0, 1, 2, \ldots$, is defined as

$$r_x(k) = \sum_{m=-\infty}^{\infty} x(m)x(m+k) \quad \text{where} \quad k \in (-\infty, \infty)$$

Find the Z-transform of the sequence $r_x(k)$ as a function of the Z-transform of the sequence $x(k)$.

8. The population of an animal colony increases during the period from kT to $(k+1)T$ by 10% (births − natural deaths). However, predation over the same period reduces the population by R animals. Estimate R, given that the number of animals is reduced from 1000 to 500 within a 12 period interval. What is the permissible rate of predation which maintains a constant population?

2.7 References

Books

[1] J.A. Cadzow, *Discrete-Time Systems, An Introduction with Interdisciplinary Applications*, Prentice Hall, Englewood Cliffs, New Jersey, 1973.

[2] G. Doetsch, *Guide to the Applications of the Laplace and Z Transforms*, Van Nostrand Reinhold, New York, 1971.

[3] E.I. Jury, *Theory and Application of the Z-Transform Method*, John Wiley, New York, 1964.

3

Description and analysis of discrete-time and sampled-data systems

3.1 Introduction

The term **discrete-time systems** covers systems which operate directly with discrete-time signals. In this case, the input, as well as the output, of the system are obviously discrete-time signals (Figure 3.1). A well-known discrete-time system is the digital computer. In this case the signals $u(k)$ and $y(k)$ are number sequences (usually 0 and 1). These types of system, as we shall see, are described by difference equations.

The term **sampled-data systems** [8] covers the usual analog (continuous-time) systems, having the following distinct characteristic: the input $u(t)$ and the output $y(t)$ are piecewise constant signals, i.e. they are constant over each interval between two consecutive sampling points (Figure 3.2). The piecewise constant signal $u(t)$ is derived from the discrete-time signal $v(kT)$ using a hold circuit (see Section 3.3.2). The output $s(t)$ of the system is a continuous-time function. Let $y(t)$ be the output of the system having a piecewise constant form with $y(t) = s(t)$ at the sampling points. Then, the system having $u(t)$ as input and $y(t)$ as output, where both signals are piecewise constant in each interval, is a sampled-data system and may be described, as will be shown in Sections 3.3.3 and 3.3.4, by difference equations. This means that sampled-data systems may be described and subsequently studied similarly to discrete-time systems. This fact is of particular importance since it **unifies** the study of hybrid systems, which consist of continuous-time and discrete-time subsystems, using one method, namely difference equations. For this reason, and for reasons of simplicity, sampled-data systems are usually addressed in the literature (and in this book) as **discrete-time systems**. It is noted that **sampled-data systems** are also called **discretized systems**.

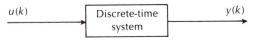

Figure 3.1 Block diagram of a discrete-time system.

32

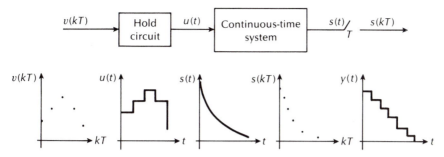

Figure 3.2   Block diagram of a sampled-data system.

In a discrete-time system, various combinations of discrete-time and continuous-time subsystems may appear. In a closed-loop discrete-time system, the following three configurations are usually encountered:

1. The discrete-time subsystem (here, a digital computer) is placed in a feedback loop (Figure 3.3). In this position, the computer may be used to control other systems simultaneously (time sharing).
2. The digital computer is programmed to operate as a controller for the closed-loop system (Figure 3.4).
3. In the closed-loop system, all signals are discrete since all subsystems are also discrete (Figure 3.5).

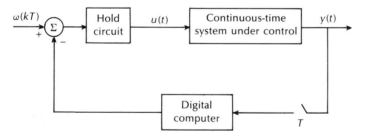

Figure 3.3   Closed-loop system with a digital computer in the feedback loop.

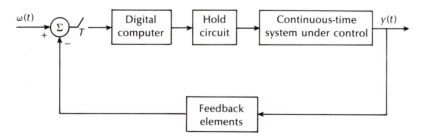

Figure 3.4   Closed-loop system with the digital computer in the feedforward path.

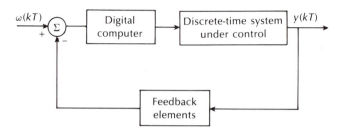

Figure 3.5 Closed-loop discrete-time system.

The purpose of this chapter is to present ways and methods for the description and analysis of discrete-time and sampled-data systems. Since the description and analysis of sampled-data systems is very similar to that of discrete-time systems, we will first present (Section 3.2) some basic notions about discrete-time systems and subsequently (Section 3.3) extend these notions to sampled-data systems. For more on the subject see the books [1–15] cited in the references.

3.2 Description and analysis of discrete-time systems

3.2.1 Properties of discrete-time systems

From a mathematical point of view discrete-time system description implies the determination of a law which assigns an output sequence $y(k)$ to a given input sequence $u(k)$ (Figure 3.6). The specific law connecting the input and output sequences $u(k)$ and $y(k)$ constitutes the mathematical model of the discrete-time system. Symbolically, this relation can be written as follows:

$$y(k) = Q[u(k)]$$

where Q is a discrete operator.

Discrete-time systems have a number of properties, some of which are of special interest and are presented below.

1. Linearity

A discrete-time system is linear if the following relation holds true:

$$Q[c_1 u_1(k) + c_2 u_2(k)] = c_1 Q[u_1(k)] + c_2 Q[u_2(k)] = c_1 y_1(k) + c_2 y_2(k) \tag{3.1}$$

for every c_1, c_2, $u_1(k)$ and $u_2(k)$, where c_1, c_2 are constants and $y_1(k) = Q[u_1(k)]$ is

$u(k)$ → [Q] → $y(k)$

Figure 3.6 Block diagram of a discrete-time system.

the output of the system with input $u_1(k)$ and $y_2(k) = Q[u_2(k)]$ is the output of the system with input $u_2(k)$.

2. Time-invariant system

A discrete-time system is time invariant if the following holds true:

$$Q[u(k - k_0)] = y(k - k_0) \tag{3.2}$$

for every k_0. Equation (3.2) shows that when the input to the system is shifted by k_0 units, the output of the system is also shifted by k_0 units.

3. Causality

A discrete-time system is called causal if the output $y(k) = 0$ for $k < k_0$, when the input $u(k) = 0$ for $k < k_0$. Since a discrete-time signal is called causal if it is zero for $k < k_0$, we may state that a system is causal if every causal excitation produces a causal response.

3.2.2 Description of linear time-invariant discrete-time systems

A linear time-invariant causal discrete-time system involves the following elements: summation units, amplification units and delay units. The block diagram of all three elements is shown in Figure 3.7. The delay unit is symbolized as z^{-1}, meaning that the output is identical to the input delayed by a time unit.

When these three elements are suitably interconnected, then one has a discrete-time system, as, for example, the discrete-time system shown in Figure 3.8. Adding the three signals at the summation point Σ, we arrive at the equation

$$y(k) + a_1 y(k - 1) = b_0 u(k) + b_1 u(k - 1) \tag{3.3a}$$

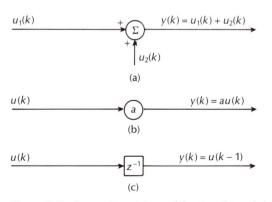

(a)

(b)

(c)

Figure 3.7 Summation (a), amplification (b) and delay (c) units.

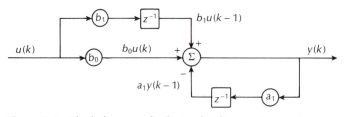

Figure 3.8 Block diagram of a first-order discrete-time system.

Similarly, for the discrete-time system shown in Figure 3.9, we obtain the equation

$$y(k) + a_1 y(k-1) + a_2 y(k-2) = b_0 u(k) + b_1 u(k-1) + b_2 u(k-2) \qquad (3.3b)$$

Obviously, equations (3.3a and b) are mathematical models describing the discrete-time systems shown in Figures 3.8 and 3.9, respectively. Equations (3.3a and b) are difference equations.

There are many ways to describe discrete-time systems, as is also the case for continuous-time systems. The most popular ones are the following: the difference equations, as in equations (3.3a and b), the transfer function, the impulse response or weight function, and the state-space equations.

In presenting these four methods, certain similarities and dissimilarities between continuous-time and discrete-time systems will be revealed. There are three basic differences, going from continuous-time to discrete-time systems: differential equations are now difference equations; the Laplace transform gives way to the Z-transform; and the integration procedure is replaced by summation.

1. Difference equations

The general form of a difference equation is as follows:

$$y(k) + a_1 y(k-1) + \cdots + a_n y(k-n) = b_0 u(k) + b_1 u(k-1) + \cdots + b_m u(k-m) \quad (3.4)$$

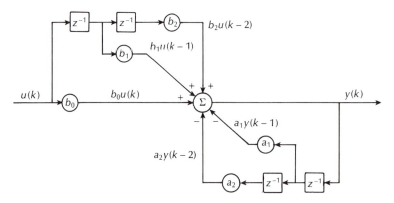

Figure 3.9 Block diagram of a second-order discrete-time system.

with initial conditions $y(-1)$, $y(-2)$, ..., $y(-n)$. The solution of equation (3.4) may be found either in the time domain (using methods similar to those for solving a differential equation in the time domain) or in the complex frequency or z-domain using the Z-transform.

2. Transfer function

The transfer function of a discrete-time system is denoted by $H(z)$ and is defined as the ratio of the Z-transform of the output $y(k)$ divided by the Z-transform of the input $u(k)$, under the condition that $u(k) = y(k) = 0$, for all negative values of k. That is,

$$H(z) = \frac{\mathcal{L}[y(k)]}{\mathcal{L}[u(k)]} = \frac{Y(z)}{U(z)} \quad \text{where} \quad u(k) = y(k) = 0 \quad \text{for} \quad k < 0 \tag{3.5}$$

The transfer function of a system described by the difference equation (3.4), with $y(k) = u(k) = 0$, for $k < 0$, is determined as follows: multiply both sides of (3.4) by the term z^{-k} and add for $k = 0, 1, 2, \ldots$, to yield:

$$\sum_{k=0}^{\infty} y(k)z^{-k} + a_1 \sum_{k=0}^{\infty} y(k-1)z^{-k} + a_2 \sum_{k=0}^{\infty} y(k-2)z^{-k} + \ldots + a_n \sum_{k=0}^{\infty} y(k-n)z^{-k}$$

$$= b_0 \sum_{k=0}^{\infty} u(k)z^{-k} + b_1 \sum_{k=0}^{\infty} u(k-1)z^{-k} + \ldots + b_m \sum_{k=0}^{\infty} u(k-m)z^{-k}$$

Using the time-shifting property (2.19a) and the assumption that $u(k)$ and $y(k)$ are zero for negative values of k, the equation above can be simplified as follows:

$$Y(z) + a_1 z^{-1} Y(z) + \cdots + a_n z^{-n} Y(z) = b_0 U(z) + b_1 z^{-1} U(z) + \cdots + b_m z^{-m} U(z)$$

Hence, using definition (3.5), we arrive at the following rational polynomial form for $H(z)$:

$$H(z) = \frac{Y(z)}{U(z)} = \frac{b_0 + b_1 z^{-1} + \ldots + b_m z^{-m}}{1 + a_1 z^{-1} + \ldots + a_n z^{-n}} \tag{3.6}$$

3. Impulse response or weight function

The impulse response (or weight function) of a system is denoted by $h(k)$ and is defined as the output of a system when its input is the impulse sequence $\delta(k)$ (see Figure 2.1), under the constraint that the initial conditions $y(-1)$, $y(-2)$, ..., $y(-n)$ of the system are zero. The block diagram definition of the impulse response is shown in Figure 3.10. The transfer function $H(z)$ and the weight function $h(k)$ are related by the following equation:

$$H(z) = \mathcal{L}[h(k)] \tag{3.7}$$

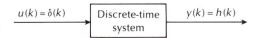

$u(k) = \delta(k)$ → Discrete-time system → $y(k) = h(k)$

Figure 3.10 Block diagram definition of the impulse response.

4. State-space equations

State-space equations are a set of first-order difference equations for describing high-order systems and have the form

$$\mathbf{x}(k+1) = \mathbf{Ax}(k) + \mathbf{Bu}(k) \tag{3.8a}$$

$$\mathbf{y}(k) = \mathbf{Cx}(k) + \mathbf{Du}(k) \tag{3.8b}$$

where $\mathbf{u}(k) \in \mathbb{R}^m$, $\mathbf{x}(k) \in \mathbb{R}^n$ and $\mathbf{y}(k) \in \mathbb{R}^p$ are the input, state and output vectors, respectively, and $\mathbf{A}, \mathbf{B}, \mathbf{C}, \mathbf{D}$ are constant matrices of appropriate dimensions.

Let $\mathbf{Y}(z) = \mathscr{L}[\mathbf{y}(k)]$ and $\mathbf{U}(z) = \mathscr{L}[\mathbf{u}(k)]$. Then, the transfer function matrix $\mathbf{H}(z)$ of (3.8) is

$$\mathbf{H}(z) = \mathbf{C}(z\mathbf{I}_n - \mathbf{A})^{-1}\mathbf{B} + \mathbf{D} \tag{3.9}$$

The impulse response matrix $\mathbf{H}(k)$ of (3.8) is

$$\mathbf{H}(k) = \mathscr{L}^{-1}[\mathbf{H}(z)] = \begin{cases} \mathbf{D} & \text{for } k = 0 \\ \mathbf{CA}^{k-1}\mathbf{B} & \text{for } k > 0 \end{cases} \tag{3.10}$$

3.2.3 *Analysis of linear time-invariant discrete-time systems*

The problem of the analysis of linear time-invariant discrete-time systems will be solved using four different methods, where each method corresponds to one of the four description models presented in Section 3.2.2.

1. Analysis based on the difference equation

We present the following two introductory examples.

■ **EXAMPLE 3.2.1** A discrete-time system is described by the difference equation

$$y(k) = u(k) + ay(k-1)$$

with the initial condition $y(-1)$. Solve the difference equation, i.e. determine $y(k)$.

■ **SOLUTION**   The difference equation may be solved to determine $y(k)$, using the Z-transform, as follows. Take the Z-transform of both sides of the equation to yield

$$\mathcal{L}[y(k)] = \mathcal{L}[u(k)] + a\mathcal{L}[y(k-1)]$$

or

$$Y(z) = U(z) + a[z^{-1}Y(z) + y(-1)]$$

and thus

$$Y(z) = \frac{z[U(z) + ay(-1)]}{z-a} = \frac{ay(-1)z}{z-a} + \frac{U(z)z}{z-a}$$

Suppose that the excitation $u(k)$ is the unit step sequence $\beta(k)$ (see Figure 2.2). In this case

$$U(z) = \mathcal{L}[\beta(k)] = \frac{z}{z-1}$$

Then, the output $Y(z)$ becomes

$$Y(z) = \frac{ay(-1)z}{z-a} + \frac{z^2}{(z-a)(z-1)} = \frac{ay(-1)z}{z-a} + \left(\frac{1}{1-a}\right)\left(\frac{z}{z-1}\right) - \left(\frac{a}{1-a}\right)\left(\frac{z}{z-a}\right)$$

and hence

$$y(k) = a^{k+1}y(-1) + \frac{1}{1-a} - \frac{1}{1-a}a^{k+1}$$

The expression for the output $y(k)$ clearly converges for $|a| < 1$. The initial condition $y(-1)$ contributes only during the transient period. The output $y(k)$, in the steady state, takes on the form

$$y_{ss}(k) = \frac{1}{1-a}$$

where $y_{ss}(k)$ denotes the steady state value of $y(k)$. Figure 3.11 shows $y(k)$ when the initial condition $y(-1) = 0$.

■ **EXAMPLE 3.2.2**   Consider a second-order discrete-time system described by the difference equation

$$y(k) = 2\zeta\omega_0 y(k-1) - \omega_0^2 y(k-2) + (1 - 2\zeta\omega_0 + \omega_0^2)u(k-1)$$

The parameters ζ and ω_0 are called the damping radio and natural frequency, respectively, and are real numbers, with $\zeta \geq 0$. The initial conditions are assumed to be zero. Solve the difference equation.

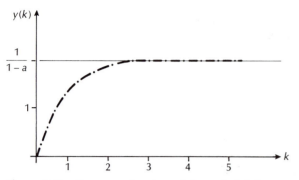

Figure 3.11   Response of system of Example 3.2.1.

■ **SOLUTION**   The difference equation may be solved for $y(k)$ using the Z-transform as follows. Taking the Z-transform of both sides of the equation yields

$$\mathscr{L}[y(k)] = \mathscr{L}[2\zeta\omega_0 y(k-1)] - \mathscr{L}[\omega_0^2 y(k-2)] + \mathscr{L}[(1 - 2\zeta\omega_0 + \omega_0^2)u(k-1)]$$

or

$$Y(z) = 2\zeta\omega_0 z^{-1}Y(z) - \omega_0^2 z^{-2}Y(z) + (1 - 2\zeta\omega_0 + \omega_0^2)z^{-1}U(z)$$

and therefore

$$Y(z) = \left(\frac{(1 - 2\zeta\omega_0 + \omega_0^2)z}{z^2 - 2\zeta\omega_0 z + \omega_0^2}\right)U(z)$$

Assuming that the excitation $u(k)$ is the unit step sequence $\beta(k)$, in which case

$$U(z) = \mathscr{L}[\beta(k)] = \frac{z}{z-1}$$

then $Y(z)$ becomes

$$Y(z) = \frac{(1 - 2\zeta\omega_0 + \omega_0^2)z^2}{(z^2 - 2\zeta\omega_0 z + \omega_0^2)(z-1)}$$

The poles of the system, i.e. the <u>roots</u> of the characteristic polynomial $z^2 - 2\zeta\omega_0^2 + \omega_0 z = 0$, are $z_{1,2} = \zeta\omega_0 \pm \omega_0\sqrt{\zeta^2 - 1}$. Choosing any positive value for ζ we obtain three interesting cases, summarized in Table 3.1. The unit step responses are different for each of the three cases and are examined separately, below.

(a) Critically damped unit step response ($\zeta = 1$)
In this case the system has a double pole at $z = \omega_0$. The output $Y(z)$ becomes

$$Y(z) = \left(\frac{(1 - \omega_0)^2 z}{(z - \omega_0)^2}\right)\left(\frac{z}{z-1}\right)$$

Table 3.1   Characterization of system behaviour for $\zeta \geqslant 0$.

ζ	System's poles	Characterization of system's behaviour
$0 \leqslant \zeta < 1$	Two complex conjugates	Underdamped
$\zeta = 1$	A double real	Critically damped
$\zeta > 1$	Two real distinct	Overdamped

which can be written as follows:

$$Y(z) = \frac{z}{z-1} - (1 - \omega_0)\left(\frac{z^2}{(z - \omega_0)^2}\right) - \omega_0\left(\frac{z}{z - \omega_0}\right)$$

Taking the inverse $\mathscr{Z}$-transform, we arrive at the following expression for the critically damped unit step response:

$$y(k) = 1 - [1 + (1 - \omega_0)k]\omega_0^k \quad \text{for } k = 0,\, 1,\, 2,\, \ldots$$

The foregoing expression for $y(k)$ converges for $|\omega_0| < 1$. The response $y(k)$ converges to the value 1 with a rate which depends on how fast the term ω_0^k converges to zero. Figure 3.12 shows the critically damped unit step response of the system for two different values of ω_0. The smaller the value of ω_0, the faster $y(k)$ converges to 1.

(b) Underdamped unit step response $(0 \leqslant \zeta < 1)$
In this case, the transfer function of the system has two complex conjugate poles, which can be written in polar form as follows:

$$z_{1,2} = \omega_0 e^{\pm j\theta} \quad \text{where } \zeta = \cos\theta$$

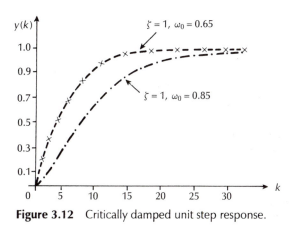

Figure 3.12   Critically damped unit step response.

The unit step response is

$$Y(z) = \left(\frac{(1 - 2\zeta\omega_0 + \omega_0^2)z}{(z - \omega_0\, e^{j\theta})\,(z - \omega_0\, e^{-j\theta})} \right)\left(\frac{z}{z-1} \right)$$

$$= \frac{z}{z-1} - \frac{(1 - 2\zeta\omega_0 + \omega_0^2)}{2j\sin\theta}\left(\frac{e^{j\theta}}{1 - \omega_0\, e^{j\theta}}\, \frac{z}{z - \omega_0\, e^{j\theta}} - \frac{e^{-j\theta}}{1 - \omega_0\, e^{-j\theta}}\, \frac{z}{z - \omega_0\, e^{-j\theta}} \right)$$

Therefore

$$y(k) = \mathcal{Z}^{-1}[Y(z)]$$

$$= 1 - \frac{(1 - 2\zeta\omega_0 + \omega_0^2)\omega_0^k}{2j\sin\theta}\left(\frac{1}{1 - \omega_0\, e^{j\theta}}\, e^{j(k+1)\theta} - \frac{1}{1 - \omega_0\, e^{-j\theta}}\, e^{-j(k+1)\theta} \right) \text{ for } k = 0, 1, 2, \ldots$$

The term $1/(1 - \omega_0 e^{j\theta})$ can be expressed in polar form, as follows:

$$\frac{1}{1 - \omega_0\, e^{j\theta}} = \frac{1}{(1 - \omega_0\cos\theta) - j\omega_0\sin\theta} = r\, e^{j\phi}$$

where

$$r = \frac{1}{\sqrt{1 - 2\omega_0\cos\theta + \omega_0^2}} \quad \text{and} \quad \tan\phi = \frac{\omega_0\sin\theta}{1 - \omega_0\cos\theta}$$

Hence, the output $y(k)$ may be written as

$$y(k) = 1 - \frac{(1 - 2\zeta\omega_0 + \omega_0^2)\omega_0^k}{2j\sin\theta}\, (r\, e^{j[(k+1)\theta + \phi]} - r\, e^{-j[(k+1)\theta + \phi]})$$

Using the relation $e^{j\phi} = \cos\phi + j\sin\phi$ and that $\zeta = \cos\theta$, we arrive at the final form

$$y(k) = 1 - \frac{\sqrt{1 - 2\zeta\omega_0 + \omega_0^2}}{\sin\theta}\, \omega_0^k \sin[(k+1)\theta + \phi] \quad \text{for} \quad k = 0, 1, 2, \ldots$$

The foregoing expression for $y(k)$ converges for $|\omega_0| < 1$. As in the previous case, the rate with which $y(k)$ converges to the value 1 depends on the rate with which ω_0^k approaches zero. For ω_0 close to $+1$ or -1, the term ω_0^k converges to zero slowly, whereas for ω_0 close to zero, the term ω_0^k converges to zero faster. Figure 3.13 shows the underdamped unit step response of the system for two different values of ω_0. These responses are characterized by a decaying sinusoidal oscillation in the transient state.

(c) Overdamped unit step response ($\zeta > 1$)
In this case, the system has two distinct real poles. Following the same procedure as

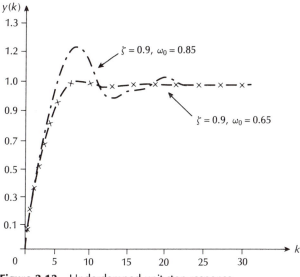

Figure 3.13   Underdamped unit step response.

in cases (a) and (b), we arrive at the following expression for the overdamped unit response:

$$y(k) = 1 + \frac{1}{2\omega_0\sqrt{\zeta^2 - 1}} \; [(\zeta\omega_0 - 1 - \omega_0\sqrt{\zeta^2 - 1})(\zeta\omega_0 + \omega_0\sqrt{\zeta^2 - 1})^{k+1}$$

$$- (\zeta\omega_0 - 1 + \omega_0\sqrt{\zeta^2 - 1})(\zeta\omega_0 - \omega_0\sqrt{\zeta^2 - 1})^{k+1}] \quad \text{for} \quad k = 0, 1, 2, \dots$$

For the convergence of $y(k)$, the parameters ζ and ω_0 must satisfy the following conditions:

$$|\zeta\omega_0 - \omega_0\sqrt{\zeta^2 - 1}| < 1 \quad \text{and} \quad |\zeta\omega_0 + \omega_0\sqrt{\zeta^2 - 1}| < 1$$

To fulfil the above requirements for a given ζ, the parameter ω_0 must be chosen such that

$$|\omega_0| < \frac{1}{\zeta + \sqrt{\zeta^2 - 1}}$$

Figure 3.14 shows the overdamped unit step response of the system for $\zeta = 2$ and $\omega_0 = 0.25$.

The results presented for the first- and second-order difference equations in Examples 3.2.1 and 3.2.2 can easily be extended to the case of an nth-order difference equation of the form

$$y(k) = \sum_{i=0}^{m} b_i u(k - i) - \sum_{i=1}^{n} a_i y(k - i) \qquad (3.11)$$

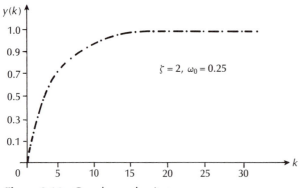

Figure 3.14   Overdamped unit step response.

with initial conditions $y(-1)$, $y(-2), ..., y(-n)$ and under the assumption that $u(k) = 0$ for $k < 0$. Take the Z-transform of equation (3.11) to yield

$$Y(z) = \sum_{i=0}^{m} b_i z^{-i} U(z) - \sum_{i=1}^{n} a_i [z^{-i} Y(z) + y(-i) + y(-i + 1)z^{-1} + ... + y(-1)z^{-(i-1)}]$$

(3.12)

Equation (3.12) can be used for the determination of $y(k)$ (for more examples see Section 2.5).

2. Analysis based on the transfer function

The input $U(z)$, the output $Y(z)$ and the transfer function $H(z)$ are related by the equation

$$Y(z) = H(z)U(z)$$

(3.13)

Hence

$$y(k) = \mathscr{L}^{-1}[Y(z)] = \mathscr{L}^{-1}[H(z)U(z)]$$

(3.14)

3. Analysis based on the impulse response

The input $u(k)$, the output $y(k)$ and the impulse response $h(k)$ are related via the following convolution relation:

$$y(k) = u(k) * h(k) = \sum_{i=-\infty}^{\infty} u(i)h(k - i)$$

(3.15)

If the system is causal, i.e. if $h(k) = 0$ for $k < 0$, then relation (3.15) becomes

$$y(k) = \sum_{i=0}^{\infty} u(i)h(k-i) = \sum_{i=0}^{\infty} u(k-i)h(i) \tag{3.16}$$

If both the system and the input signal are causal, i.e. if $h(k) = 0$ and $u(k) = 0$ for $k < 0$, then equation (3.16) becomes

$$y(k) = \sum_{i=0}^{k} h(i)u(k-i) = \sum_{i=0}^{k} h(k-i)u(i) \tag{3.17}$$

The values $y(0)$, $y(1)$, ... of the output $y(k)$ can be calculated from equation (3.17) as follows:

$$y(0) = h(0)u(0)$$
$$y(1) = h(0)u(1) + h(1)u(0)$$
$$y(2) = h(0)u(2) + h(1)u(1) + h(2)u(0)$$
$$\vdots$$
$$y(k) = h(0)u(k) + h(1)u(k-1) + \cdots + h(k)u(0)$$

or more compactly as

$$\mathbf{y} = \mathbf{Hu} = \mathbf{Uh} \tag{3.18}$$

where

$$\mathbf{y} = \begin{bmatrix} y(0) \\ y(1) \\ \vdots \\ y(k) \end{bmatrix}, \quad \mathbf{u} = \begin{bmatrix} u(0) \\ u(1) \\ \vdots \\ u(k) \end{bmatrix} \quad \text{and} \quad \mathbf{h} = \begin{bmatrix} h(0) \\ h(1) \\ \vdots \\ h(k) \end{bmatrix}$$

$$\mathbf{H} = \begin{bmatrix} h(0) & 0 & \cdots & 0 \\ h(1) & h(0) & \cdots & 0 \\ \vdots & \vdots & & \vdots \\ h(k) & h(k-1) & \cdots & h(0) \end{bmatrix} \quad \text{and} \quad \mathbf{U} = \begin{bmatrix} u(0) & 0 & \cdots & 0 \\ u(1) & u(0) & \cdots & 0 \\ \vdots & \vdots & & \vdots \\ u(k) & u(k-1) & \cdots & u(0) \end{bmatrix}$$

■ **REMARK 3.2.1**   Equation (3.18) can be used for the determination of the impulse response $h(k)$ based on the input $u(k)$ and the output $y(k)$. Indeed, from equation (3.18) we have that

$$\mathbf{h} = \mathbf{U}^{-1}\mathbf{y} \quad \text{if} \quad u(0) \neq 0 \tag{3.19}$$

The above procedure is called **deconvolution** (since it is the reverse of convolution) and constitutes a simple identification method for a discrete-time system. For more on the issue of identification see Chapter 8.

---

4. Analysis based on the state-space equations

Consider the state-space equations (3.8). From the first equation, i.e. from the equation

$$\mathbf{x}(k+1) = \mathbf{Ax}(k) + \mathbf{Bu}(k)$$

we have that

for $k = 0$: $\mathbf{x}(1) = \mathbf{Ax}(0) + \mathbf{Bu}(0)$

for $k = 1$: $\mathbf{x}(2) = \mathbf{Ax}(1) + \mathbf{Bu}(1) = \mathbf{A}[\mathbf{Ax}(0) + \mathbf{Bu}(0)] + \mathbf{Bu}(1)$
$$= \mathbf{A}^2\mathbf{x}(0) + \mathbf{ABu}(0) + \mathbf{Bu}(1)$$

for $k = 2$: $\mathbf{x}(3) = \mathbf{Ax}(2) + \mathbf{Bu}(2) = \mathbf{A}[\mathbf{A}^2\mathbf{x}(0) + \mathbf{ABu}(0) + \mathbf{Bu}(1)] + \mathbf{Bu}(2)$
$$= \mathbf{A}^3\mathbf{x}(0) + \mathbf{A}^2\mathbf{Bu}(0) + \mathbf{ABu}(1) + \mathbf{Bu}(2)$$

If we continue this procedure for $k = 3, 4, 5, \ldots$ we arrive at the following general expression for $\mathbf{x}(k)$:

$$\mathbf{x}(k) = \mathbf{A}^k\mathbf{x}(0) + \mathbf{A}^{k-1}\mathbf{Bu}(0) + \mathbf{A}^{k-2}\mathbf{Bu}(1) + \cdots + \mathbf{ABu}(k-2) + \mathbf{Bu}(k-1)$$

or

$$\mathbf{x}(k) = \mathbf{A}^k\mathbf{x}(0) + \sum_{i=0}^{k-1} \mathbf{A}^{k-i-1}\mathbf{Bu}(i) \tag{3.20}$$

According to equation (3.8b), the output vector $\mathbf{y}(k)$ is

$$\mathbf{y}(k) = \mathbf{Cx}(k) + \mathbf{Du}(k)$$

or

$$\mathbf{y}(k) = \mathbf{CA}^k\mathbf{x}(0) + \mathbf{C}\sum_{i=0}^{k-1} \mathbf{A}^{k-i-1}\mathbf{Bu}(i) + \mathbf{Du}(k) \tag{3.21}$$

where use was made of (3.20).

The matrix $\mathbf{A}^k$ is called the **fundamental** or **transition** matrix of system (3.8) and it is denoted as follows:

$$\mathbf{\Phi}(k) = \mathbf{A}^k \tag{3.22}$$

The matrix $\mathbf{\Phi}(k)$ is analogous to the matrix $\mathbf{\Phi}(t)$ of continuous-time systems (see Table 3.2).

The state vector $\mathbf{x}(k)$ may also be calculated from equation (3.8a) using the Z-transform as follows. Take the Z-transform of both sides of the equation to yield

$$z\mathbf{X}(z) - z\mathbf{x}(0) = \mathbf{AX}(z) + \mathbf{BU}(z)$$

or

$$\mathbf{X}(z) = z(z\mathbf{I} - \mathbf{A})^{-1}\mathbf{x}(0) + (z\mathbf{I} - \mathbf{A})^{-1}\mathbf{BU}(z)$$

Table 3.2 Comparison of the description methods between continuous-time and discrete-time systems.

Description method	Continuous-time system	Discrete-time system
State-space equations	$\dot{\mathbf{x}}(t) = \mathbf{F}\mathbf{x}(t) + \mathbf{G}\mathbf{u}(t)$ $\mathbf{y}(t) = \mathbf{C}\mathbf{x}(t) + \mathbf{D}\mathbf{u}(t)$	$\mathbf{x}(k+1) = \mathbf{A}\mathbf{x}(k) + \mathbf{B}\mathbf{u}(k)$ $\mathbf{y}(k) = \mathbf{C}\mathbf{x}(k) + \mathbf{D}\mathbf{u}(k)$
Transition matrix	$\boldsymbol{\Phi}(t) = e^{\mathbf{F}t}$	$\boldsymbol{\Phi}(k) = \mathbf{A}^k$
L/Z-transform of transition matrix	$\boldsymbol{\Phi}(s) = (s\mathbf{I} - \mathbf{F})^{-1}$	$\boldsymbol{\Phi}(z) = z(z\mathbf{I} - \mathbf{A})^{-1}$
Transfer function matrix	$\mathbf{H}(s) = \mathbf{C}\boldsymbol{\Phi}(s)\mathbf{G} + \mathbf{D}$	$\mathbf{H}(z) = z^{-1}\mathbf{C}\boldsymbol{\Phi}(z)\mathbf{B} + \mathbf{D}$
Impulse response matrix	$\mathbf{H}(t) = \mathbf{C}\boldsymbol{\Phi}(t)\mathbf{G} + \mathbf{D}\delta(t)$	$\mathbf{H}(k) = \mathbf{C}\boldsymbol{\Phi}(k-1)\mathbf{B}$ for $k \geq$ $\mathbf{H}(k) = \mathbf{D}$ for $k = 0$

Taking the inverse Z-transform, we have

$$\mathbf{x}(k) = \mathbf{A}^k\mathbf{x}(0) + \mathbf{A}^{k-1} * \mathbf{B}\mathbf{u}(k) = \mathbf{A}^k\mathbf{x}(0) + \sum_{i=0}^{k-1} \mathbf{A}^{k-i-1}\mathbf{B}\mathbf{u}(i) \tag{3.23}$$

Equation (3.23) is in agreement with equation (3.20), as expected. It is evident that the state transition matrix can also be expressed as

$$\boldsymbol{\Phi}(k) = \mathbf{A}^k = \mathscr{Z}^{-1}[z(z\mathbf{I} - \mathbf{A})^{-1}] \tag{3.24}$$

A comparison between the description methods used for continuous-time and discrete-time systems is shown in Table 3.2.

■ **EXAMPLE 3.2.3** Find the transition matrix, the state and the output vector of a discrete-time system with zero initial conditions with $u(k) = \beta(k)$ and

$$\mathbf{A} = \begin{bmatrix} 0 & 1 \\ -2 & 3 \end{bmatrix}, \quad \mathbf{b} = \begin{bmatrix} 0 \\ 1 \end{bmatrix} \quad \text{and} \quad \mathbf{c} = \begin{bmatrix} 1 \\ 1 \end{bmatrix}$$

■ **SOLUTION** We have

$$\boldsymbol{\Phi}(z) = \mathscr{Z}[\boldsymbol{\Phi}(k)] = z(z\mathbf{I} - \mathbf{A})^{-1} = \frac{1}{z^2 - 3z + 2} \begin{bmatrix} z(z-3) & z \\ -2z & z^2 \end{bmatrix}$$

Hence

$$\Phi(k) = \begin{bmatrix} \mathscr{L}^{-1}\left[\dfrac{z(z-3)}{z^2-3z+2}\right] & \mathscr{L}^{-1}\left[\dfrac{z}{z^2-3z+2}\right] \\[3mm] \mathscr{L}^{-1}\left[\dfrac{-2z}{z^2-3z+2}\right] & \mathscr{L}^{-1}\left[\dfrac{z^2}{z^2-3z+2}\right] \end{bmatrix}$$

Using the Z-transform pairs given in Appendix B, we obtain

$$\Phi(k) = \begin{bmatrix} 2-2^k & 2^k-1 \\ 2(1-2^k) & 2^{k+1}-1 \end{bmatrix}$$

We can check these results as follows. Since $\Phi(k) = \mathbf{A}^k$, it follows that $\Phi(0) = \mathbf{I}$ and $\Phi(1) = \mathbf{A}$. Moreover, from the initial value theorem it follows that $\lim_{z \to \infty} \Phi(z) = \Phi(0)$. Indeed

$$\Phi(0) = \lim_{z \to \infty} \Phi(z) = \begin{bmatrix} 1 & 0 \\ 0 & 1 \end{bmatrix} = \mathbf{I}$$

Since the initial conditions are zero, the state vector may be calculated as follows:

$$\mathbf{X}(z) = (z\mathbf{I} - \mathbf{A})^{-1}\mathbf{B}U(z) = \frac{\Phi(z)}{z}\mathbf{B}U(z) = \frac{1}{(z-1)^2(z-2)}\begin{bmatrix} z \\ z^2 \end{bmatrix}$$

Using the Z-transform pairs given in Appendix B, we obtain

$$\mathbf{x}(k) = \mathscr{L}^{-1}[\mathbf{X}(z)] = \begin{bmatrix} 2^k-k-1 \\ 2(2^k-1)-k \end{bmatrix}$$

From this relation, it follows that $\mathbf{x}(0) = \mathbf{0}$. Finally, the output of the system is given by:

$$y(k) = \mathbf{c}^\mathrm{T}\mathbf{x}(k) = 2^k - k - 1 + (2)2^k - 2 - k = (3)2^k - 2k - 3$$

■ **REMARK 3.2.2** When the initial conditions hold for $k = k_0$ the above results take on the following general forms:

$$\Phi(k, k_0) = \mathbf{A}^{k-k_0} \tag{3.25a}$$

$$\mathbf{x}(k) = \Phi(k - k_0)\mathbf{x}(k_0) + \sum_{i=k_0}^{k-1} \Phi(k - i - 1)\mathbf{B}\mathbf{u}(i) \tag{3.25b}$$

$$\mathbf{y}(k) = \mathbf{C}\Phi(k - k_0)\mathbf{x}(k_0) + \sum_{i=k_0}^{k-1} \mathbf{C}\Phi(k - i - 1)\mathbf{B}\mathbf{u}(i) + \mathbf{D}\mathbf{u}(k) \tag{3.25c}$$

3.2.4 *Description and analysis of linear time-varying discrete-time systems*

When the discrete-time system is linear and time varying, then the state-space equations have the form

$$\mathbf{x}(k+1) = \mathbf{A}(k)\mathbf{x}(k) + \mathbf{B}(k)\mathbf{u}(k) \tag{3.26a}$$

$$\mathbf{y}(k) = \mathbf{C}(k)\mathbf{x}(k) + \mathbf{D}(k)\mathbf{u}(k) \tag{3.26b}$$

where $\mathbf{A}(k)$, $\mathbf{B}(k)$ and $\mathbf{C}(k)$, $\mathbf{D}(k)$ are matrices which vary with k. The transition matrix of equations (3.26) is denoted as $\mathbf{\Phi}(k, k_0)$ and is given by the following equation:

$$\mathbf{\Phi}(k, k_0) = \begin{cases} \mathbf{I} & \text{for } k = k_0 \\ \prod\limits_{i=k_0}^{k-1} \mathbf{A}(i) & \text{for } k > k_0 \end{cases} \tag{3.27}$$

The solution of equation (3.26a) is given by the following relation:

$$\mathbf{x}(k) = \mathbf{\Phi}(k, k_0)\mathbf{x}(k_0) + \sum_{i=k_0}^{k-1} \mathbf{\Phi}(k, i+1)\mathbf{B}(i)\mathbf{u}(i) \tag{3.28a}$$

The output vector $\mathbf{y}(k)$ is then

$$\mathbf{y}(k) = \mathbf{C}(k)\mathbf{\Phi}(k, k_0)\mathbf{x}(k_0) + \sum_{i=k_0}^{k-1} \mathbf{C}(k)\mathbf{\Phi}(k, i+1)\mathbf{B}(i)\mathbf{u}(i) + \mathbf{D}(k)\mathbf{u}(k) \tag{3.28b}$$

It is noted that equations (3.25b and c) can be derived from equations (3.28a and b) if, instead of (k, k_0), we substitute the difference $(k - k_0)$. Similar results can be derived for the case of continuous-time systems [12].

3.3 Description and analysis of sampled-data systems

3.3.1 *Introduction to D/A and A/D converters*

As we have already noted in Chapter 1, in modern control systems a continuous-time system is usually controlled using a digital computer (Figure 1.1). As a result, the closed-loop system involves continuous-time, as well as discrete-time, subsystems. In order to have a common base for the study of the closed-loop system, it is logical to use the same mathematical model for both continuous-time and discrete-time systems. The mathematical model uses difference equations. This approach **unifies** the study of closed-loop systems. Furthermore, it **facilitates** the study of closed-loop systems, since well-known methods and results, such as transfer functions, stability criteria, controller design techniques, etc., may be extended to cover this case.

A practical problem which we come across in such systems is that the output of a discrete-time system, which is a discrete-time signal, may be the input to a continuous-time system (in which case, of course, the input ought to be a continuous-time signal). And, vice versa, the output of a continuous-time system, which is a continuous-time signal, could be the input to a discrete-time system (which, of course, ought to be a discrete-time signal). This problem is dealt with by special devices called converters. There are two types of converters: D/A converters which convert the discrete-time signals to continuous-time signals (Figure 3.15a) and A/D converters which convert continuous-time signals to discrete-time signals (Figure 3.15b). The constant T is the sampling period. It is noted that before the discrete-time signal $y(kT)$ of the A/D converter is fed into a digital computer, it is first converted to a digital signal with the help of a device called a quantizer. The digital signal is a sequence of 0 and 1 digits (see also Figure 1.1).

3.3.2 Hold circuits

A D/A converter is actually a hold circuit whose output $y(t)$ is a piecewise constant function. Specifically, the operation of the hold circuit (Figure 3.15a) is described by the following equations:

$$y(t) = u(kT) \quad \text{for } kT \leqslant t < (k+1)T \tag{3.29a}$$

or

$$y(kT + \xi) = u(kT) \quad \text{for } 0 \leqslant \xi < T \tag{3.29b}$$

We will show that the operation of a hold circuit, described by equations (3.29a and b), may be equivalently described (from a mathematical point of view) by the

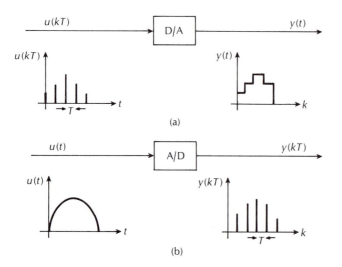

Figure 3.15   The operation of (a) D/A and (b) A/D converters.

idealized hold circuit shown in Figure 3.16(b). Here, it is assumed that the signal $u(t)$ goes through an ideal sampler δ_T, whose output $u^*(t)$ is given by the relation

$$u^*(t) = \sum_{k=0}^{\infty} u(kT)\delta(t - kT) \tag{3.30}$$

Clearly, $u^*(t)$ is a sequence of impulse functions. Applying the Laplace transform on $u^*(t)$ we obtain

$$U^*(s) = \sum_{k=0}^{\infty} u(kT)e^{-skT} \tag{3.31}$$

Therefore, the output $Y(s)$ is

$$Y(s) = \left(\frac{1 - e^{-sT}}{s}\right) \sum_{k=0}^{\infty} u(kT)e^{-skT} \tag{3.32}$$

We would have arrived at the same result as in equation (3.32) if we had calculated $Y(s)$ for the hold circuit shown in Figure 3.16(a). Indeed, since the impulse response $h(t)$ of the hold circuit is a gate function (see Figure 2.3), i.e. $h(t) = \beta(t) - \beta(t - T)$, the output $y(t)$ should be the convolution of $u(kT)$ and $h(t)$, i.e.

$$y(t) = u(kT) * h(t) = \sum_{k=0}^{\infty} u(kT)[\beta(t - kT) - \beta(t - kT - T)] \tag{3.33}$$

Taking the Laplace transform of equation (3.33) we have

$$Y(s) = \sum_{k=0}^{\infty} u(kT)\left(\frac{e^{-skT} - e^{-s(k+1)T}}{s}\right) = \left(\frac{1 - e^{-sT}}{s}\right) \sum_{k=0}^{\infty} u(kT)e^{-skT} \tag{3.34}$$

Equations (3.34) and (3.32) are identical. Therefore, both configurations in Figures 3.16(a) and (b) are equivalent with regard to the output $y(t)$. It is clear that the device shown in Figure 3.16(b) cannot be realized in practice and is used only because it facilitates the mathematical description of the hold circuit.

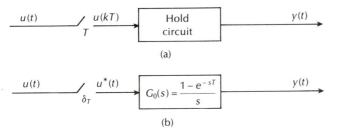

Figure 3.16 Hold circuit block diagram: (a) hold circuit; (b) idealized hold circuit.

It is noted that the foregoing hold circuit is a zero-order hold circuit. There are hold circuits of higher order (first, second, etc.), which are sometimes but rarely used in practice, and we will not deal with them here. We will only refer briefly to the first-order hold circuit. The output of this circuit is not piecewise constant, as in the case of the zero-order hold circuit (Figure 3.17a), but piecewise linear (Figure 3.17b). The slope of each line is given by $\{u(kT) - u[(k-1)T]\}/T$. The first-order hold circuit can be described by the following relation:

$$y(kT + \xi) = u(kT) + \left(\frac{u(kT) - u[(k-1)T]}{T}\right)\xi \quad \text{for } 0 \leq \xi \leq T$$

It has the following transfer function:

$$G_1(s) = \left(\frac{Ts+1}{s}\right)\left(\frac{1-e^{-sT}}{s}\right)^2$$

In this book, only the zero-order hold circuit will be used and for convenience its transfer function will be denoted by $G_h(s)$.

3.3.3 Conversion of G(s) to G(z)

In order to convert a continuous-time transfer function $G(s)$ to the transfer function $G(z)$ of the respective discretized system, various techniques have been proposed. In what follows, we briefly present some of the most popular techniques.

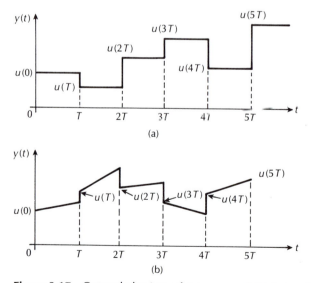

Figure 3.17 Output behaviour of (a) zero- and (b) first-order hold circuits.

1. The backward difference method

For simplicity, we consider the case of a first-order system described by the transfer function

$$G(s) = \frac{Y(s)}{U(s)} = \frac{a}{s+a} \tag{3.35}$$

whose system's differential equation is

$$y^{(1)} = -ay + au \tag{3.36}$$

Integrating both sides of the differential equation from 0 to t, we obtain

$$\int_0^t \frac{dy}{dt}\,dt = -a\int_0^t y\,dt + a\int_0^t u\,dt$$

Suppose that we want to determine the values of the output $y(t)$ at the sampling instants, i.e. at the points where $t = kT$. Then, the integral equation above becomes

$$\int_0^{kT} \frac{dy}{dt}\,dt = -a\int_0^{kT} y\,dt + a\int_0^{kT} u\,dt$$

Hence, we have

$$y(kT) - y(0) = -a\int_0^{kT} y\,dt + a\int_0^{kT} u\,dt \tag{3.37}$$

Substituting kT by $(k-1)T$ in this equation, we obtain

$$y[(k-1)T] - y(0) = -a\int_0^{(k-1)T} y\,dt + a\int_0^{(k-1)T} u\,dt \tag{3.38}$$

Subtracting equation (3.38) from (3.37) we further obtain

$$y(kT) - y[(k-1)T] = -a\int_{(k-1)T}^{kT} y\,dt + a\int_{(k-1)T}^{kT} u\,dt \tag{3.39}$$

Both terms on the right-hand side of (3.39) may be calculated approximately in various ways. If the approximation is done as shown in Figure 3.18(a), i.e. by applying the backward difference method, then equation (3.39) takes the form

$$y(kT) = y[(k-1)T] - aT[y(kT) - u(kT)] \tag{3.40}$$

Obviously, equation (3.40) is the equivalent difference equation of the differential equation (3.36). To find $G(z)$, we only need to take the Z-transform of equation (3.40) to yield

$$G(z) = \frac{Y(z)}{U(z)} = \frac{a}{[(1 - z^{-1})/T] + a} \tag{3.41}$$

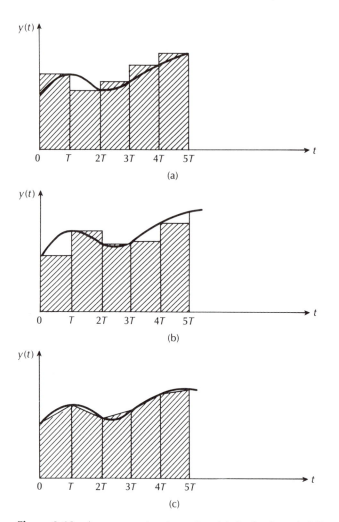

Figure 3.18 Area approximation using (a) the backward difference method, (b) the forward difference method and (c) the trapezoidal rule.

Extending the results of the above example to the general case we arrive at the following procedure for discretizing $G(s)$ using the backward difference method:

$$G(z) = G(s)\big|_{s=(1-z^{-1})/T} \tag{3.42}$$

2. The forward difference method

In this case, the approximation of the two terms on the right-hand side of equation (3.39) is done as shown in Figure 3.18(b). Working in the same way as in the previous case, we arrive at the following result for discretizing $G(s)$ using the forward difference method:

$$G(z) = G(s)\big|_{s = (1 - z^{-1})/Tz^{-1}} \tag{3.43}$$

3. The bilinear transformation method or Tustin transformation method

The Tustin transformation is based on the approximation of the two terms on the right-hand side of equation (3.39) using the trapezoidal rule as shown in Figure 3.18(c). This leads to the following result for discretizing $G(s)$:

$$G(z) = G(s)\big|_{s = (2/T)[(1 - z^{-1})/(1 + z^{-1})]} \tag{3.44}$$

4. The invariant impulse response method

In this case, both $G(s)$ and $G(z)$ present a common characteristic in that their respective impulse functions, $g(t)$ and $g(kT)$, are equal for $t = kT$. This is achieved when

$$G(z) = \mathscr{Z}[g(kT)] \quad \text{where } g(kT) = [\mathscr{L}^{-1}G(s)]_{t = kT} \tag{3.45}$$

where $\mathscr{L}^{-1}$ denotes the inverse Laplace transform.

5. The invariant step response method

In this case, both $G(s)$ and $G(z)$ present a common characteristic in that their step responses, i.e. the response $y(t)$ produced by the excitation $u(t) = 1$ and the response $y(kT)$ produced by the excitation $u(kT) = 1$, are equal for $t = kT$. This is achieved when

$$\mathscr{Z}^{-1}\left[G(z) \frac{1}{1 - z^{-1}}\right] = \left[\mathscr{L}^{-1}\left[G(s) \frac{1}{s}\right]\right]_{t = kT} \tag{3.46}$$

where, obviously, the left-hand side of equation (3.46) is equal to $y(kT)$, whereas the right-hand side is equal to $y(t)$ at $t = kT$. Applying the Z-transform to equation (3.46) we obtain

$$G(z) = (1 - z^{-1})\mathscr{Z}\left[\frac{G(s)}{s}\right] = \mathscr{Z}\left[\frac{1 - e^{-Ts}}{s} G(s)\right] = \mathscr{Z}[G_h(s)G(s)] \tag{3.47}$$

where $G_h(s)$ is the transfer function of the zero-order hold circuit presented in Section 3.3.2.

The invariant step response method is the preferred and perhaps the most correct method, when using a computer and a zero-order hold circuit.

6. Pole–zero matching method

Consider the transfer function

$$G(s) = K_s \frac{(s + \mu_1)(s + \mu_2)\dots(s + \mu_m)}{(s + \pi_1)(s + \pi_2)\dots(s + \pi_n)} \quad \text{where} \quad m \leqslant n \tag{3.48}$$

Then, the pole–zero matching method assumes that $G(z)$ has the general form

$$G(z) = K_z \frac{(z+1)^{n-m}(z+z_1)(z+z_2)...(z+z_m)}{(z+p_1)(z+p_2)...(z+p_n)} \qquad (3.49)$$

where the z_i and the p_i are 'matched' to the respective μ_i and π_i according to the following relations:

$$z_i = -e^{-\mu_i T} \quad \text{and} \quad p_i = -e^{-\pi_i T} \qquad (3.50)$$

The $n-m$ multiple zeros $(z+1)^{n-m}$ which appear in $G(z)$ represent the order difference between the polynomial of the numerator and that of the denominator in equation (3.48). The constant K_z is calculated so as to satisfy particular requirements. For example, when we are interested in the behaviour of a system at low frequencies (and this is the usual case in control systems), K_z is chosen so that $G(s)$ and $G(z)$ are equal for $s=0$ and $z=1$, respectively; that is, the following relation should hold:

$$G(z)|_{z=1} = K_z 2^{n-m} \frac{(1+z_1)(1+z_2)...(1+z_m)}{(1+p_1)(1+p_2)...(1+p_n)} = G(s)|_{s=0} = K_s \frac{\mu_1 \mu_2 ... \mu_m}{\pi_1 \pi_2 ... \pi_n} \qquad (3.51)$$

Using equation (3.51), one can easily determine K_z.

■ **EXAMPLE 3.3.1** For the system with transfer function $G(s) = a/(s+a)$, find all the above six equivalent descriptions of $G(z)$.

■ **SOLUTION** After several simple algebraic calculations, all six descriptions of $G(z)$ were found and are summarized in Table 3.3.

■ **EXAMPLE 3.3.2** Consider a second-order continuous-time system having the following transfer function:

$$H(s) = \frac{b}{s(s+a)}$$

Find $H(z)$ using the invariant impulse response method.

■ **SOLUTION** We have

$$h(t) = \mathcal{L}^{-1} H(s) = \frac{b}{a} \mathcal{L}^{-1}\left[\frac{1}{s} - \frac{1}{(s+a)}\right] = \frac{b}{a}(1 - e^{-at})$$

Hence

$$h(kT) = h(t)|_{t=kT} = \frac{b}{a}(1 - e^{-akT})$$

Table 3.3 Equivalent discrete-time transfer functions $G(z)$ of the continuous-time first-order transfer function $G(s) = a/(s + a)$.

Method	Conversion of $G(s)$ to $G(z)$	Equivalent discrete-time transfer function $G(z)$
Backward difference method	$s = \dfrac{1 - z^{-1}}{T}$	$G(z) = \dfrac{a}{[(1 - z^{-1})/T] + a}$
Forward difference method	$s = \dfrac{1 - z^{-1}}{Tz^{-1}}$	$G(z) = \dfrac{a}{[(1 - z^{-1})/Tz^{-1}] + a}$
Tustin transformation	$s = \dfrac{2}{T}\left(\dfrac{1 - z^{-1}}{1 + z^{-1}}\right)$	$G(z) = \dfrac{a}{\{(2/T)[(1 - z^{-1})/(1 + z^{-1})]\} + a}$
Invariant impulse response method	$G(z) = \mathscr{L}[g(t)]$ where $g(t) = \mathscr{L}^{-1}G(s)$	$G(z) = \dfrac{a}{1 - e^{-aT}z^{-1}}$
Invariant step response method	$G(z) = \mathscr{L}\left[\dfrac{1 - e^{Ts}}{s}G(s)\right]$	$G(z) = \dfrac{(1 - e^{-aT})z^{-1}}{1 - e^{-aT}z^{-1}}$
Pole–zero matching method	$G(z) = K_z\left(\dfrac{z + 1}{z + p_1}\right)$	$G(z) = \left(\dfrac{1 - e^{-aT}}{2}\right)\left(\dfrac{1 + z^{-1}}{1 - e^{-aT}z^{-1}}\right)$

The Z-transform of $h(kT)$ is

$$H(z) = \mathscr{L}[h(kT)] = \frac{b}{a}\sum_{k=0}^{\infty}(1 - e^{-akT})z^{-k} = \frac{b/a}{1 - z^{-1}} - \frac{b/a}{1 - e^{-aT}z^{-1}}$$

or

$$H(z) = \frac{bz^{-1}(1 - e^{-aT})}{a(1 - z^{-1})(1 - e^{-aT}z^{-1})}$$

■ **EXAMPLE 3.3.3** Consider a second-order continuous-time system having the following transfer function:

$$H(s) = \frac{2}{(s + 1)(s + 2)}$$

Find $H(z)$ using the invariant step response method.

■ **SOLUTION**   We have

$$H(z) = \mathcal{L}[G_h(s)H(s)] = \mathcal{L}\left[\left[\frac{1 - e^{-Ts}}{s}\right]H(s)\right] = (1 - z^{-1})\mathcal{L}\left[\frac{H(s)}{s}\right]$$

Using the relation

$$\frac{H(s)}{s} = \frac{2}{s(s + 1)(s + 2)} = \frac{1}{s} - \frac{2}{s + 1} + \frac{1}{s + 2}$$

and the Z-transform tables of Appendix B, we obtain

$$\mathcal{L}\left[\frac{H(s)}{s}\right] = \frac{1}{1 - z^{-1}} - \frac{2}{1 - e^{-T}z^{-1}} + \frac{1}{1 - e^{-2T}z^{-1}}$$

Hence

$$H(z) = (1 - z^{-1})\mathcal{L}\left[\frac{H(s)}{s}\right] = (1 - z^{-1})\left(\frac{1}{1 - z^{-1}} - \frac{2}{1 - e^{-T}z^{-1}} + \frac{1}{1 - e^{-2T}z^{-1}}\right)$$

3.3.4 Conversion of differential state-space equations to difference state-space equations

Consider the continuous-time multi-input multi-output open-loop system shown in Figure 3.19. Let this system be described in state space by the equations

$$\dot{\mathbf{x}}(t) = \mathbf{F}\mathbf{x}(t) + \mathbf{G}\mathbf{m}(t) \tag{3.52a}$$

$$\mathbf{y}(t) = \mathbf{C}\mathbf{x}(t) + \mathbf{D}\mathbf{m}(t) \tag{3.52b}$$

We will show that equations (3.52a and b) may be approximated by difference equations. To this end, consider the piecewise constant excitation vector $\mathbf{m}(t)$ described as

$$\mathbf{m}(t) = \mathbf{u}(kT) \quad \text{for } kT \leqslant t < (k + 1)T \tag{3.53}$$

which is a reasonable assumption, where zero-order hold elements are used. Then, solving equation (3.52a) for $\mathbf{x}(t)$, we have that

$$\mathbf{x}(t) = e^{\mathbf{F}t}\mathbf{x}(0) + \int_0^t e^{\mathbf{F}(t - \xi)}\mathbf{G}\mathbf{m}(\xi) \, d\xi \tag{3.54}$$

According to equation (3.53), $\mathbf{m}(0) = \mathbf{u}(0)$ for $0 \leqslant t < T$ and hence equation (3.54) becomes

$$\mathbf{x}(t) = e^{\mathbf{F}t}\mathbf{x}(0) + \int_0^t e^{\mathbf{F}(t - \xi)}\mathbf{G}\mathbf{u}(0) \, d\xi \quad \text{for} \quad 0 \leqslant t < T \tag{3.55}$$

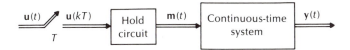

Figure 3.19 Open-loop system with a sampler and a hold circuit.

The state vector $\mathbf{x}(t)$, for $t = T$, will be

$$\mathbf{x}(T) = e^{\mathbf{F}T}\mathbf{x}(0) + \int_0^T e^{\mathbf{F}(T - \xi)}\mathbf{G}\mathbf{u}(0)\,d\xi \tag{3.56}$$

Define

$$\mathbf{A}(T) = e^{\mathbf{F}T} \tag{3.57a}$$

$$\mathbf{B}(T) = \int_0^T e^{\mathbf{F}(T - \xi)}\mathbf{G}\,d\xi = \int_0^T e^{\mathbf{F}\lambda}\mathbf{G}\,d\lambda \quad \text{where } \lambda = T - \xi \tag{3.57b}$$

Then, for $t = T$, equation (3.55) can be simplified as follows:

$$\mathbf{x}(T) = \mathbf{A}(T)\mathbf{x}(0) + \mathbf{B}(T)\mathbf{u}(0) \tag{3.58}$$

Repeating the above procedure for $T \leqslant t < 2T$, $2T \leqslant t < 3T$, etc., we arrive at the following general formula:

$$\mathbf{x}[(k + 1)T] = \mathbf{A}(T)\mathbf{x}(kT) + \mathbf{B}(T)\mathbf{u}(kT)$$

The output equation (3.52b) may therefore be written as

$$\mathbf{y}(kT) = \mathbf{C}\mathbf{x}(kT) + \mathbf{D}\mathbf{u}(kT)$$

Hence, the state-space differential equations (3.52) can be approximated as a system of difference equations as follows:

$$\mathbf{x}[(k + 1)T] = \mathbf{A}(T)\mathbf{x}(kT) + \mathbf{B}(T)\mathbf{u}(kT) \tag{3.59a}$$

$$\mathbf{y}(kT) = \mathbf{C}\mathbf{x}(kT) + \mathbf{D}\mathbf{u}(kT) \tag{3.59b}$$

The state-space equations (3.52) and (3.59) are equivalent only for the time instants $t = kT$ under the constraint that the input vectors $\mathbf{m}(t)$ and $\mathbf{u}(t)$ satisfy the condition (3.53).

■ **REMARK 3.3.1**   The transfer function matrix of the continuous-time system (3.52) is $\mathbf{H}(s) = \mathbf{C}(s\mathbf{I} - \mathbf{F})^{-1}\mathbf{G} + \mathbf{D}$ and the transfer function matrix of the equivalent discrete-time system (3.59) is $\mathbf{G}(z) = \mathbf{C}[z\mathbf{I} - \mathbf{A}(T)]^{-1}\mathbf{B}(T) + \mathbf{D}$. These two matrices are related as follows:

$$\mathbf{G}(z) = \mathscr{L}[G_h(s)\mathbf{H}(s)] = \mathscr{L}\left[\left[\frac{1 - e^{-sT}}{s}\right]\mathbf{H}(s)\right] \tag{3.60}$$

This means that the matrix $\mathbf{G}(z)$ is equivalent to the matrix $\mathbf{H}(z)$ in the sense of the invariant step response method (see equation (3.47)).

■ **REMARK 3.3.2**   Going from the continuous-time state-space description (3.52) to the discrete-time (sampled-data) state-space description (3.59), we need to determine the matrices $\mathbf{A}(T)$ and $\mathbf{B}(T)$ using the definitions (3.57). For the determination of $\mathbf{A}(T)$ we note that it can be easily carried out using the following relation:

$$\mathbf{A}(T) = e^{\mathbf{F}t}\big|_{t=T} = [\mathcal{L}^{-1}[s\mathbf{I} - \mathbf{F}]^{-1}]_{t=T} \tag{3.61a}$$

Equation (3.61a) facilitates the determination of $\mathbf{B}(T)$, since, according to the definition (3.57b) of $\mathbf{B}(T)$, we have that

$$\mathbf{B}(T) = \int_0^T [\mathcal{L}^{-1}[s\mathbf{I} - \mathbf{F}]^{-1}]_{t=\lambda}\,\mathbf{G}\,d\lambda = \int_0^T \mathbf{A}(\lambda)\mathbf{G}\,d\lambda \tag{3.61b}$$

■ **EXAMPLE 3.3.4**   Consider the continuous-time system

$$\dot{\mathbf{x}}(t) = \mathbf{F}\mathbf{x}(t) + \mathbf{g}m(t)$$
$$y(t) = \mathbf{c}^{\mathsf{T}}\mathbf{x}(t)$$

where

$$\mathbf{F} = \begin{bmatrix} -1 & 0 \\ 1 & 0 \end{bmatrix}, \quad \mathbf{g} = \begin{bmatrix} 2 \\ 1 \end{bmatrix} \quad \text{and} \quad \mathbf{c} = \begin{bmatrix} 0 \\ 1 \end{bmatrix}$$

Find the equivalent discrete-time (sampled-data) system, i.e. find the matrix $\mathbf{A}(T)$ and the vector $\mathbf{b}(T)$.

■ **SOLUTION**   We have

$$s\mathbf{I} - \mathbf{F} = \begin{bmatrix} s+1 & 0 \\ -1 & s \end{bmatrix}, \quad (s\mathbf{I} - \mathbf{F})^{-1} = \begin{bmatrix} \dfrac{1}{s+1} & 0 \\ \dfrac{1}{s(s+1)} & \dfrac{1}{s} \end{bmatrix} \quad \text{and} \quad \mathcal{L}^{-1}[s\mathbf{I} - \mathbf{F}]^{-1} = \begin{bmatrix} e^{-t} & 0 \\ 1 - e^{-t} & 1 \end{bmatrix}$$

Therefore, from equation (3.61a) we obtain

$$\mathbf{A}(T) = [\mathcal{L}^{-1}[s\mathbf{I} - \mathbf{F}]^{-1}]_{t=1} - e^{\mathbf{F}T} = \begin{bmatrix} e^{-T} & 0 \\ 1 - e^{-T} & 1 \end{bmatrix}$$

Moreover, from equation (3.61b) we obtain

$$\mathbf{b}(T) = \int_0^T e^{\mathbf{F}\lambda}\mathbf{g}\,d\lambda = \int_0^T \begin{bmatrix} e^{-\lambda} & 0 \\ 1 - e^{-\lambda} & 1 \end{bmatrix}\begin{bmatrix} 2 \\ 1 \end{bmatrix} d\lambda = \int_0^T \begin{bmatrix} 2e^{-\lambda} \\ 3 - 2e^{-\lambda} \end{bmatrix} d\lambda = \begin{bmatrix} 2(1 - e^{-T}) \\ 3T - 2(1 - e^{-T}) \end{bmatrix}$$

■ **EXAMPLE 3.3.5**   Consider an oscillator with transfer function

$$G(s) = \frac{Y(s)}{U(s)} = \frac{\omega^2}{s^2 + \omega^2}$$

A state-space description of the oscillator is of the form $\dot{\mathbf{x}} = \mathbf{F}\mathbf{x} + \mathbf{g}u$, $y = \mathbf{c}^T\mathbf{x}$, where

$$\mathbf{x} = \begin{bmatrix} x_1 \\ x_2 \end{bmatrix} = \begin{bmatrix} y \\ \omega^{-1}y^{(1)} \end{bmatrix}, \quad \mathbf{F} = \begin{bmatrix} 0 & \omega \\ -\omega & 0 \end{bmatrix}, \quad \mathbf{g} = \begin{bmatrix} 0 \\ \omega \end{bmatrix} \quad \text{and} \quad \mathbf{c} = \begin{bmatrix} 1 \\ 0 \end{bmatrix}$$

Find the equivalent discrete-time (sampled-data) system of the form (3.59), i.e. find the matrix $\mathbf{A}(T)$ and the vector $\mathbf{b}(T)$.

■ **SOLUTION**   We have

$$(s\mathbf{I} - \mathbf{F})^{-1} = \begin{bmatrix} \dfrac{s}{s^2 + \omega^2} & \dfrac{\omega}{s^2 + \omega^2} \\[2mm] \dfrac{-\omega}{s^2 + \omega^2} & \dfrac{s}{s^2 + \omega^2} \end{bmatrix} \quad \text{and} \quad e^{\mathbf{F}t} = \begin{bmatrix} \cos\omega t & \sin\omega t \\ -\sin\omega t & \cos\omega t \end{bmatrix}$$

Therefore

$$\mathbf{A}(T) = [\mathscr{L}^{-1}[s\mathbf{I} - \mathbf{F}]^{-1}]_{t=T} = \begin{bmatrix} \cos\omega T & \sin\omega T \\ -\sin\omega T & \cos\omega T \end{bmatrix}$$

$$\mathbf{b}(T) = \int_0^T e^{\mathbf{F}\lambda}\mathbf{g}\,d\lambda = \begin{bmatrix} 1 - \cos\omega t \\ \sin\omega t \end{bmatrix}$$

3.3.5   *Analysis of sampled-data systems*

1. Analysis based on the state-space equations

In order to solve (3.59), we take advantage of the results of Section 3.2.3, since they differ only by the constant T in equations (3.59). We therefore have that the general solution of equation (3.59a) is given by

$$\mathbf{x}(kT) = \mathbf{\Phi}[(k - k_0)T]\mathbf{x}(k_0T) + \sum_{i=k_0}^{k-1} \mathbf{\Phi}[(k - i - 1)T]\mathbf{B}(T)\mathbf{u}(iT) \qquad (3.62a)$$

and the general solution of (3.59b) by

$$\mathbf{y}(kT) = \mathbf{C}\mathbf{\Phi}[(k - k_0)T]\mathbf{x}(k_0T) + \mathbf{C}\sum_{i=k_0}^{k-1} \mathbf{\Phi}[(k - i - 1)T]\mathbf{B}(T)\mathbf{u}(iT) + \mathbf{D}\mathbf{u}(kT) \quad (3.62b)$$

where $\mathbf{\Phi}[(k - k_0)T]$ is the transition matrix, given by the relation

$$\mathbf{\Phi}[(k - k_0)T] = [\mathbf{A}(T)]^{k - k_0}$$

Clearly, if we set $T = 1$ in equations (3.62a and b), then we obtain the formulae (3.20) and (3.21), respectively.

In the case of time-varying systems we have the following state-space description:

$$\mathbf{x}[(k+1)T] = \mathbf{A}(kT)\mathbf{x}(kT) + \mathbf{B}(kT)\mathbf{u}(kT) \tag{3.63a}$$

$$\mathbf{y}(kT) = \mathbf{C}(kT)\mathbf{x}(kT) + \mathbf{D}(kT)\mathbf{u}(kT) \tag{3.63b}$$

The transition matrix $\boldsymbol{\Phi}(k, k_0)$ is given by the equation

$$\boldsymbol{\Phi}[(k, k_0)T] = \begin{cases} \mathbf{I} & \text{for } k = k_0 \\ \displaystyle\prod_{i=k_0}^{k-1} \mathbf{A}(iT) & \text{for } k > k_0 \end{cases} \tag{3.64}$$

The following expression gives the solution of equation (3.63a):

$$\mathbf{x}(kT) = \boldsymbol{\Phi}[(k, k_0)T]\mathbf{x}(k_0 T) + \sum_{i=k_0}^{k-1} \boldsymbol{\Phi}[(k, i+1)T]\mathbf{B}(iT)\mathbf{u}(iT) \tag{3.65a}$$

Consequently, the output vector $\mathbf{y}(kT)$ will be

$$\mathbf{y}(kT) = \mathbf{C}(kT)\boldsymbol{\Phi}[(k, k_0)T]\mathbf{x}(k_0 T) + \sum_{i=k_0}^{k-1} \mathbf{C}(kT)\boldsymbol{\Phi}[(k, i+1)T]\mathbf{B}(iT)\mathbf{u}(iT) + \mathbf{D}(kT)\mathbf{u}(kT) \tag{3.65b}$$

The transition matrix $\boldsymbol{\Phi}[(k, k_0)T]$ has a number of properties. For example,

$$\boldsymbol{\Phi}[(k_0, k_0)T] = \mathbf{I} \tag{3.66a}$$

$$\boldsymbol{\Phi}[(k_2, k_1)T]\boldsymbol{\Phi}[(k_1, k_0)T] = \boldsymbol{\Phi}[(k_2, k_0)T] \tag{3.66b}$$

$$\boldsymbol{\Phi}[(k_1, k_2)T] = \boldsymbol{\Phi}^{-1}[(k_2, k_1)T] \tag{3.66c}$$

For the special case of time-invariant systems the following relations are true:

$$\boldsymbol{\Phi}[(k+n)] = \boldsymbol{\Phi}(k)\boldsymbol{\Phi}(n) \tag{3.67a}$$

$$\boldsymbol{\Phi}(k) = \boldsymbol{\Phi}^{-1}(-k) \tag{3.67b}$$

2. Analysis based on $\mathbf{H}(kT)$

Consider the continuous-time system shown in Figure 3.20, where the two samplers are synchronized. The output vector $\mathbf{y}(kT)$, i.e. the vector $\mathbf{y}(t)$ at the sampling instant $t = kT$, is

$$\mathbf{y}(kT) = \sum_{i=0}^{\infty} \mathbf{H}(kT - iT)\mathbf{u}(iT) \tag{3.68}$$

Equation (3.68) represents, as is already known, a convolution. If we set $T = 1$, then equation (3.68) has the form of the scalar convolution (3.16).

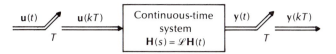

Figure 3.20   A continuous-time system with input and output samplers.

3. Analysis based on **H**(z)

If we take the Z-transform of equation (3.68) we obtain the following expression for the output vector:

$$\mathbf{Y}(z) = \mathbf{H}(z)\mathbf{U}(z) \tag{3.69}$$

where

$$\mathbf{Y}(z) = \mathscr{Z}[\mathbf{y}(kT)], \quad \mathbf{U}(z) = \mathscr{Z}[\mathbf{u}(kT)] \quad \text{and} \quad \mathbf{H}(z) = \mathscr{Z}[\mathbf{H}(kT)] \tag{3.70}$$

■ **EXAMPLE 3.3.6**   For the system of Example 3.3.5, for $T = 1$ and $\omega = \pi/2$, find:
(a) The transition matrix $\mathbf{\Phi}(kT)$.
(b) The state-space vector $\mathbf{x}(kT)$.
(c) The output $y(kT)$.

■ **SOLUTION**   For the given values of ω and T, then **A**, **b** and **c** become

$$\mathbf{A} = \begin{bmatrix} 0 & 1 \\ -1 & 0 \end{bmatrix}, \quad \mathbf{b} = \begin{bmatrix} 1 \\ 1 \end{bmatrix} \quad \text{and} \quad \mathbf{c} = \begin{bmatrix} 1 \\ 0 \end{bmatrix}$$

(a) $\mathbf{\Phi}(z) = \mathscr{Z}[\mathbf{\Phi}(kT)] = z(z\mathbf{I} - \mathbf{A})^{-1} = \dfrac{1}{z^2 + 1}\begin{bmatrix} z^2 & z \\ -z & z^2 \end{bmatrix}$

Therefore

$$\mathbf{\Phi}(kT) = \begin{bmatrix} \mathscr{Z}^{-1}\left[\dfrac{z^2}{z^2 + 1}\right] & \mathscr{Z}^{-1}\left[\dfrac{z}{z^2 + 1}\right] \\[2ex] \mathscr{Z}^{-1}\left[\dfrac{-z}{z^2 + 1}\right] & \mathscr{Z}^{-1}\left[\dfrac{z^2}{z^2 + 1}\right] \end{bmatrix}$$

Using the tables of Z-transform pairs given in Appendix B, we have

$$\mathbf{\Phi}(kT) = \begin{bmatrix} \cos\left(\dfrac{k\pi T}{2}\right) & \sin\left(\dfrac{k\pi T}{2}\right) \\[2ex] -\sin\left(\dfrac{k\pi T}{2}\right) & \cos\left(\dfrac{k\pi T}{2}\right) \end{bmatrix}$$

(b) $\mathbf{X}(z) = (z\mathbf{I} - \mathbf{A})^{-1}\mathbf{B}\mathbf{U}(z) = \dfrac{1}{z^2 + 1}\begin{bmatrix} z & 1 \\ -z & z \end{bmatrix}\begin{bmatrix} 1 \\ 1 \end{bmatrix}\begin{bmatrix} z \\ z - 1 \end{bmatrix} = \dfrac{1}{(z^2 + 1)(z - 1)}\begin{bmatrix} z(z + 1) \\ z(z - 1) \end{bmatrix}$

Hence

$$\mathbf{x}(kT) = \mathbf{Z}^{-1}[X(z)] = \begin{bmatrix} 1 - \cos\left(\dfrac{k\pi T}{2}\right) \\ \sin\left(\dfrac{k\pi T}{2}\right) \end{bmatrix}$$

(c) Finally, the output of the system is given by $\mathbf{y}(kT) = \mathbf{c}^T\mathbf{x}(kT) = 1 - \cos\left(\dfrac{k\pi T}{2}\right)$

3.4 Determination of closed-loop system transfer functions

In a control system where both continuous-time and discrete-time signals are involved, the transfer function may be determined in the usual way. However, care must be taken in handling the continuous-time and discrete-time systems and signals appearing in various points of the overall system. The following examples demonstrate some basic steps in calculating the transfer function of such open- and closed-loop systems.

■ **EXAMPLE 3.4.1**   Find the transfer function of the open-loop system of Figure 3.21.

■ **SOLUTION**   The output $Q(z)$ of the digital computer is $Q(z) = D(z)\Omega(z)$, where $D(z) = \mathscr{L}[d(kT)]$, $Q(z) = \mathscr{L}[q(kT)]$ and $\Omega(z) = \mathscr{L}[\omega(kT)]$. The output $Y(z)$ of the entire system would be $Y(z) = \mathscr{L}[G(s)G_h(s)]Q(z)$ or $Y(z) = \mathscr{L}[G(s)G_h(s)]D(z)\Omega(z)$, where the relation $Q(z) = D(z)\Omega(z)$ was used. Therefore, the transfer function $H(z)$ of the open-loop system will be $H(z) = \mathscr{L}[G(s)G_h(s)]D(z)$

■ **EXAMPLE 3.4.2**   Find the transfer function of the closed-loop system of Figure 3.22.

■ **SOLUTION**   The output $Y(z)$ and the error $E(z)$ are related, according to the results of Example 3.4.1, by the relation $Y(z) = \mathscr{L}[G(s)G_h(s)]D(z)E(z)$. The error $e(t)$ would be $e(t) = \omega(t) - b(t)$ or $e(kT) = \omega(kT) - b(kT)$ at the instants where $t = kT$. The Z-transform of $e(kT)$ is $E(z) = \Omega(z) - B(z)$, where $\Omega(z) = \mathscr{L}[\omega(kT)]$ and

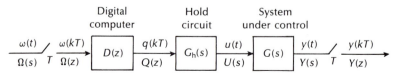

Figure 3.21   The open-loop system of Example 3.4.1.

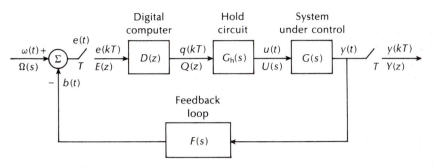

Figure 3.22 The closed-loop system of Example 3.4.2.

$B(z) = \mathcal{L}[b(kT)]$ and $\omega(kT)$ and $b(kT)$ are the input and feedback signals at the sampling points. The signals $B(z)$ and $E(z)$ are related by the equation

$$B(z) = D(z)\mathcal{L}[G(s)G_h(s)F(s)]E(z)$$

Hence $E(z) = \Omega(z) - B(z) = \Omega(z) - D(z)\mathcal{L}[G(s)G_h(s)F(s)]E(z)$ and therefore

$$E(z) = \frac{\Omega(z)}{1 + D(z)\mathcal{L}[G(s)G_h(s)F(s)]}$$

Finally, the transfer function $H(z)$ of the closed system is

$$H(z) = \frac{Y(z)}{\Omega(z)} = \frac{\mathcal{L}[G(s)G_h(s)]D(z)}{1 + D(z)\mathcal{L}[G(s)G_h(s)F(s)]}$$

■ **EXAMPLE 3.4.3** Find the transfer function of the closed-loop system of Figure 3.22 when

$$G_h(s) = \frac{1 - e^{-sT}}{s}, \quad G(s) = \frac{a}{s+a}, \quad D(z) = 1 \quad \text{and} \quad F(s) = 1$$

■ **SOLUTION** Substituting the data of the present example into the formula of $H(z)$ found in the previous example, we arrive at the following result:

$$H(z) = \frac{\mathcal{L}[G(s)G_h(s)]}{1 + \mathcal{L}[G(s)G_h(s)]}$$

For the determination of $\mathcal{L}[G(s)G_h(s)]$ we apply the results of the fourth technique in Section 3.3.3. For the sake of simplicity, we define $\hat{G}(s) = G(s)G_h(s)$ and hence

$$\hat{G}(s) = \left(\frac{1 - e^{-sT}}{s}\right)\left(\frac{a}{s+a}\right) = \frac{1}{s} - \frac{1}{s+a} - \left(\frac{1}{s} - \frac{1}{s+a}\right)e^{-sT}$$

Taking the inverse Laplace transform of $\hat{G}(s)$ yields

$$\hat{g}(t) = \mathcal{L}^{-1}\hat{G}(s) = (1 - e^{-at})\beta(t) - [\beta(t-T) - e^{-a(t-T)}]\beta(t-T)$$

Table 3.4 Block diagrams of closed-loop systems with their respective outputs in the z-domain. At the points where a hold circuit is needed, it is assumed that it is included in the transfer function of the system that follows these points.

Closed-loop system	System output $Y(z)$

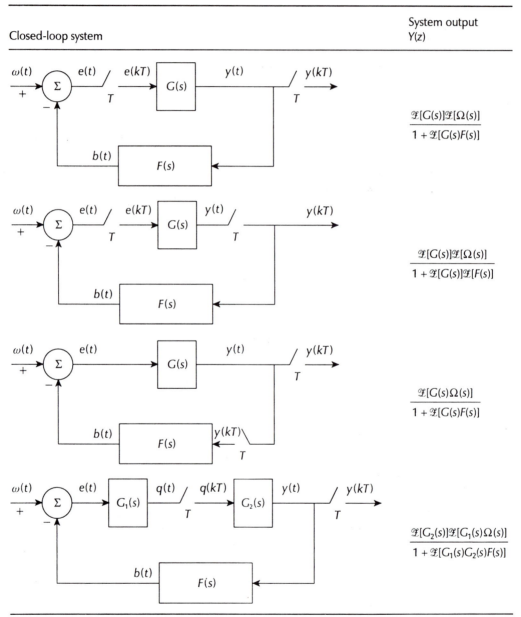

$$\frac{\mathscr{Z}[G(s)]\mathscr{Z}[\Omega(s)]}{1 + \mathscr{Z}[G(s)F(s)]}$$

$$\frac{\mathscr{Z}[G(s)]\mathscr{Z}[\Omega(s)]}{1 + \mathscr{Z}[G(s)]\mathscr{Z}[F(s)]}$$

$$\frac{\mathscr{Z}[G(s)\Omega(s)]}{1 + \mathscr{Z}[G(s)F(s)]}$$

$$\frac{\mathscr{Z}[G_2(s)]\mathscr{Z}[G_1(s)\Omega(s)]}{1 + \mathscr{Z}[G_1(s)G_2(s)F(s)]}$$

At the instants where $t = kT$, the function $\hat{g}(kT)$ becomes

$$\hat{g}(kT) = (1 - e^{-akT})\beta(kT) - [\beta(kT - T) - e^{-a(k-1)T}]\beta(kT - T)$$

Applying equation (2.19b) (i.e. the formula $\mathscr{L}[f(kT - \sigma T)\beta(kT - \sigma T)] = z^{-\sigma}\mathscr{L}[f(kT)]$) to the calculation of $\hat{G}(z) = \mathscr{L}[\hat{g}(kT)]$, we obtain

$$\hat{G}(z) = \mathscr{L}[\hat{g}(kT)] = \mathscr{L}[(1 - e^{-akT})\beta(kT)] - z^{-1}\mathscr{L}[(1 - e^{-akT})\beta(kT)]$$

Since

$$\mathscr{L}[(1 - e^{-akT})\beta(kT)] = \frac{1}{1 - z^{-1}} - \frac{1}{1 - e^{-aT}z^{-1}}$$

we finally have that

$$\hat{G}(z) = \mathscr{L}[\hat{g}(kT)] = \left(\frac{1}{1 - z^{-1}} - \frac{1}{1 - e^{-aT}z^{-1}}\right)(1 - z^{-1}) = \frac{(1 - e^{-aT})z^{-1}}{1 - e^{-aT}z^{-1}}$$

Therefore, the transfer function of the closed-loop system becomes

$$H(z) = \frac{\hat{G}(z)}{1 + \hat{G}(z)} = \frac{(1 - e^{-aT})z^{-1}}{1 + (1 - 2e^{-aT})z^{-1}}$$

We close Section 3.4 with Table 3.4, in which the block diagrams and their respective outputs of some typical closed-loop systems are given. At the points where a hold circuit is needed, we assume that it is included in the transfer function of the system that follows these points.

3.5 Problems

1. Solve the difference equations:

(a) $y(k + 1) - ay(k) = 0$
(b) $y(k + 2) - 0.3y(k + 1) - 0.1y(k) = 0$
(c) $y(k + 2) - y(k + 1) + 0.25y(k) = 0$
(d) $y(k + 2) - y(k + 1) + 0.5y(k) = 0$
(e) $y(k + 1) - 0.6y(k) = 0.4\beta(k)$

with zero initial conditions, where $\beta(k)$ is the unit step sequence.

2. A discrete-time system is described by the state-space equations (3.8), where

$$\mathbf{A} = \begin{bmatrix} 0 & 1 \\ -6 & -5 \end{bmatrix}, \quad \mathbf{b} = \begin{bmatrix} 0 \\ 1 \end{bmatrix}, \quad \mathbf{c} = \begin{bmatrix} 1 \\ 1 \end{bmatrix} \quad \mathbf{D} = 0$$

Find the transition matrix, the transfer function and the response of the system when the input is the unit step sequence.

3. Find the transfer functions of the systems described by the difference equations:

(a) $y(k+2) + 5y(k+1) + 6y(k) = u(k+1)$

(b) $y(k+2) - y(k) = u(k)$

4. Find the values of the responses $y(0)$, $y(1)$, $y(2)$ and $y(3)$ using the convolution method when

(a) $h(k) = 1 - e^{-k}$ and $u(k) = \beta(k)$

(b) $h(k) = 1 - e^{-k}$ and $u(k) = k\beta(k)$

(c) $h(k) = \sin k$ and $u(k) = \beta(k)$

(d) $h(k) = e^{-k}$ and $u(k) = e^{-2k}$

5. Consider the continuous-time system (3.52), where

$$\mathbf{F} = \begin{bmatrix} -1 & 0 \\ 0 & -2 \end{bmatrix}, \quad \mathbf{g} = \begin{bmatrix} 1 \\ -1 \end{bmatrix}, \quad \mathbf{c}^{\mathrm{T}} = \begin{bmatrix} 0 \\ 1 \end{bmatrix} \quad \text{and} \quad \mathbf{D} = 0$$

Discretize the system, i.e. find the difference equations of the system in state space, when the sampling frequency $T = 0.1$ s.

6. Consider the discrete-time filter shown in Figure 3.23. Find the transfer function $Y(z)/U(z)$.

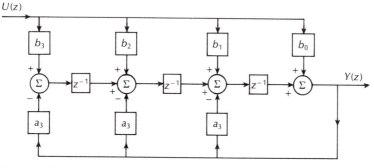

Figure 3.23 Discrete-time filter.

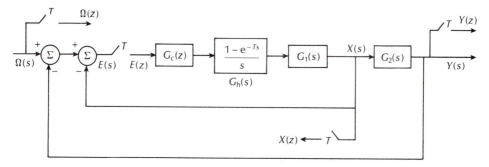

Figure 3.24 A discrete-time control system.

7. Find the transfer function $Y(z)/\Omega(z)$ for the closed-loop system given in Figure 3.24. What is the transfer function relating $X(z)$ and $\Omega(z)$?

8. Find the transfer function $Y(z)/\Omega(z)$ for the system given in Figure 3.25.

9. Consider the systems represented by the block diagrams of Figure 3.26. Find the corresponding discrete-time transfer functions $Y_i(z)/\Omega_i(z)$, $i = 1, 2$. Compare the results.

10. Find the equivalent discrete-time transfer function $G(z)$ of the continuous-time transfer function $G(s) = 1/s(s + 1)$ preceded by a zero-order hold described by $G_h(s) = (1 - e^{-Ts})/s$. Use two approaches, one utilizing the Z-transform tables and the other via a time domain analysis.

11. The block diagram of a digital space-vehicle control system is shown in Figure 3.27, where G and F are constants and J is the vehicle moment of inertia (all in appropriate units). Find the discrete-time transfer function of the closed-loop system.

12. A continuous-time process described by the transfer function K/s is controlled by a digital computer as shown in Figure 3.28.

(a) Find the closed-loop transfer function $Y(z)/\Omega(z)$, the disturbance-to-output transfer function $Y(z)/D(z)$, and the open-loop transfer function $Y(z)/E(z)$.
(b) Obtain the steady state characteristics of the system using the final value theorem.

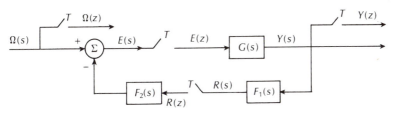

Figure 3.25   A system with continuous- and discrete-time signals.

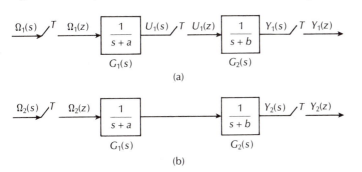

Figure 3.26   Discretized systems: (a) system with sampler between $G_1(s)$ and $G_2(s)$; (b) system without sampler between $G_1(s)$ and $G_2(s)$.

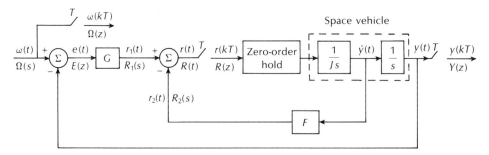

Figure 3.27 A digital space-vehicle control system.

(c) Find the unit step response of the system for $K = 10$, $F = 4$, $K_c = 1$ and $T = 0.01$ s.

13. A simplified state-space model for the attitude control (roll control) of a spacecraft is

$$\begin{bmatrix} \dot{x}_1(t) \\ \dot{x}_2(t) \end{bmatrix} = \begin{bmatrix} 0 & 1 \\ 0 & 0 \end{bmatrix} \begin{bmatrix} x_1(t) \\ x_2(t) \end{bmatrix} + \begin{bmatrix} 0 \\ 1/J \end{bmatrix} u(t) = \mathbf{F}\mathbf{x}(t) + \mathbf{g}u(t)$$

$$y(t) = [1 \ 0] \begin{bmatrix} x_1(t) \\ x_2(t) \end{bmatrix} = \mathbf{c}^T\mathbf{x}(t)$$

where

x_1 = the roll of the spacecraft in rad,
x_2 = the roll rate in rad/s,
u = the control torque about the roll axis produced by the thrusters in N m,
J = the moment of inertia of the vehicle about the roll axis at the vehicle centre of mass in kg m^2.

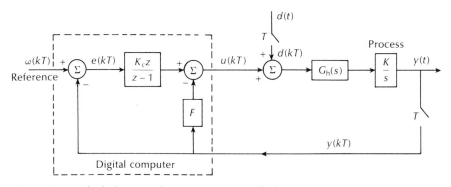

Figure 3.28 Block diagram of a computer-controlled process.

The transfer function relating the roll of the spacecraft to the torque input is

$$G(s) = \frac{Y(s)}{U(s)} = \frac{1}{Js^2}$$

Find the equivalent discrete-time description of the system with sampling period T.

14. Consider the continuous-time system

$$\dot{\mathbf{x}}(t) = \mathbf{A}\mathbf{x}(t) + \mathbf{B}\mathbf{u}(t)$$

$$\mathbf{y}(t) = \mathbf{C}\mathbf{x}(t) + \mathbf{D}\mathbf{u}(t)$$

where $\mathbf{u}(t)$ is the input, $\mathbf{x}(t)$ is the state and $\mathbf{y}(t)$ is the output of the system. Find the discrete-time state description using:

1. the forward difference method;
2. the backward difference method;
3. the Tustin transformation method.

3.6 References

Books

[1] J. Ackermann, *Sampled Data Control Systems*, Springer Verlag, New York, 1985.
[2] K.J. Åström and B. Wittenmark, *Computer Controlled Systems*, Prentice Hall, Englewood Cliffs, New Jersey, 1984.
[3] D.M. Auslander and P. Sagues, *Microprocessors for Measurement and Control*, Osborne/McGraw-Hill, Berkeley, California, 1981.
[4] J.A. Cadzow, *Discrete-Time Systems, An Introduction with Interdisciplinary Applications*, Prentice Hall, Englewood Cliffs, New Jersey, 1973.
[5] J.A. Cadzow and H.R. Martens, *Discrete-Time Systems and Computer Control Systems*, Prentice Hall, Englewood Cliffs, New Jersey, 1970.
[6] G.F. Franklin, J.D. Powell and M.L. Workman, *Digital Control of Dynamic Systems*, Addison-Wesley, London, 1990 (Second Edition).
[7] C.H. Houpis and G.B. Lamont, *Digital Control Systems*, McGraw-Hill, New York, 1985.
[8] E.I. Jury, *Sampled Data Control Systems*, John Wiley, New York, 1958, Robert E. Krieger, Huntington, New York, 1973 (Second Edition).
[9] P. Katz, *Digital Control Using Microprocessors*, Prentice Hall, London, 1981.
[10] B.C. Kuo, *Digital Control Systems*, Holt-Saunders, Tokyo, 1980.
[11] K. Ogata, *Discrete-Time Control Systems*, Prentice Hall, Englewood Cliffs, New Jersey, 1987.
[12] P.N. Paraskevopoulos, *Introduction to Automatic Control Systems*, in preparation.
[13] G.H. Perdikaris, *Computer Controlled Systems, Theory and Applications*, Kluwer, London, 1991.

[14] C.L. Phillips and H.T. Nagle Jr, *Digital Control System Analysis and Design*, Prentice Hall, Englewood Cliffs, New Jersey, 1984.

[15] J.R. Ragazzini and G.F. Franklin, *Sampled Data Control Systems*, McGraw-Hill, Holland, 1980.

4

Stability, controllability and observability

4.1 Introduction

In this chapter we study certain fundamental structural properties of system theory, namely stability, controllability and observability. These properties constitute a basis for understanding the behaviour of a system and are of great theoretical and practical importance.

4.2 Definitions and basic theorems of stability

Stability is the most important structural system property. In designing a control system, our primary goal is to guarantee stability. As soon as this is done, then one seeks to satisfy other design requirements; for example, speed of response, settling time, bandwidth, steady state error, etc. If the system is unstable, then the amplitude of at least one of its state and/or output variables tends to infinity as time increases, even though the input is bounded. This usually results in driving the system beyond its model's limits (i.e. to saturation) and in certain cases the consequences may be even more undesirable and the system may suffer severe damage such as burn-out, breakdown, explosion, etc. For these and other reasons, securing the stability of a control system is our first and most important concern.

Consider the discrete-time system

$$\mathbf{x}(kT + T) = f[\mathbf{x}(kT), kT, \mathbf{u}(kT)] \quad \text{with} \quad \mathbf{x}(k_0 T) = \mathbf{x}_0 \tag{4.1}$$

Let $\mathbf{u}(kT) = \mathbf{0}$, for $k \geq k_0$. Moreover, let $\mathbf{x}(kT)$ and $\tilde{\mathbf{x}}(kT)$ be the solutions of equation (4.1) when the initial conditions are $\mathbf{x}(k_0 T)$ and $\tilde{\mathbf{x}}(k_0 T)$, respectively. Also let the symbol $\| \cdot \|$ represent the Euclidean norm

$$\| \mathbf{x} \| = (x_1^2 + x_2^2 + \cdots + x_n^2)^{1/2}$$

We give the following definitions of stability.

■ **DEFINITION 4.2.1 Stability** The solution $\mathbf{x}(kT)$ of equation (4.1) is stable if for a given $\varepsilon > 0$ there exists a $\delta(\varepsilon, k_0) > 0$ such that all solutions satisfying $\| \mathbf{x}(k_0 T) - \tilde{\mathbf{x}}(k_0 T) \| < \varepsilon$ imply that $\| \mathbf{x}(kT) - \tilde{\mathbf{x}}(kT) \| < \delta$ for all $k \geq k_0$.

■ **DEFINITION 4.2.2 Asymptotic stability** The solution $\mathbf{x}(kT)$ of equation (4.1) is asymptotically stable if it is stable and if $\| \mathbf{x}(kT) - \tilde{\mathbf{x}}(kT) \| \to 0$ as $k \to +\infty$, under the constraint that $\| \mathbf{x}(k_0 T) - \tilde{\mathbf{x}}(k_0 T) \|$ is sufficiently small.

In the case where the system (4.1) is stable in accordance with Definition 4.2.1, the point $\mathbf{x}(k_0 T)$ is called the **equilibrium point**. In the case where the system (4.1) is asymptotically stable, the equilibrium point is the origin $\mathbf{0}$.

4.2.1 Stability of linear time-invariant discrete-time systems

Consider the linear time-invariant discrete-time system

$$\mathbf{x}(kT + T) = \mathbf{A}\mathbf{x}(kT) + \mathbf{B}\mathbf{u}(kT) \quad \text{with} \quad \mathbf{x}(k_0 T) = \mathbf{x}_0$$
$$\mathbf{y}(kT) = \mathbf{C}\mathbf{x}(kT) + \mathbf{D}\mathbf{u}(kT) \tag{4.2}$$

Applying Definition 4.2.1 to the system (4.2) we have the following theorem.

■ **THEOREM 4.2.1** System (4.2) is stable, according to Definition 4.2.1, if and only if the eigenvalues λ_i of the matrix $\mathbf{A}$, i.e. the roots of the polynomial equation $| \lambda \mathbf{I} - \mathbf{A} | = 0$, lie inside the unit circle (i.e. $| \lambda_i | < 1$), or the matrix $\mathbf{A}$ has eigenvalues on the unit circle (i.e. $| \lambda_i | = 1$) of multiplicity 1.

Applying Definition 4.2.2 to the system (4.2) we have the following theorem.

■ **THEOREM 4.2.2** System (4.2) is asymptotically stable if and only if $\lim_{k \to \infty} \mathbf{x}(kT) = \mathbf{0}$ for every $\mathbf{x}(k_0 T)$, when $\mathbf{u}(kT) = \mathbf{0}$ $(k \geq k_0)$.

On the basis of Theorem 4.2.2 we prove the following theorem.

■ **THEOREM 4.2.3** System (4.2) is asymptotically stable if and only if the eigenvalues λ_i of $\mathbf{A}$ are inside the unit circle.

■ **PROOF** The theorem will be proved for the special case where the matrix $\mathbf{A}$ has distinct eigenvalues. The proof of the case where certain or all eigenvalues are repeated is left as an exercise. When $\mathbf{u}(kT) = \mathbf{0}$ $(k \geq k_0)$ the state vector $\mathbf{x}(kT)$ is given by

$$\mathbf{x}(kT) = \mathbf{A}^{k-k_0}\mathbf{x}(k_0 T) \tag{4.3}$$

Let the eigenvalues $\lambda_1, \lambda_2, \ldots, \lambda_n$ of the matrix $\mathbf{A}$ be distinct. Then, according to the Sylvester theorem (see Appendix A), the matrix $\mathbf{A}^k$ can be written as

$$\mathbf{A}^k = \sum_{i=1}^{n} \mathbf{A}_i \lambda_i^k \tag{4.4}$$

where $\mathbf{A}_i$ are special matrices which depend only on $\mathbf{A}$. Substituting equation (4.4) in equation (4.3), we obtain

$$\mathbf{x}(kT) = \sum_{i=1}^{n} \mathbf{A}_i \lambda_i^{k-k_0} \mathbf{x}(k_0 T) \tag{4.5}$$

Hence

$$\lim_{k \to \infty} \mathbf{x}(kT) = \left[\sum_{i=1}^{n} \mathbf{A}_i \mathbf{x}(k_0 T) \left(\lim_{k \to \infty} \lambda_i^{k-k_0} \right) \right] \tag{4.6}$$

From equation (4.6) it is obvious that $\lim_{k \to \infty} \mathbf{x}(kT) = \mathbf{0} \; \forall \mathbf{x}(k_0 T)$, if and only if $\lim_{k \to \infty} \lambda_i^{k-k_0} = 0, \quad \forall i = 1, 2, \ldots, n$, which is true if and only if $|\lambda_i| < 1$ $\forall i = 1, 2, \ldots, n$, where $|\cdot|$ stands for the magnitude of a complex number.

Table 4.1 A comparison of asymptotic stability between continuous-time and disrete-time systems.

	Continuous-time systems	Discrete-time systems
State-space equations	$\dot{\mathbf{x}}(t) = \mathbf{F}\mathbf{x}(t) + \mathbf{G}\mathbf{u}(t)$ $\mathbf{y}(t) = \mathbf{C}\mathbf{x}(t) + \mathbf{D}\mathbf{u}(t)$	$\mathbf{x}[(k+1)T] = \mathbf{A}\mathbf{x}(kT) + \mathbf{B}\mathbf{u}(kT)$ $\mathbf{y}(kT) = \mathbf{C}\mathbf{x}(kT) + \mathbf{D}\mathbf{u}(kT)$
Characteristic equation	$\|\lambda\mathbf{I} - \mathbf{F}\| = 0$	$\|\lambda\mathbf{I} - \mathbf{A}\| = 0$
Stability definition	$\mathrm{Re}\{\lambda_i\} < 0$	$\|\lambda_i\| < 1$
Stability description		

A comparison between linear time-invariant continuous-time systems and linear time-invariant discrete-time systems with respect to asymptotic stability is shown in Table 4.1 (p. 75).

4.2.2 *Bounded-input bounded-output stability*

■ **DEFINITION 4.2.3**   A linear time-invariant system is bounded-input bounded-output (BIBO) stable if for any bounded input a bounded output is produced for every initial condition.

Applying Definition 4.2.3 to system (4.2) we have the following theorem.

■ **THEOREM 4.2.4**   The linear time-invariant system (4.2) is BIBO stable if and only if the poles of the transfer function $\mathbf{H}(z) = \mathbf{C}(z\mathbf{I} - \mathbf{A})^{-1}\mathbf{B} + \mathbf{D}$, before any pole–zero cancellation, are inside the unit circle.

From Definition 4.2.3 we may conclude that asymptotic stability is the strongest, since it implies both stability and BIBO stability. It is easy to give examples showing that stability does not imply BIBO stability and vice versa.

■ **EXAMPLE 4.2.1**   Consider the system (see Example 3.3.5)

$$\mathbf{x}(kT + T) = \begin{bmatrix} \cos \omega T & \sin \omega T \\ -\sin \omega T & \cos \omega T \end{bmatrix} \mathbf{x}(kT) + \begin{bmatrix} 1 - \cos \omega T \\ \sin \omega T \end{bmatrix} u(kT)$$

$$y(kT) = [1 \ 0]\mathbf{x}(kT)$$

Examine the stability of the system according to Definition 4.2.1 and the BIBO stability of the system.

■ **SOLUTION**   The eigenvalues of the system are $\lambda_{1,2} = \cos \omega T \pm \mathrm{j} \sin \omega T$. Here the magnitudes of the eigenvalues are $|\lambda_1| = |\lambda_2| = 1$. When $u(kT) = 0$, then $\|\mathbf{x}(kT)\| = \|\mathbf{A}^k\mathbf{x}(0)\| \le \|\mathbf{A}^k\| \|\mathbf{x}(0)\| \le \|\mathbf{x}(0)\|$. Thus, the system is stable inside the circle $\|\mathbf{x}(0)\|$. To study the BIBO stability of the system, excite the system with a square waveform of frequency ω. The input and output waveforms of the system, when $\omega = 1$ and $T = 0.5$, are shown in Figure 4.1. From Figure 4.1 we observe that the output amplitude increases with time. Although the system accepts a bounded amplitude signal at its input, it produces an unbounded amplitude signal at its output. Therefore, the system is not BIBO stable [1].

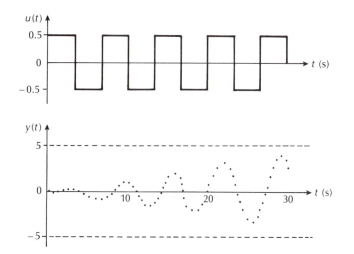

Figure 4.1 Input and output waveforms of the system presented in Example 4.2.1 for zero initial conditions and $\omega = 1$ and $T = 0.5$ [1].

In the material that follows, whenever the term stability is used without further clarification, it refers to asymptotic stability.

4.3 Stability criteria

Various techniques for testing stability have been proposed. The most popular techniques for determining the stability of a discrete-time system are the following [1], [2], [4], [7], [9]:

1. The Routh criterion, using the Möbius transformation
2. The Jury criterion
3. The Lyapunov method
4. The root-locus method
5. The Bode and Nyquist criteria

Techniques 1, 2, and 3 are algebraic and will be presented in this chapter, whereas techniques 4 and 5 are graphical and will be presented in Sections 5.3 and 5.4 of Chapter 5, respectively.

4.3.1 *The Routh criterion using the Möbius transformation*

Consider the polynomial

$$a(z) = a_0 z^n + a_1 z^{n-1} + \cdots + a_n \tag{4.7}$$

The roots of the polynomial are the roots of the equation

$$a(z) = a_0 z^n + a_1 z^{n-1} + \cdots + a_n = 0 \tag{4.8}$$

As stated in Theorem 4.2.3, asymptotic stability is secured if all the roots of the characteristic polynomial lie inside the unit circle. The well-known Routh criterion for continuous-time systems is a simple method for determining if all the roots of an arbitrary polynomial lie in the left-half complex plane without requiring determination of the values of the roots [8]. The Möbius bilinear transformation

$$w = \frac{z+1}{z-1} \quad \text{or} \quad z = \frac{w+1}{w-1} \tag{4.9}$$

maps the unit circle of the z-plane into the left w-plane. Consequently, if the Möbius transformation is applied to equation (4.8), then the Routh criterion may be applied as in the case of continuous-time systems.

■ **EXAMPLE 4.3.1** The characteristic polynomial $a(z)$ of a system is given by $a(z) = z^2 + 0.7z + 0.1$. Investigate the stability of the system.

■ **SOLUTION** Applying the transformation (4.9) to $a(z)$ yields

$$a(w) = \left(\frac{w+1}{w-1}\right)^2 + 0.7\left(\frac{w+1}{w-1}\right) + 0.1 = \frac{(w+1)^2 + 0.7(w^2 - 1) + 0.1(w-1)^2}{(w-1)^2}$$

$$= \frac{1.8w^2 + 1.8w + 0.4}{(w-1)^2}$$

The numerator of $a(w)$ is called the 'auxiliary' characteristic polynomial to which the well-known Routh criterion will be applied. For the present example, the Routh array is

w^2	1.8	0.4
w^1	1.8	0
w^0	0.4	0

The coefficients of the first column have the same sign and according to Routh's criterion the system is stable. We can reach the same result if we factorize $a(z)$ into a product of terms, i.e. $a(z) = (z + 0.5)(z + 0.2)$. The two roots of $a(z)$ are -0.5 and -0.2, which are both inside the unit circle, and hence the system is stable.

4.3.2 The Jury criterion

It is useful to establish criteria which can directly show whether a polynomial $a(z)$ has all its roots inside the unit circle instead of determining its eigenvalues. Such a criterion, which is equivalent to the Routh criterion for continuous-time systems, has been developed by Schur, Cohn and Jury. This criterion is usually called the Jury criterion and is described in detail below.

First, the Jury table is formed for the polynomial $a(z)$, given by equation (4.7), as shown in Table 4.2. The first two rows of the table are the coefficients of the polynomial $a(z)$ presented in forward and reverse order, respectively. The third row is formed by multiplying the second row by $\beta_n = a_n/a_0$ and subtracting the result from the first row. Note that the last element of the third row becomes zero. The fourth row is identical to the third row, but in reverse order. The above procedure is repeated until the $(2n + 1)$th row is reached. The last row consists of only a single element. The following theorem holds.

■ **THEOREM 4.3.1 The Jury stability criterion**   If $a_0 > 0$, then the polynomial $a(z)$ has all its roots inside the unit circle, if and only if all a_0^k, $k = 0, 1, \ldots, n-1$, are positive. If all coefficients a_0^k differ from zero, then the number of negative coefficients a_0^k is equal to the number of roots which lie outside the unit circle.

■ **REMARK 4.3.1**   If all coefficients a_0^k, $k = 1, 2, \ldots, n$, are positive then it can be shown that the condition $a_0^0 > 0$ is equivalent to the following two conditions:

$$a(1) > 0 \tag{4.10a}$$

$$(-1)^n a(-1) > 0 \tag{4.10b}$$

Relations (4.10a and b) present necessary conditions for stability and may therefore be used to check for stability, prior to construction of the Jury table.

■ **EXAMPLE 4.3.2**   Consider the characteristic polynomial $a(z) = z^3 - 1.3z^2 - 0.8z + 1$. Investigate the stability of the system.

■ **SOLUTION**   Condition (4.10a) is examined first. Here $a(1) = 1 - 1.3 - 0.8 + 1 = -0.1$ and since $a(1) < 0$, the necessary condition (4.10a) is not satisfied. Therefore, one or

Table 4.2   The Jury table.

a_0	a_1	$\cdots$	a_{n-1}	a_n	
a_n	a_{n-1}	$\cdots$	a_1	a_0	$\beta_n = a_n/a_0$
a_0^{n-1}	a_1^{n-1}	$\cdots$	a_{n-1}^{n-1}		
a_{n-1}^{n-1}	a_{n-2}^{n-1}	$\cdots$	a_0^{n-1}		$\beta_{n-1} = a_{n-1}^{n-1}/a_0^{n-1}$
$\vdots$					
a_0^0	where $a_i^{k-1} = a_i^k - \beta_k a_{k-i}^k$ and $\beta_k = a_k^k/a_0^k$				

Table 4.3 The Jury table for Example 4.3.3.

1	a_1	a_2	
a_2	a_1	1	
			$\beta_2 = a_2$
$1 - a_2^2$	$a_1(1 - a_2)$		
$a_1(1 - a_2)$	$1 - a_2^2$		
			$\beta_1 = a_1/(1 + a_2)$
$1 - a_2^2 - \dfrac{a_1^2(1 - a_2)}{1 + a_2}$			

more roots of the characteristic polynomial lie outside the unit circle. It is immediately concluded that the system under consideration must be unstable.

■ **EXAMPLE 4.3.3** Consider the second-order characteristic polynomial $a(z) = z^2 + a_1 z + a_2$. Investigate the stability of the system.

■ **SOLUTION** The Jury table is formed as shown in Table 4.3. All the roots of the characteristic polynomial are inside the unit circle if

$$1 - a_2^2 > 0 \quad \text{and} \quad \left(\frac{1 - a_2}{1 + a_2}\right)[(1 + a_2)^2 - a_1^2] > 0$$

which lead to the conditions $-1 < a_2 < 1$, $a_2 > -1 + a_1$ and $a_2 > -1 - a_1$. The stability region of the second-order characteristic polynomial is shown in Figure 4.2.

■ **EXAMPLE 4.3.4** Consider the characteristic polynomial $a(z) = z^3 + Kz^2 + 0.5z + 2$. Find the values of the constant K for which all the roots of the polynomial $a(z)$ lie inside the unit circle.

■ **SOLUTION** The Jury table is formed as shown in Table 4.4. According to Theorem 4.3.1, all a_0^k elements are positive, except for the last element which may become

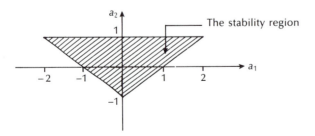

Figure 4.2 The stability region for Example 4.3.3.

Table 4.4   The Jury table for Example 4.3.4.

1	K	0.5	0.5
1	2	K	1

$\beta_3 = 0.5$

$1 - 0.25$	$K - 1$	$0.5 - 0.5K$
$0.5 - 0.5K$	$K - 1$	0.75

$\beta_2 = \frac{2}{3}(1 - K)$

$\frac{1}{3}(1 - K)^2$	$\frac{2}{3}(1 - K)(K - 1) - (K - 1)$	0
$\frac{2}{3}(1 - K)(K - 1) - (K - 1)$	$\frac{1}{3}(1 - K)^2$	

$\beta_1 = 2K + 1$

$\frac{1}{3}(1 - K)^2 - \frac{1}{3}(2K + 1)^2(1 - K)$

positive for certain values or for a range of values of K. This last element in the Jury table may be written as

$$\frac{1}{3}(1 - K)^2 - \frac{1}{3}(2K + 1)^2(1 - K) = \frac{1}{3}(1 - K)[1 - K - (2K + 1)^2] = \frac{1}{3}(1 - K)(-4K^2 - 5K)$$
$$= \frac{1}{3}(1 - K)(-K)(4K + 5)$$

The last product of terms becomes positive if the following inequalities are true: $K > -\frac{5}{4}$ and $K < 0$. Hence, when $-\frac{5}{4} < K < 0$, the characteristic polynomial has all its roots inside the unit circle and the system under consideration is stable.

■ **EXAMPLE 4.3.5**   Find the range of values of K which stabilizes the closed-loop system shown in Figure 4.3.

■ **SOLUTION**   The transfer function of the open system is

$$G(s) = \frac{K}{s(s + 1)} = K\left(\frac{1}{s} - \frac{1}{s + 1}\right)$$

To derive $G(z)$ from $G(s)$, we apply the method of the invariant pulse response (Section 3.3.3). As a result, we have $g(t) = K - Ke^{-t}$, and hence $g(kT) = K - Ke^{-kT}$. Consequently

$$G(z) = \mathscr{L}[g(kT)] = K\left(\frac{z(1 - e^{-T})}{z^2 - (1 + e^{-T})z + e^{-T}}\right)$$

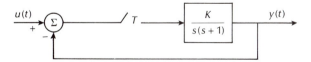

Figure 4.3   The closed-loop system of Example 4.3.5.

The poles of the closed-loop system are the roots of the equation $1 + G(z) = 0$, i.e. they are the roots of the equation

$$a(z) = z^2 - [e^{-T}(1 + K) + 1 - K]z + e^{-T} = 0$$

Method 1: Routh criterion

If we apply the transformation (4.9) to $a(z)$, then the closed-loop characteristic polynomial becomes

$$a(w) = \frac{w^2[K(1 - e^{-T})] + w[2(1 - e^{-T})] + 2(1 + e^{-T}) - K(1 - e^{-T})}{(w - 1)^2}$$

The numerator of $a(w)$ is now the 'auxiliary' characteristic polynomial on which the Routh criterion will be applied. The Routh table for $a(w)$ is

w^2	$K(1 - e^{-T})$	$2(1 + e^{-T}) - K(1 - e^{-T})$
w^1	$2(1 - e^{-T})$	0
w^0	$2(1 + e^{-T}) - K(1 - e^{-T})$	0

Applying the Routh criterion we obtain the inequalities

$$K(1 - e^{-T}) > 0$$

$$2(1 - e^{-T}) > 0$$

$$2(1 + e^{-T}) - K(1 - e^{-T}) > 0$$

Clearly, the second inequality is always true. From the first and third inequalities we deduce the following restrictions on K:

$$K > 0 \quad \text{and} \quad \frac{K}{2} < \frac{(1 + e^{-T})}{(1 - e^{-T})} = \cot\left(\frac{T}{2}\right)$$

These two restrictions, combined together, yield the following relation which K must satisfy so that the closed-loop system is stable:

$$0 < K < 2 \cot(T/2)$$

Method 2: Jury criterion

We form the Jury table for $a(z)$, as shown in Table 4.5. According to the Jury criterion, the conditions which must hold true for the closed-loop system to be stable are

$$1 > 0$$

$$1 - e^{-2T} > 0$$

$$1 - e^{-2T} > \frac{(e^{-T} - 1)^2}{1 - e^{-2T}} [e^{-T}(1 + K) + 1 - K]^2$$

Table 4.5 The Jury table for Example 4.3.5.

1		$-e^{-T}(1 + K) - 1 + K$	e^{-T}
e^{-T}		$-e^{-T}(1 + K) - 1 + K$	1
$1 - e^{-2T}$		$(e^{-T} - 1)[e^{-T}(1 + K) + 1 - K]$	0
$(e^{-T} - 1)[e^{-T}(1 + K) + 1 - K]$		$1 - e^{-2T}$	
$1 - e^{-2T} - \dfrac{(e^{-T} - 1)^2}{1 - e^{-2T}} [e^{-T}(1 + K) + 1 - K]^2$			

From these three inequalities, the first two are always true, whereas the third one becomes

$$(1 - e^{-T})K^2 - 2K(1 - e^{-2T}) < 0$$

This inequality is true if K lies between the roots of the polynomial of the left-hand side of the inequality, which are 0 and $2(1 + e^{-T})/(1 - e^{-T}) = 2 \cot(T/2)$. Thus K must satisfy the condition

$$0 < K < 2 \cot(T/2)$$

As expected, both the Routh and Jury methods lead to the same results.

4.3.3 *Stability in the sense of Lyapunov*

The technique for determining the stability of a system using the Lyapunov method is based on the following theorem. This theorem covers the general case for non-linear systems. The special case of linear time-invariant systems is presented in Theorem 4.3.3 and is illustrated in the examples that follow.

■ **THEOREM 4.3.2** Consider the non-linear system

$$\mathbf{x}[(k + 1)T] = \mathbf{f}[\mathbf{x}(kT)] \tag{4.11}$$

where $\mathbf{x}, \mathbf{f} \in \mathbb{R}^n$. Suppose that there exists a scalar energy-like function $V(\mathbf{x})$ which satisfies the following conditions:

1. $V(\mathbf{x}) > 0, \quad \text{for } \mathbf{x} \neq \mathbf{0}$ \hfill (4.12a)

2. $\Delta V(\mathbf{x}) < 0, \quad \text{for } \mathbf{x} \neq \mathbf{0}$ \hfill (4.12b)

   where $\Delta V[\mathbf{x}(kT)] = V[\mathbf{x}[(k + 1)]T] - V[\mathbf{x}(kT)]$

3. $V(\mathbf{0}) = 0$ (4.12c)

4. $V(\mathbf{x}) \rightarrow +\infty$, as $\|\mathbf{x}\| \rightarrow +\infty$ (4.12d)

Then, system (4.11) is asymptotically stable with equilibrium state $\mathbf{x} = \mathbf{0}$ and $V(\mathbf{x})$ is a **Lyapunov function**.

For the special case of linear time-invariant systems, equation (4.11) takes on the form

$$\mathbf{x}(k+1) = \mathbf{A}\mathbf{x}(k) \tag{4.13}$$

Choose as a Lyapunov function the quadratic (energy-like) form

$$\mathbf{V}[\mathbf{x}(k)] = \mathbf{x}^{\mathrm{T}}(k)\mathbf{P}\mathbf{x}(k) \tag{4.14}$$

where $\mathbf{P}$ is a positive definite Hermitian matrix, i.e. $\mathbf{P}$ is a positive definite real symmetric matrix (see Appendix A). Then

$$\Delta V[\mathbf{x}(k)] = V[\mathbf{x}(k+1)] - V[\mathbf{x}(k)] = \mathbf{x}^{\mathrm{T}}(k+1)\mathbf{P}\mathbf{x}(k+1) - \mathbf{x}^{\mathrm{T}}(k)\mathbf{P}\mathbf{x}(k)$$
$$= [\mathbf{A}\mathbf{x}(k)]^{\mathrm{T}}\mathbf{P}[\mathbf{A}\mathbf{x}(k)] - \mathbf{x}^{\mathrm{T}}(k)\mathbf{P}\mathbf{x}(k) = \mathbf{x}^{\mathrm{T}}(k)[\mathbf{A}^{\mathrm{T}}\mathbf{P}\mathbf{A} - \mathbf{P}]\mathbf{x}(k) \tag{4.15}$$

According to relation (4.12a) presented in Theorem 4.3.2, we choose the Lyapunov function $V[\mathbf{x}(k)]$ to be positive definite. Thus, for system (4.13) to be stable, the change $\Delta V[\mathbf{x}(k)]$ must be negative definite. That is, it must be

$$\Delta V[\mathbf{x}(k)] = -\mathbf{x}^{\mathrm{T}}(k)\mathbf{Q}\mathbf{x}(k) \tag{4.16}$$

where $\mathbf{Q}$ is a positive definite real symmetric matrix. Comparing equations (4.15) and (4.16), we arrive at the relation

$$\mathbf{A}^{\mathrm{T}}\mathbf{P}\mathbf{A} - \mathbf{P} = -\mathbf{Q} \tag{4.17}$$

For linear time-invariant systems, the following theorem holds true.

■ **THEOREM 4.3.3**   The necessary and sufficient condition for the equilibrium state $\mathbf{x} = \mathbf{0}$ of the system (4.13) to be asymptotically stable is the following: for any positive definite Hermitian matrix $\mathbf{Q}$ there must exist a positive definite Hermitian matrix $\mathbf{P}$ which satisfies equation (4.17). The function $\mathbf{x}^{\mathrm{T}}\mathbf{P}\mathbf{x}$ is the Lyapunov function of system (4.13).

It is noted that when asymptotic stability is required, it is sufficient that the matrix $\mathbf{Q}$ is positive definite. Moreover, in the above theorem, we usually choose the matrix $\mathbf{Q}$ to be real and symmetric, so as to facilitate the determination of the matrix $\mathbf{P}$.

■ **EXAMPLE 4.3.6**   Consider the system (4.13), where

$$\mathbf{A} = \begin{bmatrix} 0 & 1 \\ -0.5 & -1 \end{bmatrix}$$

Investigate the stability of the system using the Lyapunov method.

■ **SOLUTION**   For simplicity, choose $\mathbf{Q} = \mathbf{I}_2$. Then equation (4.17) becomes

$$\begin{bmatrix} 0 & -0.5 \\ 1 & -1 \end{bmatrix}\begin{bmatrix} p_{11} & p_{12} \\ p_{21} & p_{22} \end{bmatrix}\begin{bmatrix} 0 & 1 \\ -0.5 & -1 \end{bmatrix} - \begin{bmatrix} p_{11} & p_{12} \\ p_{21} & p_{22} \end{bmatrix} = \begin{bmatrix} -1 & 0 \\ 0 & -1 \end{bmatrix}$$

Solving this equation for p_{11}, $p_{12} = p_{21}$ and p_{22} we obtain

$$\mathbf{P} = \begin{bmatrix} p_{11} & p_{12} \\ p_{21} & p_{22} \end{bmatrix} = \begin{bmatrix} \frac{11}{5} & \frac{8}{5} \\ \frac{8}{5} & \frac{24}{5} \end{bmatrix}$$

Applying the Sylvester criterion (Appendix A), we find that the matrix $\mathbf{P}$ is positive definite. Hence, the system under consideration is stable. The Lyapunov function is

$$V(\mathbf{x}) = \mathbf{x}^T \mathbf{P} \mathbf{x} = \frac{11}{5} x_1^2 + \frac{16}{5} x_1 x_2 + \frac{24}{5} x_2^2$$

■ **EXAMPLE 4.3.7**   Consider the system (4.13), where

$$\mathbf{A} = \begin{bmatrix} 0 & 1 \\ -2 & -3 \end{bmatrix}$$

Investigate the stability of the system using the Lyapunov method.

■ **SOLUTION**   Choose $\mathbf{Q} = \mathbf{I}_2$. Then, equation (4.17) becomes

$$\begin{bmatrix} 0 & -2 \\ 1 & -3 \end{bmatrix}\begin{bmatrix} p_{11} & p_{12} \\ p_{12} & p_{22} \end{bmatrix}\begin{bmatrix} 0 & 1 \\ -2 & -3 \end{bmatrix} - \begin{bmatrix} p_{11} & p_{12} \\ p_{12} & p_{22} \end{bmatrix} = \begin{bmatrix} -1 & 0 \\ 0 & -1 \end{bmatrix}$$

where we have set $p_{21} = p_{12}$ since we have assumed that the matrix $\mathbf{P}$ is symmetric. Owing to the symmetry in the matrices $\mathbf{P}$ and $\mathbf{Q}$, this expression yields the following $n(n + 1)/2 = 3$ algebraic equations:

$$4p_{22} - p_{11} = -1$$

$$-3p_{12} + 6p_{22} = 0$$

$$p_{11} - 6p_{12} + 8p_{22} = -1$$

Clearly, there cannot be a unique solution for p_{11}, p_{12} and p_{22}. Hence, we conclude that the system is unstable. We would have reached the same conclusion if we had examined the eigenvalues of the matrix $\mathbf{A}$, which are the roots of the characteristic equation $\lambda^2 + 3\lambda + 2 = 0$, i.e. $\lambda_1 = -1$ and $\lambda_2 = -2$. The eigenvalue λ_2 lies outside the unit circle and therefore the system is unstable.

4.3.4 Influence of pole position on the transient response

The position of the poles of the transfer function on the z-plane has a decisive influence on the system's transient response. We distinguish three cases:

1. When a pole lies inside the unit circle, the corresponding transient response converges to zero as time increases.
2. When a pole lies on the unit circle, the corresponding amplitude of the transient response remains constant with time.
3. When a pole lies outside the unit circle, the corresponding transient response increases with time.

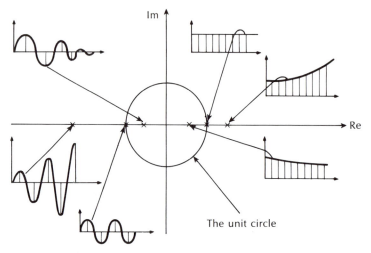

Figure 4.4   Influence of real poles on the transient response.

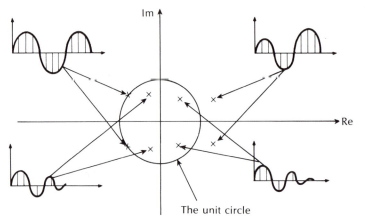

Figure 4.5   Influence of complex poles on the transient response.

These observations are illustrated in Figures 4.4 and 4.5. Figure 4.4 refers to transient responses corresponding to a real pole, whereas Figure 4.5 refers to transient responses corresponding to a pair of complex conjugate poles.

4.4 Controllability and observability

4.4.1 *Controllability*

The concept of controllability was initially introduced by Kalman in 1960 [11–21] and since then it has played a significant theoretical, as well as a practical, role in modern control systems. Gilbert [10] was also one of the earlier contributors on the subject. Controllability is a necessary condition for many control problems to have a solution, such as arbitrary eigenvalue assignment, optimal control, system identification, etc.

Simply speaking, controllability is a property of a system which is strongly related to the ability of the system to go from a given initial state to a desired final state within a finite time. Consider the system

$$\mathbf{x}(k+1) = \mathbf{A}\mathbf{x}(k) + \mathbf{B}\mathbf{u}(k) \quad \text{with} \quad \mathbf{x}(0) = \mathbf{x}_0 \tag{4.18a}$$

$$\mathbf{y}(k) = \mathbf{C}\mathbf{x}(k) \tag{4.18b}$$

The state $\mathbf{x}(k)$ at time k is given by equation (3.20), which can be rewritten as follows:

$$\mathbf{x}(k) = \mathbf{A}^k\mathbf{x}(0) + [\mathbf{B} \mid \mathbf{AB} \mid \ldots \mid \mathbf{A}^{k-1}\mathbf{B}] \begin{bmatrix} \mathbf{u}(k-1) \\ \mathbf{u}(k-2) \\ \vdots \\ \mathbf{u}(0) \end{bmatrix} \tag{4.19}$$

■ **DEFINITION 4.4.1**   Assume that $|\mathbf{A}| \neq 0$. Then system (4.18a) is controllable if it is possible to find a control sequence $\{\mathbf{u}(0), \mathbf{u}(1), \ldots, \mathbf{u}(q-1)\}$ which allows the system to reach an arbitrary final state $\mathbf{x}(q) = \boldsymbol{\xi}$, $\boldsymbol{\xi} \in \mathbb{R}^n$, within a finite time, say q, from any initial state $\mathbf{x}(0)$.

According to Definition 4.4.1, equation (4.19) takes the form

$$\boldsymbol{\xi} - \mathbf{A}^q\mathbf{x}(0) = [\mathbf{B} \mid \mathbf{AB} \mid \ldots \mid \mathbf{A}^{q-1}\mathbf{B}] \begin{bmatrix} \mathbf{u}(q-1) \\ \mathbf{u}(q-2) \\ \vdots \\ \mathbf{u}(0) \end{bmatrix} \tag{4.20}$$

This relation is an inhomogeneous algebraic system of equations, having the control sequence $\{\mathbf{u}(0), \mathbf{u}(1), \dots, \mathbf{u}(q-1)\}$ and the parameter q as unknowns. From linear algebra, it is known that this equation has a solution if and only if

$$\text{rank}[\mathbf{B} \vdots \cdots \vdots \mathbf{A}^{q-1}\mathbf{B} \mid \boldsymbol{\xi} - \mathbf{A}^q\mathbf{x}(0)] = \text{rank}[\mathbf{B} \vdots \cdots \vdots \mathbf{A}^{q-1}\mathbf{B}]$$

This condition, for every arbitrary final state $\mathbf{x}(q) = \boldsymbol{\xi}$, holds true if and only if

$$\text{rank}[\mathbf{B} \vdots \mathbf{AB} \vdots \cdots \vdots \mathbf{A}^{q-1}\mathbf{B}] = n \quad \text{where} \quad q \in \mathbb{N} \tag{4.21}$$

Clearly, an increase in time q improves the possibility of satisfying condition (4.21). However, the Cayley–Hamilton theorem (Appendix A) states that the terms $\mathbf{A}^j\mathbf{B}$, for $j \geqslant n$, are linearly dependent on the first n terms (i.e. on the terms $\mathbf{B}$, $\mathbf{AB}, \dots, \mathbf{A}^{n-1}\mathbf{B}$). Thus, condition (4.21) holds true if and only if $q = n$, i.e. if

$$\text{rank}[\mathbf{B} \vdots \mathbf{AB} \vdots \cdots \vdots \mathbf{A}^{n-1}\mathbf{B}] = n \tag{4.22}$$

Therefore, the following theorem has been proved.

■ **THEOREM 4.4.1**   System (4.18a) is controllable if and only if

$$\text{rank } \mathbf{S} = n \quad \text{where } \mathbf{S} = [\mathbf{B} \vdots \mathbf{AB} \vdots \cdots \vdots \mathbf{A}^{n-1}\mathbf{B}] \tag{4.23}$$

Here the $n \times nm$ matrix $\mathbf{S}$ is called the **controllability matrix**.

■ **REMARK 4.4.1**   For a controllable system of order n, n time units are sufficient for the system to reach any final state $\boldsymbol{\xi} = \mathbf{x}(n)$.

■ **EXAMPLE 4.4.1**   Consider the system (4.18a), where

$$\mathbf{A} = \begin{bmatrix} 1 & 0 \\ 0 & 1 \end{bmatrix} \quad \text{and} \quad \mathbf{b} = \begin{bmatrix} 1 \\ 1 \end{bmatrix}$$

Investigate the controllability of the system.

■ **SOLUTION**   The controllability matrix is

$$\mathbf{S} = [\mathbf{b} \vdots \mathbf{Ab}] = \begin{bmatrix} 1 & 1 \\ 1 & 1 \end{bmatrix}$$

Since $|\mathbf{S}| = 0$ and rank $\mathbf{S} < 2$, it follows that the system is uncontrollable.

■ **EXAMPLE 4.4.2**   Consider the system (4.18a), where

$$\mathbf{A} = \begin{bmatrix} 1 & 1 \\ -0.25 & 0 \end{bmatrix}, \quad \mathbf{b} = \begin{bmatrix} 1 \\ -0.5 \end{bmatrix} \quad \text{and} \quad \mathbf{x}(0) = \begin{bmatrix} 2 \\ 2 \end{bmatrix}$$

Is it possible to find a control sequence such that $\mathbf{x}^{\mathsf{T}}(2) = [-0.5 \ \ 1]$?

■ **SOLUTION** From equation (4.19), for $k = 2$, we have

$$\mathbf{x}(2) = \mathbf{A}^2\mathbf{x}(0) + \mathbf{A}\mathbf{b}u(0) + \mathbf{b}u(1)$$

or

$$\begin{bmatrix} -0.5 \\ 1 \end{bmatrix} = \begin{bmatrix} 3.5 \\ -1 \end{bmatrix} + \begin{bmatrix} 1 \\ -0.5 \end{bmatrix} [0.5u(0) + u(1)]$$

The equation above leads to the scalar equation $0.5u(0) + u(1) = -4$. A possible control sequence would be $u(0) = -2$ and $u(1) = -3$. Assuming that $\mathbf{x}^T(2) = [0.5 \ 1]$ we obtain the following system of equations:

$$\begin{bmatrix} 0.5 \\ 1 \end{bmatrix} = \begin{bmatrix} 3.5 \\ -1 \end{bmatrix} + \begin{bmatrix} 1 \\ -0.5 \end{bmatrix} [0.5u(0) + u(1)]$$

which does not possess a solution. This occurs because the system is uncontrollable, since rank $\mathbf{S} = 1$, where

$$\mathbf{S} = [\mathbf{b} \,\vdots\, \mathbf{Ab}] = \begin{bmatrix} 1 & 0.5 \\ -0.5 & -0.25 \end{bmatrix}$$

■ **EXAMPLE 4.4.3** Consider the system (4.18a), where

$$\mathbf{A} = \begin{bmatrix} 1 & 1 \\ 0 & 1 \end{bmatrix}, \quad \mathbf{b} = \begin{bmatrix} 0 \\ 1 \end{bmatrix} \quad \text{and} \quad \mathbf{x}(0) = \begin{bmatrix} 0 \\ 0 \end{bmatrix}$$

Find a control sequence, if it exists, which can drive the system to the desired final state $\boldsymbol{\xi} = [1 \ 1.2]^T$.

■ **SOLUTION** The controllability matrix of the system is

$$\mathbf{S} = [\mathbf{b} \,\vdots\, \mathbf{Ab}] = \begin{bmatrix} 0 & 1 \\ 1 & 1 \end{bmatrix}$$

Here, $|\mathbf{S}| \neq 0$. Hence, the system is controllable and therefore there exists a control sequence which can drive the system to the desired final state $\boldsymbol{\xi} = [1 \ 1.2]^T$. The response of the system at time $k = 2$ is

$$\mathbf{x}(2) = \mathbf{b}u(1) + \mathbf{A}\mathbf{b}u(0) = \begin{bmatrix} 0 \\ 1 \end{bmatrix} u(1) + \begin{bmatrix} 1 \\ 1 \end{bmatrix} u(0) = \begin{bmatrix} u(0) \\ u(1) + u(0) \end{bmatrix}$$

For $\boldsymbol{\xi} = \mathbf{x}(2)$, we obtain $u(0) = 1$ and $u(1) = 0.2$. Thus, the desired control sequence is $\{u(0), u(1)\} = \{1, 0.2\}$.

4.4.2 Observability

The concept of observability was introduced in parallel to controllability by Kalman [11–21]. Simply stated, observability is a property of a system which is strongly related to our ability to determine the initial state $\mathbf{x}(0)$ on the basis of the input–output data. Because of the close resemblance between controllability and observability, they are frequently referred to as dual concepts.

■ **DEFINITION 4.4.2**   System (4.18) is observable if there exists a finite time q such that, on the basis of the input sequence $\{\mathbf{u}(0), \mathbf{u}(1), ..., \mathbf{u}(q-1)\}$ and the output sequence $\{\mathbf{y}(0), \mathbf{y}(1), ..., \mathbf{y}(q-1)\}$, the initial state $\mathbf{x}(0)$ of the system may be uniquely determined.

Consider the system (4.18). The influence of the input signal $\mathbf{u}(k)$ on the behaviour of the system can always be determined. Therefore, without loss of generality, we can assume that $\mathbf{u}(k) = \mathbf{0}$. We also assume that the output sequence $\{\mathbf{y}(0), \mathbf{y}(1), ..., \mathbf{y}(q-1)\}$ is known (for a certain q). This leads to the following system of equations:

$$\begin{bmatrix} \mathbf{C} \\ \mathbf{CA} \\ \vdots \\ \mathbf{CA}^{q-1} \end{bmatrix} \mathbf{x}(0) = \begin{bmatrix} \mathbf{y}(0) \\ \mathbf{y}(1) \\ \vdots \\ \mathbf{y}(q-1) \end{bmatrix} \tag{4.24}$$

where use was made of (3.21) with $\mathbf{u}(k) = \mathbf{0}$. Equation (4.24) is an inhomogeneous linear algebraic system of equations with $\mathbf{x}(0)$ unknown. This equation has a unique solution for $\mathbf{x}(0)$ (as is required from Definition 4.4.2) if and only if there exists a finite q such that

$$\operatorname{rank} \begin{bmatrix} \mathbf{C} \\ \mathbf{CA} \\ \vdots \\ \mathbf{CA}^{q-1} \end{bmatrix} = n \tag{4.25}$$

Clearly, an increase in time q improves the possibility of satisfying condition (4.25). However, the Cayley–Hamilton theorem (Appendix A) states that the terms $\mathbf{CA}^j$, for $j \geqslant n$, are linearly dependent on the first n terms (i.e. on the terms $\mathbf{C}$, $\mathbf{CA}, ..., \mathbf{CA}^{n-1}$). Thus, condition (4.25) holds true if and only if $q = n$, i.e. if

$$\operatorname{rank} \begin{bmatrix} \mathbf{C} \\ \mathbf{CA} \\ \vdots \\ \mathbf{CA}^{n-1} \end{bmatrix} = n \tag{4.26}$$

Therefore, the following theorem has been proved.

■ **THEOREM 4.4.2** System (4.18) is observable if and only if

$$\text{rank } \mathbf{R} = n \quad \text{where} \quad \mathbf{R} = \begin{bmatrix} \mathbf{C} \\ \mathbf{CA} \\ \vdots \\ \mathbf{CA}^{n-1} \end{bmatrix}$$

Here the $np \times n$ matrix $\mathbf{R}$ is called the **observability matrix**.

■ **REMARK 4.4.2** In an observable system of order n, knowledge of the first n output vectors $\{\mathbf{y}(0), \mathbf{y}(1), \ldots, \mathbf{y}(n-1)\}$ is sufficient to determine the initial condition $\mathbf{x}(0)$ of the system uniquely.

■ **EXAMPLE 4.4.4** Consider the system (4.18), where

$$\mathbf{A} = \begin{bmatrix} 1.1 & -0.3 \\ 1 & 0 \end{bmatrix} \quad \text{and} \quad \mathbf{c} = \begin{bmatrix} 1 \\ -0.5 \end{bmatrix}$$

Investigate the observability of the system.

■ **SOLUTION** The observability matrix of the system is

$$\mathbf{R} = \begin{bmatrix} \mathbf{c}^{\mathrm{T}} \\ \mathbf{c}^{\mathrm{T}}\mathbf{A} \end{bmatrix} = \begin{bmatrix} 1 & -0.5 \\ 0.6 & -0.3 \end{bmatrix}$$

The rank of $\mathbf{R}$ is one and hence the system is unobservable.

■ **EXAMPLE 4.4.5** Consider the system (4.18), where $\mathbf{u}(kT) = \mathbf{0} \ \forall \ k$, and

$$\mathbf{A} = \begin{bmatrix} 1 & 0 \\ 1 & 1 \end{bmatrix} \quad \text{and} \quad \mathbf{c} = \begin{bmatrix} 0 \\ 1 \end{bmatrix}$$

The output sequence of the system is $\{y(0), y(1)\} = \{1, 1.2\}$. Find the initial state $\mathbf{x}(0)$ of the system.

■ **SOLUTION** The observability matrix $\mathbf{R}$ of the system is

$$\mathbf{R} = \begin{bmatrix} \mathbf{c}^{\mathrm{T}} \\ \mathbf{c}^{\mathrm{T}}\mathbf{A} \end{bmatrix} = \begin{bmatrix} 0 & 1 \\ 1 & 1 \end{bmatrix}$$

which has a non-zero determinant. Hence, the system is observable and the initial conditions may be determined from equation (4.24) which, for the present example, is

$$\begin{bmatrix} 0 & 1 \\ 1 & 1 \end{bmatrix} \begin{bmatrix} x_1(0) \\ x_2(0) \end{bmatrix} = \begin{bmatrix} 1 \\ 1.2 \end{bmatrix}$$

From this equation we obtain $x_1(0) = 0.2$ and $x_2(0) = 1$.

4.4.3 *Loss of controllability and observability due to sampling*

As we already know from Section 3.3, when sampling a continuous-time system, the resulting discrete-time system matrices depend on the sampling period T. How does this sampling period T affect the controllability and the observability of the discretized system?

For a discretized system to be controllable, it is necessary that the initial continuous-time system be controllable. This is due to the fact that the control signals of the sampled-data system are only a subset of the control signals of the continuous-time system. However, the controllability may be lost for certain values of the sampling period. Hence, the initial continuous-time system may be controllable, but the equivalent discrete-time system may not. Similar problems occur for the observability of the system.

■ **EXAMPLE 4.4.6**   The state-space equations of a harmonic oscillator with transfer function $H(s) = \omega^2/(s^2 + \omega^2)$ are

$$\dot{\mathbf{x}}(t) = \begin{bmatrix} 0 & \omega \\ -\omega & 0 \end{bmatrix} \mathbf{x}(t) + \begin{bmatrix} 0 \\ \omega \end{bmatrix} u(t)$$

$$y(t) = [1 \ 0]\mathbf{x}(t)$$

Investigate the controllability and the observability of the sampled-data (discrete-time) system whose states are sampled with a sampling period T.

■ **SOLUTION**   The discrete-time model of the harmonic oscillator is (see Example 3.3.5)

$$\mathbf{x}(kT + T) = \begin{bmatrix} \cos \omega T & \sin \omega T \\ -\sin \omega T & \cos \omega T \end{bmatrix} \mathbf{x}(kT) + \begin{bmatrix} 1 - \cos \omega T \\ \sin \omega T \end{bmatrix} u(kT)$$

$$y(kT) = [1 \ 0]\mathbf{x}(kT)$$

One can easily calculate the determinants of the controllability and observability matrices to yield $|\mathbf{S}| = -\sin \omega T(1 - \cos \omega T)$ and $|\mathbf{R}| = \sin \omega T$, respectively. We observe that the controllability and observability of the discrete-time system is lost when $\omega T = q\pi$, where q is an integer, although the respective continuous-time system is both controllable and observable.

4.4.4 *Geometric considerations of controllability*

■ **DEFINITION 4.4.3 Controllable subspace** $\mathscr{E}$ The set of all state vectors $\boldsymbol{\xi} \in \mathbb{R}^n$ which the system (4.18) can reach, within a finite time q, when subjected to a suitable input sequence and starting from the origin ($\mathbf{x}(0) = \mathbf{0}$), constitutes the controllable subspace of system (4.18).

■ **THEOREM 4.4.3**   The controllable subspace $\mathscr{E}$ of the linear time-invariant system (4.18) is a vector subspace of $\mathbb{R}^n$ and is defined by the relation

$$\mathscr{E} = \text{Im}[\mathbf{S}] \quad \text{where} \quad \mathbf{S} = [\mathbf{B} \mid \mathbf{AB} \mid \cdots \mid \mathbf{A}^{n-1}\mathbf{B}] \tag{4.27}$$

where the term $\text{Im}[\mathbf{S}]$ (image of the $n \times nm$ matrix $\mathbf{S}$) is the space produced by all the $(n \times 1)$-dimensional vectors $\boldsymbol{\omega}$ for which $\boldsymbol{\omega} = \mathbf{Sv}$, where $\mathbf{v}$ is any vector in $\mathbb{R}^{nm}$. That is, the controllable subspace $\mathscr{E}$ consists of all the linearly independent columns of $\mathbf{S}$.

■ **PROOF**   From equation (4.19) and for zero initial conditions, i.e. for $\mathbf{x}(0) = \mathbf{0}$, we obtain

$$\boldsymbol{\xi} = \mathbf{x}(q) = [\mathbf{B} \mid \mathbf{AB} \mid \ldots \mid \mathbf{A}^{q-1}\mathbf{B}] \begin{bmatrix} \mathbf{u}(q-1) \\ \mathbf{u}(q-2) \\ \vdots \\ \mathbf{u}(0) \end{bmatrix} \tag{4.28}$$

Let the set of all controllable states $\boldsymbol{\xi}$ at the time instant q be denoted by $\mathscr{E}_q$. Then, according to equation (4.28), the following relation holds true:

$$\mathscr{E}_q = \text{Im}[\mathbf{B} \mid \mathbf{AB} \mid \cdots \mid \mathbf{A}^{q-1}\mathbf{B}] \tag{4.29}$$

Thus, the controllable subspace $\mathscr{E}$ of system (4.18) is given by the relation

$$\mathscr{E} = \bigcup_{q=1}^{\infty} \mathscr{E}_q = \bigcup_{q=1}^{\infty} \text{Im}[\mathbf{B} \mid \mathbf{AB} \mid \ldots \mid \mathbf{A}^{q-1}\mathbf{B}] \quad \forall q \in \mathbb{N} \tag{4.30}$$

From Definition 4.4.3 we have that

$$\text{Im}[\mathbf{B} \mid \mathbf{AB} \mid \cdots \mid \mathbf{A}^{q-1}\mathbf{B}] \subseteq \text{Im}[\mathbf{B} \mid \mathbf{AB} \mid \cdots \mid \mathbf{A}^{n-1}\mathbf{B}] \quad \forall q < n \tag{4.31}$$

Using the Cayley–Hamilton theorem (Appendix A) we obtain

$$\text{Im}[\mathbf{B} \mid \mathbf{AB} \mid \cdots \mid \mathbf{A}^{q-1}\mathbf{B}] = \text{Im}[\mathbf{B} \mid \mathbf{AB} \mid \cdots \mid \mathbf{A}^{n-1}\mathbf{B}] \quad \forall q \geq n \tag{4.32}$$

From equations (4.30), (4.31) and (4.32) the following relation is derived:

$$\mathscr{E} = \text{Im}[\mathbf{B} \mid \mathbf{AB} \mid \cdots \mid \mathbf{A}^{n-1}\mathbf{B}] = \text{Im}[\mathbf{S}] \tag{4.33}$$

From this equation, it is concluded that $\dim \mathscr{E} = \text{rank } \mathbf{S}$ and that a basis of $\mathscr{E}$ consists of all the linearly independent columns of $\mathbf{S}$, where $\dim$ stands for the dimension of a vector subspace [3].

The combination of Theorems 4.4.1 and 4.4.3 leads to the following proposition.

■ **PROPOSITION 4.4.1**   System (4.18) is controllable if and only if $\mathscr{C} = \mathbb{R}^n$.

■ **EXAMPLE 4.4.7**   Consider the system (4.18), where

$$\mathbf{A} = \begin{bmatrix} 1 & 0 \\ 0 & 1 \end{bmatrix} \quad \text{and} \quad \mathbf{b} = \begin{bmatrix} 0 \\ 1 \end{bmatrix}$$

Find the controllable subspace of the system.

■ **SOLUTION**   According to Theorem 4.4.3, the controllable subspace is

$$\mathscr{C} = \mathrm{Im}[\mathbf{b} \mid \mathbf{Ab}] = \mathrm{Im} \begin{bmatrix} 0 & 0 \\ 1 & 1 \end{bmatrix}$$

The dimension of $\mathscr{C}$ is one and its basis vector $[0 \; 1]^T$. Thus, the controllable states are $\boldsymbol{\xi} = [\xi_1 \; \xi_2]^T$, where $\xi_1 = 0$ and ξ_2 is arbitrary. The representation of the controllable subspace $\mathscr{C}$ on $\mathbb{R}^2$ is the vertical axis, as shown in Figure 4.6.

4.4.5 *Geometric considerations of observability*

■ **DEFINITION 4.4.4 The unobservable subspace** $\mathscr{N}$ The set of all initial conditions $\mathbf{x}(0)$ which do not affect the output of the system (4.18) form the unobservable subspace of system (4.18).

■ **THEOREM 4.4.4**   The unobservable subspace $\mathscr{N}$ of the linear time-invariant system (4.18) is a vector subspace of $\mathbb{R}^n$ and is defined by the relation

$$\mathscr{N} = \mathrm{Ker}[\mathbf{R}] \quad \text{where} \quad \mathbf{R} = \begin{bmatrix} \mathbf{C} \\ \mathbf{CA} \\ \vdots \\ \mathbf{CA}^{n-1} \end{bmatrix} \tag{4.34}$$

The term $\mathrm{Ker}[\mathbf{R}]$ (kernel of the $np \times n$ matrix $\mathbf{R}$) is the space of all $(n \times 1)$-dimensional vectors $\boldsymbol{\omega}$, for which $\mathbf{R}\boldsymbol{\omega} = \mathbf{0}$. That is, the base of the unobservable subspace $\mathscr{N}$ consists of the linearly independent vectors which are orthogonal to the rows of $\mathbf{R}$.

■ **PROOF**   From equation (4.24), the initial condition vectors $\mathbf{x}(0)$ which do not appear in the output of the system at the time instant q are the solutions of the

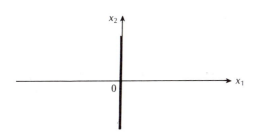

Figure 4.6 The controllable subspace of the system of Example 4.4.7.

respective homogeneous equation, i.e. the solutions of the equation

$$
\begin{bmatrix} \mathbf{C} \\ \mathbf{CA} \\ \vdots \\ \mathbf{CA}^{q-1} \end{bmatrix} \mathbf{x}(0) = \mathbf{0}
\tag{4.35}
$$

Let the set of all solutions of equation (4.35) be denoted as $\mathcal{N}_q$. Then, according to equation (4.35), $\mathcal{N}_q$ is given by

$$
\mathcal{N}_q = \mathrm{Ker} \begin{bmatrix} \mathbf{C} \\ \mathbf{CA} \\ \vdots \\ \mathbf{CA}^{q-1} \end{bmatrix}
\tag{4.36}
$$

Since the unobservable subspace refers to the initial conditions whose effect is not present at the output at any point of time, it follows that

$$
\mathcal{N} = \bigcap_{q=1}^{\infty} \mathcal{N}_q
\tag{4.37}
$$

From Definition 4.4.4, it follows that

$$
\mathrm{Ker} \begin{bmatrix} \mathbf{C} \\ \mathbf{CA} \\ \vdots \\ \mathbf{CA}^{q-1} \end{bmatrix} \supseteq \mathrm{Ker} \begin{bmatrix} \mathbf{C} \\ \mathbf{CA} \\ \vdots \\ \mathbf{CA}^{n-1} \end{bmatrix} \quad \forall \, q < n
\tag{4.38}
$$

Using the Cayley–Hamilton theorem we obtain

$$
\mathrm{Ker} \begin{bmatrix} \mathbf{C} \\ \mathbf{CA} \\ \vdots \\ \mathbf{CA}^{q-1} \end{bmatrix} = \mathrm{Ker} \begin{bmatrix} \mathbf{C} \\ \mathbf{CA} \\ \vdots \\ \mathbf{CA}^{n-1} \end{bmatrix} \quad \forall \, q \geq n
\tag{4.39}
$$

Finally, from relations (4.37), (4.38) and (4.39) it follows that

$$\mathcal{N} = \text{Ker}[\mathbf{R}] \tag{4.40}$$

For an $r \times n$ matrix $\mathbf{A}$ the dimension of the subspace $\mathcal{A} = \text{Ker}[\mathbf{A}]$ is equal to the number of linearly independent vectors $\mathbf{x}$ in $\mathbb{R}^n$ which are orthogonal to the matrix $\mathbf{A}$, i.e. $\mathbf{Ax} = \mathbf{0}$ [3]. Thus, $\dim \mathcal{A} = n - \text{rank } \mathbf{A}$. Hence, from equation (4.40) it is concluded that $\dim \mathcal{N} = n - \text{rank } \mathbf{R}$ and a basis of $\mathcal{N}$ consists of all the linearly independent vectors which are orthogonal to the rows of $\mathbf{R}$.

The combination of Theorems 4.4.2 and 4.4.4 leads to the following proposition.

■ **PROPOSITION 4.4.2**   System (4.18) is observable if and only if $\mathcal{N} = \varnothing$.

■ **EXAMPLE 4.4.8**   Consider a system of the form (4.18), where

$$\mathbf{A} = \begin{bmatrix} 1 & 0 \\ 0 & 1 \end{bmatrix} \quad \text{and} \quad \mathbf{c} = \begin{bmatrix} 0 \\ 1 \end{bmatrix}$$

Find the unobservable subspace of the system.

■ **SOLUTION**   According to Theorem 4.4.4, the unobservable subspace is

$$\mathcal{N} = \text{Ker}\begin{bmatrix} \mathbf{c}^T \\ \mathbf{c}^T\mathbf{A} \end{bmatrix} = \text{Ker}\begin{bmatrix} 0 & 1 \\ 0 & 1 \end{bmatrix}$$

The dimension of $\mathcal{N}$ is one and one of its basis vectors is $[1 \ 0]^T$. Hence, the unobservable initial conditions are $\mathbf{x}(0) = [x_1(0) \ x_2(0)]^T$, where $x_1(0)$ is arbitrary and $x_2(0) = 0$. The image of the unobservable subspace in $\mathbb{R}^2$ is the horizontal axis, as shown in Figure 4.7.

4.4.6   *State vector transformations – state-space canonical forms*

1. General remarks

Consider the linear transformation of the state-space vector $\mathbf{x}(t)$ of system (4.18) having the form [5], [6], [8]

$$\mathbf{x} = \mathbf{Tx}^* \tag{4.41}$$

where $\mathbf{T}$ is an $n \times n$ non-singular transformation matrix and $\mathbf{x}^*$ is the new n-dimensional state-space vector. Substituting the transformation (4.41) in equation

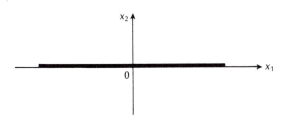

Figure 4.7 The unobservable subspace of the system of Example 4.4.8.

(4.18) we obtain

$$\mathbf{Tx}^*(k+1) = \mathbf{ATx}^*(k) + \mathbf{Bu}(k) \quad \text{with} \quad \mathbf{x}(0) = \mathbf{Tx}^*(0)$$
$$\mathbf{y} = \mathbf{CTx}^*(k)$$

or

$$\mathbf{x}^*(k+1) = \mathbf{A}^*\mathbf{x}^*(k) + \mathbf{B}^*\mathbf{u}(k) \quad \text{with} \quad \mathbf{x}^*(0) = \mathbf{x}^* \tag{4.42a}$$
$$\mathbf{y} = \mathbf{C}^*\mathbf{x}^*(k) \tag{4.42b}$$

where

$$\mathbf{A}^* = \mathbf{T}^{-1}\mathbf{AT}, \quad \mathbf{B}^* = \mathbf{T}^{-1}\mathbf{B}, \quad \mathbf{C}^* = \mathbf{CT} \quad \text{and} \quad \mathbf{x}_0^* = \mathbf{T}^{-1}\mathbf{x}(0) \tag{4.43}$$

For simplicity, systems (4.18) and (4.42) are represented as $(\mathbf{A}, \mathbf{B}, \mathbf{C})_n$ and $(\mathbf{A}^*, \mathbf{B}^*, \mathbf{C}^*)_n$, respectively. System $(\mathbf{A}^*, \mathbf{B}^*, \mathbf{C}^*)_n$ is called the **transformed state-space model** of system $(\mathbf{A}, \mathbf{B}, \mathbf{C})_n$. The motivation for transforming the state-space vector $\mathbf{x}$ into the state-space vector $\mathbf{x}^*$ is to obtain a new coordinate system $x_1^*, x_2^*, \ldots, x_n^*$ which has advantages over the initial coordinate system $x_1, x_2, \ldots, x_n$. These advantages are related to the physical structure and properties of the system, as well as to the simplicity of the computational aspects of the system under consideration. Usually the problem of transforming the state-space vector $\mathbf{x}$ into $\mathbf{x}^*$ becomes one of determining an appropriate transformation matrix $\mathbf{T}$ so that the matrices $\mathbf{A}^*$, $\mathbf{B}^*$ and $\mathbf{C}^*$ of the new system (4.42) have certain desired properties. Particular attention is given to the matrix $\mathbf{A}^*$ whose form should be as simple as possible, e.g. diagonal or in phase canonical form.

2. The invariance of the characteristic polynomial and of the transfer function matrix

Regardless of the particular choice of the transformation matrix $\mathbf{T}$, there are certain properties of the initial system $(\mathbf{A}, \mathbf{B}, \mathbf{C})_n$ which remain invariant. Two such properties are the characteristic polynomial and the transfer function matrix. This is stated in the following theorem which can easily he proved.

■ **THEOREM 4.4.5**   The following hold true:

$$p(z) = p^*(z) \qquad \qquad (4.44)$$

$$\mathbf{H}(z) = \mathbf{H}^*(z) \qquad \qquad (4.45)$$

where

$p(z) = |z\mathbf{I} - \mathbf{A}|$ = the characteristic polynomial of matrix $\mathbf{A}$,
$p^*(z) = |z\mathbf{I} - \mathbf{A}^*|$ = the characteristic polynomial of the transformed matrix $\mathbf{A}^*$,
$\mathbf{H}(z) = \mathbf{C}(z\mathbf{I} - \mathbf{A})^{-1}\mathbf{B}$ = the transfer function matrix of system (4.18),
$\mathbf{H}^*(z) = \mathbf{C}^*(z\mathbf{I} - \mathbf{A}^*)^{-1}\mathbf{B}^*$ = the transfer function matrix of the transformed system (4.42).

3. The invariance of controllability and observability

The properties of controllability and observability also remain invariant under the transformation of the state-space vector. Indeed, the matrices $\mathbf{S}^*$ and $\mathbf{R}^*$ of the transformed system (4.42) are related to the matrices $\mathbf{S}$ and $\mathbf{R}$ of the initial system (4.18) as follows: for the matrix $\mathbf{S}^*$ we have

$$\begin{aligned}
\mathbf{S}^* &= [\mathbf{B}^* \vdots \mathbf{A}^*\mathbf{B}^* \vdots \cdots \vdots (\mathbf{A}^*)^{n-1}\mathbf{B}^*] \\
&= [\mathbf{T}^{-1}\mathbf{B} \vdots \mathbf{T}^{-1}\mathbf{A}\mathbf{T}\mathbf{T}^{-1}\mathbf{B} \vdots \cdots \vdots (\mathbf{T}^{-1}\mathbf{A}\mathbf{T})^{n-1}\mathbf{T}^{-1}\mathbf{B}] \\
&= [\mathbf{T}^{-1}\mathbf{B} \vdots \mathbf{T}^{-1}\mathbf{A}\mathbf{B} \vdots \cdots \vdots \mathbf{T}^{-1}\mathbf{A}^{n-1}\mathbf{B}] \\
&= \mathbf{T}^{-1}[\mathbf{B} \vdots \mathbf{A}\mathbf{B} \vdots \cdots \vdots \mathbf{A}^{n-1}\mathbf{B}] = \mathbf{T}^{-1}\mathbf{S} \qquad (4.46)
\end{aligned}$$

Also

$$\begin{aligned}
\mathbf{R}^{*\mathrm{T}} &= [\mathbf{C}^{*\mathrm{T}} \vdots \mathbf{A}^{*\mathrm{T}}\mathbf{C}^{*\mathrm{T}} \vdots \cdots \vdots (\mathbf{A}^{*\mathrm{T}})^{n-1}\mathbf{C}^{*\mathrm{T}}] \\
&= [\mathbf{T}^{\mathrm{T}}\mathbf{C}^{\mathrm{T}} \vdots \mathbf{T}^{\mathrm{T}}\mathbf{A}^{\mathrm{T}}(\mathbf{T}^{-1})^{\mathrm{T}}\mathbf{T}^{\mathrm{T}}\mathbf{C}^{\mathrm{T}} \vdots \cdots \vdots [\mathbf{T}^{\mathrm{T}}\mathbf{A}^{\mathrm{T}}(\mathbf{T}^{-1})^{\mathrm{T}}]^{n-1}\mathbf{T}^{\mathrm{T}}\mathbf{C}^{\mathrm{T}}] \\
&= [\mathbf{T}^{\mathrm{T}}\mathbf{C}^{\mathrm{T}} \vdots \mathbf{T}^{\mathrm{T}}\mathbf{A}^{\mathrm{T}}\mathbf{C}^{\mathrm{T}} \vdots \cdots \vdots \mathbf{T}^{\mathrm{T}}(\mathbf{A}^{\mathrm{T}})^{n-1}\mathbf{C}^{\mathrm{T}}] \\
&= \mathbf{T}^{\mathrm{T}}[\mathbf{C}^{\mathrm{T}} \vdots \mathbf{A}^{\mathrm{T}}\mathbf{C}^{\mathrm{T}} \vdots \cdots \vdots (\mathbf{A}^{\mathrm{T}})^{n-1}\mathbf{C}^{\mathrm{T}}] = \mathbf{T}^{\mathrm{T}}\mathbf{R}^{\mathrm{T}} \qquad (4.47)
\end{aligned}$$

From relations (4.46) and (4.47) it follows that, if the initial system is controllable and observable, so is the transformed system.

4. Transformation of state-space equations into special forms

■ **DEFINITION 4.4.5**   System $(\mathbf{A}^*, \mathbf{B}^*, \mathbf{C}^*)_n$ is in phase canonical form when the matrices $\mathbf{A}^*$ and $\mathbf{B}^*$ have the form [5], [22]

$$\mathbf{A}^* = \begin{bmatrix} \mathbf{A}_{11}^* & \mathbf{A}_{12}^* & \cdots & \mathbf{A}_{1m}^* \\ \mathbf{A}_{21}^* & \mathbf{A}_{22}^* & \cdots & \mathbf{A}_{2m}^* \\ \vdots & \vdots & & \vdots \\ \mathbf{A}_{m1}^* & \mathbf{A}_{m2}^* & \cdots & \mathbf{A}_{mm}^* \end{bmatrix} \quad \text{and} \quad \mathbf{B} = \begin{bmatrix} \mathbf{B}_1^* \\ \mathbf{B}_2^* \\ \vdots \\ \mathbf{B}_m^* \end{bmatrix} \qquad (4.48a)$$

where

$$\mathbf{A}_{ii}^* = \begin{bmatrix} 0 & 1 & 0 & \dots & 0 \\ 0 & 0 & 1 & \dots & 0 \\ \vdots & \vdots & \vdots & & \vdots \\ 0 & 0 & 0 & \dots & 1 \\ -(a_{ii}^*)_0 & -(a_{ii}^*)_1 & -(a_{ii}^*)_2 & \dots & -(a_{ii}^*)_{\sigma_i-1} \end{bmatrix} \qquad (4.48b)$$

$$\mathbf{A}_{ij}^* = \left[\begin{array}{ccccc} & & \mathbf{0} & & \\ \hline -(a_{ij}^*)_0 & -(a_{ij}^*)_1 & -(a_{ij}^*)_2 & \dots & -(a_{ij}^*)_{\sigma_j-1} \end{array} \right] \qquad (4.48c)$$

$$\mathbf{B}_i^* = \left[\begin{array}{cccccccc} & & & & \mathbf{0} & & & \\ \hline 0 & 0 & \dots & 0 & 1 & (b_i^*)_{i+1} & (b_i^*)_{i+2} & \dots & (b_i^*)_m \end{array} \right] \qquad (4.48d)$$

$$\underset{\text{— } i\text{th position}}{}$$

where $\sigma_1, \sigma_2, \dots, \sigma_m$ are positive integers satisfying the condition

$$\sum_{i=1}^m \sigma_i = n$$

■ **DEFINITION 4.4.6** Assuming that the system $(\mathbf{A}^*, \mathbf{B}^*, \mathbf{C}^*)_n$ is a single-input system, then the system can be described in phase canonical form when the matrix $\mathbf{A}^*$ and the vector $\mathbf{b}^*$ have the form

$$\mathbf{A}^* = \begin{bmatrix} 0 & 1 & 0 & \dots & 0 \\ 0 & 0 & 1 & \dots & 0 \\ \vdots & \vdots & \vdots & & \vdots \\ 0 & 0 & 0 & \dots & 1 \\ -a_0^* & -a_1^* & -a_2^* & \dots & -a_{n-1}^* \end{bmatrix} \quad \text{and} \quad \mathbf{b}^* = \begin{bmatrix} 0 \\ 0 \\ \vdots \\ 0 \\ 1 \end{bmatrix} \qquad (4.49)$$

It is obvious that Definition 4.4.6 is a special (but important) case of Definition 4.4.5.

The basic computational steps necessary to transform a system into its phase canonical form are given in the following two theorems.

■ **THEOREM 4.4.6** Let the system $(\mathbf{A}, \mathbf{B}, \mathbf{C})_n$ be a single-input system and assume that the controllability matrix

$$\mathbf{S} = [\mathbf{b} \mathbin{\vdots} \mathbf{A}\mathbf{b} \mathbin{\vdots} \mathbf{A}^2\mathbf{b} \mathbin{\vdots} \cdots \mathbin{\vdots} \mathbf{A}^{n-1}\mathbf{b}] \qquad (4.50)$$

is invertible. Then, there exists a matrix $\mathbf{T}$ which transforms the given system into its phase canonical form $(\mathbf{A}^*, \mathbf{B}^*, \mathbf{C}^*)_n$, defined by (4.49), where the matrix $\mathbf{T}$ is given by

$$\mathbf{T} = \mathbf{P}^{-1}$$

where

$$\mathbf{P} = \begin{bmatrix} \mathbf{p}_1 \\ \mathbf{p}_2 \\ \vdots \\ \mathbf{p}_n \end{bmatrix} = \begin{bmatrix} \mathbf{q} \\ \mathbf{q}\mathbf{A} \\ \vdots \\ \mathbf{q}\mathbf{A}^{n-1} \end{bmatrix} \tag{4.51}$$

and $\mathbf{q}$ is the last row of the matrix $\mathbf{S}^{-1}$.

Theorem 4.4.6 can be generalized for multi-input systems as follows.

■ **THEOREM 4.4.7** Let the system $(\mathbf{A}, \mathbf{B}, \mathbf{C})_n$ be a multi-input system and suppose that the matrix

$$\hat{\mathbf{S}} = [\mathbf{b}_1 \vdots \mathbf{A}\mathbf{b}_1 \vdots \cdots \mathbf{A}^{\sigma_1-1}\mathbf{b}_1 \mid \mathbf{b}_2 \vdots \mathbf{A}\mathbf{b}_2 \vdots \cdots \vdots \mathbf{A}^{\sigma_2-1}\mathbf{b}_2 \mid \cdots \mid \mathbf{b}_m \vdots \mathbf{A}\mathbf{b}_m \vdots \cdots \vdots \mathbf{A}^{\sigma_m-1}\mathbf{b}_m] \quad (4.52)$$

is invertible, where $\mathbf{b}_i$ is the ith column of the matrix $\mathbf{B}$ and $\sigma_1, \sigma_2, ..., \sigma_m$ are positive integers such that $\sum_{i=1}^{m} \sigma_i = n$. Necessary and sufficient conditions for the matrix $\hat{\mathbf{S}}$ to be regular (invertible) are that the system $(\mathbf{A}, \mathbf{B}, \mathbf{C})_n$ is controllable and that rank $\mathbf{B} = m$. Next, consider the transformation matrix $\mathbf{T}$, where $\mathbf{T}$ is given by the relation $\mathbf{T} = \mathbf{P}^{-1}$, where

$$\mathbf{P} = \begin{bmatrix} \begin{bmatrix} \mathbf{P}_1 \end{bmatrix} \\ \begin{bmatrix} \mathbf{P}_2 \end{bmatrix} \\ \vdots \\ \begin{bmatrix} \mathbf{P}_m \end{bmatrix} \end{bmatrix} = \begin{bmatrix} \mathbf{p}_1 \\ \mathbf{p}_2 \\ \vdots \\ \mathbf{p}_{\sigma_1} \\ \mathbf{p}_{\sigma_1+1} \\ \mathbf{p}_{\sigma_1+2} \\ \vdots \\ \mathbf{p}_{\sigma_1+\sigma_2} \\ \vdots \\ \mathbf{p}_{n-\sigma_m} \\ \mathbf{p}_{n-\sigma_m+1} \\ \vdots \\ \mathbf{p}_n \end{bmatrix} = \begin{bmatrix} \mathbf{q}_1 \\ \mathbf{q}_1\mathbf{A} \\ \vdots \\ \mathbf{q}_1\mathbf{A}^{\sigma_1-1} \\ \mathbf{q}_2 \\ \mathbf{q}_2\mathbf{A} \\ \vdots \\ \mathbf{q}_2\mathbf{A}^{\sigma_2-1} \\ \vdots \\ \mathbf{q}_m \\ \mathbf{q}_m\mathbf{A} \\ \vdots \\ \mathbf{q}_m\mathbf{A}^{\sigma_m-1} \end{bmatrix} \tag{4.53}$$

and $\mathbf{q}_k$ is the δ_kth row of the matrix $\hat{\mathbf{S}}^{-1}$, where $\delta_k = \sum_{i=1}^{k} \sigma_i$, $k = 1, 2, \ldots, m$. Then, there always exist certain integers $\sigma_1, \sigma_2, \ldots, \sigma_m$ such that the matrix $\mathbf{T}$ transforms the given system $(\mathbf{A}, \mathbf{B}, \mathbf{C})_n$ into its phase canonical form $(\mathbf{A}^*, \mathbf{B}^*, \mathbf{C}^*)_n$, defined by (4.48). The integers $\sigma_1, \sigma_2, \ldots, \sigma_m$ are called the controllability indices. A systematic method for determining these indices is given in [5].

4.4.7 *Kalman decomposition*

Kalman showed that it is possible to introduce certain coordinates, using a suitable transformation matrix $\mathbf{T}$, so that a system can be decomposed as follows:

$$\mathbf{x}(k+1) = \begin{bmatrix} \mathbf{A}_{11} & \mathbf{A}_{12} & \mathbf{0} & \mathbf{0} \\ \mathbf{0} & \mathbf{A}_{22} & \mathbf{0} & \mathbf{0} \\ \mathbf{A}_{31} & \mathbf{A}_{32} & \mathbf{A}_{33} & \mathbf{A}_{34} \\ \mathbf{0} & \mathbf{A}_{42} & \mathbf{0} & \mathbf{A}_{44} \end{bmatrix} \mathbf{x}(k) + \begin{bmatrix} \mathbf{B}_1 \\ \mathbf{0} \\ \mathbf{B}_3 \\ \mathbf{0} \end{bmatrix} \mathbf{u}(k) \tag{4.54a}$$

$$\mathbf{y}(k) = [\mathbf{C}_1 \quad \mathbf{C}_2 \quad \mathbf{0} \quad \mathbf{0}]\mathbf{x}(k) \tag{4.54b}$$

where $\mathbf{A}_{ij}$, $\mathbf{B}_i$ and $\mathbf{C}_i$ are block matrices of suitable dimensions. The state-space vector $\mathbf{x}(k)$ is accordingly decomposed into four subvectors each of which corresponds to one of the following four cases: states which are both controllable and observable; states which are uncontrollable but are observable; states which are controllable but are unobservable; and states which are both uncontrollable and unobservable.

The following theorem which is related to the results above is called **the Kalman decomposition theorem**.

■ **THEOREM 4.4.8** A system can be decomposed into four subsystems with the following properties:

Subsystem S_{OC}: the observable and controllable subsystem $(\mathbf{A}_{11}, \mathbf{B}_1, \mathbf{C}_1)$.
Subsystem $S_{\mathrm{O\bar{C}}}$: the observable but uncontrollable subsystem $(\mathbf{A}_{22}, \mathbf{0}, \mathbf{C}_2)$.
Subsystem $S_{\mathrm{\bar{O}C}}$: the unobservable but controllable subsystem $(\mathbf{A}_{33}, \mathbf{B}_3, \mathbf{0})$.
Subsystem $S_{\mathrm{\bar{O}\bar{C}}}$: the unobservable and uncontrollable subsystem $(\mathbf{A}_{44}, \mathbf{0}, \mathbf{0})$.

The transfer function matrix of the system is unique and can be determined from the subsystem which is both controllable and observable. Indeed, straightforward calculations show that the transfer function matrix of system (4.54) is given by

$$\mathbf{H}(z) = \mathbf{C}_1(z\mathbf{I} - \mathbf{A}_{11})^{-1}\mathbf{B}_1 \tag{4.55}$$

Relation (4.55) contains only the controllable and the observable part of the system.

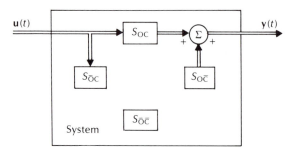

Figure 4.8 Block diagram of the Kalman decomposition.

Figure 4.8 shows the block diagram of the Kalman decomposition, involving all four subsystems and the way in which they are linked to each other. Also, it shows that the input is related to the output only through the subsystem S_{OC}.

4.5 Problems

1. Investigate the stability of a discrete-time system when its characteristic polynomial is given by:

(a) $z^2 + 0.8z + 0.2$ (b) $z^2 - 0.8z - 0.4$ (c) $z^3 + z^2 + 0.2z + 1$

(d) $z^3 + z^2 + 4z - 8$ (e) $z^2 + z + 1$ (f) $z^4 + 2z^3 + z^2 + 2z + 2$

2. Find all values of K for which the roots of the following characteristic polynomials lie inside the unit circle:

(a) $z^2 + 0.2z + K$ (b) $z^2 + Kz + 0.4$ (c) $z^2 + (K + 0.4)z + 1$

(d) $z^3 + Kz^2 + 2z + 2$ (e) $z^3 - 0.5z^2 - 0.2z + K$ (f) $z^3 + (K + 1)z^2 - 0.5z + 1$

3. A magnetic disk drive requires a motor to position a read/write head over tracks of data on a spinning disk. The motor and the head may be approximated by the transfer function

$$G(s) = \frac{1}{s(T_1 s + 1)}$$

where $T_1 > 0$. The controller takes the difference of the actual and desired positions and generates an error. This error is discretized with sampling period T, multiplied by a gain K and applied to the motor with the use of a zero-order hold of period T (see Figure 4.9).

Determine the range of values of the gain K, so that the closed-loop discrete-time system is stable. Apply the invariant impulse response method and the Routh criterion.

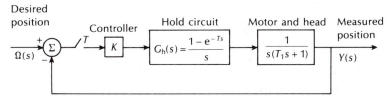

Figure 4.9 Block diagram of disk drive control of problem 3.

4. Consider the system given in Figure 4.10. Apply the Jury criterion to determine the range of values of the gain K for which the system is stable. Assume sampling period $T = 0.1$ s, 0.2 s and 1 s.

5. Consider a system described by the linear time-invariant difference equation

$$y(k+2) + a_1 y(k+1) + a_2 y(k) = 0$$

Investigate the stability of the system using the Lyapunov method.

6. Apply the method of Lyapunov to study the stability of $\mathbf{x}(k+1) = \mathbf{A}\mathbf{x}(k)$, where $\mathbf{A}$ is

(a) $\begin{bmatrix} 0 & 1 \\ -2 & -1 \end{bmatrix}$ (b) $\begin{bmatrix} 0 & 1 \\ 0 & 0 \end{bmatrix}$ (c) $\begin{bmatrix} -0.1 & 0 \\ 0 & -4 \end{bmatrix}$

(d) $\begin{bmatrix} 0 & 1 \\ 2 & 2 \end{bmatrix}$ (e) $\begin{bmatrix} -0.2 & 0 & 0 \\ 0 & 0.5 & 0 \\ 0 & 0 & -0.4 \end{bmatrix}$ (f) $\begin{bmatrix} 0 & 1 & 0 \\ 0 & 0 & 1 \\ 0.4 & -1 & 1 \end{bmatrix}$

7. Check the stability of the system described by

$$x_1(k+1) = x_2(k)$$

$$x_2(k+1) = 2.5x_1(k) + x_2(k) + u(k)$$

$$y(k) = x_1(k)$$

If the system is unstable, use the output feedback law $u(k) = -gy(k)$ to stabilize it. Determine the range of values of a suitable g.

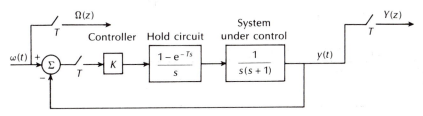

Figure 4.10 Block diagram of system of problem 4.

8. A single-input system is described by

$$\mathbf{x}[(k+1)T] = \mathbf{A}\mathbf{x}(kT) + \mathbf{b}u(kT) \quad \text{where} \quad \mathbf{x} \in \mathbb{R}^n$$

(a) Obtain the sequence $\{u(0), u(T), \ldots, u[(n-1)T]\}$ which will drive the system from a known initial state $\mathbf{x}(0)$ to a known final state $\mathbf{x}(nT)$.

(b) A controller $G_c(z)$ is to be designed having as input the difference

$$\omega(kT) - y(kT) \quad \text{with} \quad y(kT) = [1 \quad 0]\begin{bmatrix} x_1(k) \\ x_2(k) \end{bmatrix} = \mathbf{c}^T\mathbf{x}(kT)$$

as shown in Figure 4.11 so that the system state is driven from $\mathbf{x}(0)$ to $\mathbf{x}(nT)$. Suggest a formula by which the controller transfer function $G_c(z)$ may be determined.

(c) Assume that $\omega(kT) = 1$, $\mathbf{x}^T(0) = [0 \quad 0]$, the desired state is $\mathbf{x}^T(2) = [2 \quad 1]$ and that

$$\mathbf{A} = \begin{bmatrix} 0 & 0.1 \\ -0.15 & 0.8 \end{bmatrix} \quad \text{and} \quad \mathbf{b} = \begin{bmatrix} 0 \\ 1 \end{bmatrix}$$

Use the method of part (b) to design a controller $G_c(z)$ to achieve the objective.

(d) Find the closed-loop transfer function of the system and plot the poles and the zeros in the complex plane.

9. Consider a system described by the state-space equations $\mathbf{x}(k+1) = \mathbf{A}\mathbf{x}(k) + \mathbf{B}u(k)$, $\mathbf{y}(k) = \mathbf{C}\mathbf{x}(k)$ where $\mathbf{A}$, $\mathbf{B}$ and $\mathbf{C}$ are

(a) $\mathbf{A} = \begin{bmatrix} 0 & 1 & 0 & 0 \\ 0 & 0 & 1 & 0 \\ 0 & 0 & 0 & 1 \\ -4 & -2 & 1 & 0.4 \end{bmatrix}$, $\mathbf{B} = \begin{bmatrix} 0 \\ 0 \\ 0 \\ 1 \end{bmatrix}$ and $\mathbf{C} = [1 \quad -2 \quad 1 \quad 4]$

(b) $\mathbf{A} = \begin{bmatrix} 0 & 0 & 1 \\ 1 & 1 & 0 \\ -1 & 0 & 0.5 \end{bmatrix}$, $\mathbf{B} = \begin{bmatrix} 1 & 1 \\ 1 & 1 \\ -1 & -0.5 \end{bmatrix}$ and $\mathbf{C} = \begin{bmatrix} 0 & 1 & 1 \\ 1 & -1 & 1 \end{bmatrix}$

(c) $\mathbf{A} = \begin{bmatrix} 1 & 0 & 1 \\ 1 & 2 & 1 \\ -T & 0 & 0 \end{bmatrix}$, $\mathbf{B} = \begin{bmatrix} 1 & 1 \\ 0 & 1 \\ 1 & 0 \end{bmatrix}$ and $\mathbf{C} = \begin{bmatrix} 0 & 0 & 1 \\ 1 & 0 & 0 \end{bmatrix}$

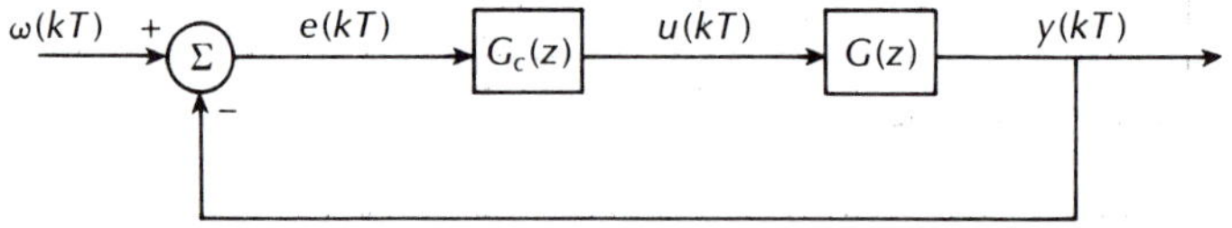

Figure 4.11 Closed-loop system of problem 8.

Investigate the controllability and observability of these systems. For case (c), find the values of the sampling time T, appearing in matrix $\mathbf{A}$, which make the system controllable and/or observable.

10. Consider the single-input single-output system:

$$\mathbf{x}(k+1) = \mathbf{A}\mathbf{x}(k) + \mathbf{b}u(k)$$

$$y(k) = \mathbf{c}^{\mathrm{T}}\mathbf{x}(k)$$

which is assumed to be observable. Obtain the coordinate transformation $\mathbf{x} = \mathbf{P}\mathbf{z}$ which transforms the state equations in the observable cononical form described by the matrix $\mathbf{A}^*$ and the vector $\mathbf{c}^*$ having the following form:

$$\mathbf{A}^* = \begin{bmatrix} -a_{n-1} & 1 & 0 & \cdots & 0 & 0 \\ -a_{n-2} & 0 & 1 & \cdots & 0 & 0 \\ & \vdots & & & \vdots & \\ -a_1 & 0 & 0 & \cdots & 0 & 1 \\ -a_0 & 0 & 0 & \cdots & 0 & 0 \end{bmatrix} \quad \text{and} \quad \mathbf{c}^{*\mathrm{T}} = [1 \quad 0 \quad \cdots \quad 0]$$

where $\lambda^n + a_{n-1}\lambda^{n-1} + \cdots + a_1\lambda + a_0 = 0$ is the characteristic equation of $\mathbf{A}$. Determine the conditions for the existence of the transformation.

11. The continuous-time system under consideration is a rotating body described by the dynamical equation:

$$\dot{\omega} = \frac{\mathrm{d}^2\theta}{\mathrm{d}t^2} = \frac{L}{J}$$

where θ is the position (angle of rotation), ω is the rate of the angle of rotation, L is the externally applied torque and J is the moment of inertia. If $x_1 = \theta$ and $x_2 = \dot{\theta} = \omega$, then the state-space description is

$$\begin{bmatrix} \dot{x}_1 \\ \dot{x}_2 \end{bmatrix} = \begin{bmatrix} 0 & 1 \\ 0 & 0 \end{bmatrix}\begin{bmatrix} x_1 \\ x_2 \end{bmatrix} + \begin{bmatrix} 0 \\ 1 \end{bmatrix}\frac{L}{J} = \mathbf{A}\mathbf{x} + \mathbf{b}u$$

Obtain the discrete-time description when using a zero-order hold and a sampling period T. If θ is taken to be the output, determine if this description is observable. What happens if the angular velocity ω is measured instead? Discuss the results in both cases.

12. A system is described by the state equations

$$\dot{\mathbf{x}}(k+1) = \mathbf{A}\mathbf{x}(k) + \mathbf{b}u(k)$$

$$y(k) = \mathbf{c}^{\mathrm{T}}\mathbf{x}(k)$$

where

$$\mathbf{A} = \begin{bmatrix} 0 & 1 \\ -2 & -3 \end{bmatrix}, \quad \mathbf{b} = \begin{bmatrix} 1 \\ 1 \end{bmatrix} \quad \text{and} \quad \mathbf{c}^T = [1 \quad 2]$$

Determine the controllability and observability of both the open-loop and the closed-loop systems when

$$u(k) = \omega(k) - \mathbf{f}^T \mathbf{x}(k)$$

where $\omega(k)$ is some reference input and $\mathbf{f} = [f_1 \ f_2]^T$.

13. Transform in phase canonical form all the systems presented in problem 9.

14. Consider the system $\mathbf{x}(k+1) = \mathbf{A}\mathbf{x}(k) + \mathbf{B}\mathbf{u}(k)$, $x \in \mathbb{R}^n$, $u \in \mathbb{R}^m$. Show that if for an eigenvector of $\mathbf{A}$, say $\mathbf{a}^T$, $\mathbf{a}^T\mathbf{B} = \mathbf{0}$ holds true, then the system is uncontrollable.

4.6   References

Books

[1] K.J. Åström and B. Wittenmark, *Computer Controlled Systems*, Prentice Hall, Englewood Cliffs, New Jersey, 1984.
[2] G.F. Franklin, J.D. Powell and M.L. Workman, *Digital Control of Dynamic Systems*, Addison-Wesley, London, 1990 (Second Edition).
[3] F.R. Gantmacher, *The Theory of Matrices*, Volumes I and II, Chelsea, New York, 1959.
[4] C.H. Houpis and G.B. Lamont, *Digital Control Systems*, McGraw-Hill, New York, 1985.
[5] T. Kailath, *Linear Systems*, Prentice Hall, Englewood Cliffs, New Jersey, 1980.
[6] K. Ogata, *State Space Analysis of Control Systems*, Prentice Hall, Englewood Cliffs, New Jersey, 1967.
[7] K. Ogata, *Discrete-Time Control Systems*, Prentice Hall, Englewood Cliffs, New Jersey, 1987.
[8] P.N. Paraskevopoulos, *Introduction to Automatic Control Systems*, in preparation.
[9] C.L. Phillips and H.T. Nagle Jr, *Digital Control System Analysis and Design*, Prentice Hall, Englewood Cliffs, New Jersey, 1984.

Papers

[10] E.G. Gilbert, 'Controllability and Observability in Multivariable Control Systems', *Journal of the SIAM – Control Series*, A, Vol. 1, No. 2, pp. 128–151, 1963.
[11] R.E. Kalman, 'A New Approach to Linear Filtering and Prediction Problems', *Transactions of the ASME (Journal of Basic Engineering)*, Vol. 82D, pp. 35–45, 1960.

[12] R.E. Kalman, 'On the General Theory of Control Systems', *Proceedings of the First International Congress, IFAC, Moscow, USSR*, pp. 481–492, 1960.

[13] R.E. Kalman, 'Contributions to the Theory of Optimal Control', *Proceedings of the 1959 Mexico City Conference on Differential Equations, Mexico City*, pp. 102–119, 1960.

[14] R.E.Kalman, 'Canonical Structure of Linear Dynamical Systems', *Proceedings of the National Academy of Science*, Vol. 48, pp. 596–600, 1962.

[15] R.E. Kalman, 'Mathematical Description of Linear Dynamic Systems', *SIAM Journal on Control*, Vol. 1, pp. 152–192, 1963.

[16] R.E. Kalman, 'When is a Linear Control System Optimal?', *Transactions of the ASME (Journal of Basic Engineering)*, Vol. 86D, pp. 51–60, 1964.

[17] R.E. Kalman and J.E. Bertram, 'A Unified Approach to the Theory of Sampling Systems', *Journal of the Franklin Institute*, pp. 405–425, 1959.

[18] R.E. Kalman and J.E. Bertram, 'Control Systems Analysis and Design via Second Method of Liapunov: II Discrete-Time Systems', *Transactions of the ASME (Journal of Basic Engineering)*, Series D, pp. 371–400, 1960.

[19] R.E. Kalman and R.S. Bucy, 'New Results in Linear Filtering and Prediction Theory', *Transactions of the ASME (Journal of Basic Engineering)*, Vol. 83D, pp. 95–108, 1961.

[20] R.E. Kalman and R.W. Koepcke, 'Optimal Synthesis of Linear Sampling Control Systems Using Generalized Performance Indices', *Transactions of the ASME*, Vol. 80, pp. 1820–1826, 1958.

[21] R.E. Kalman, Y.C. Ho and K.S. Narendra, 'Controllability of Linear Dynamic Systems', *Contributions to Differential Equations*, Vol. 1, pp. 189–213, 1963.

[22] D.G. Luenberger, 'Canonical Forms for Linear Multivariable Systems', *IEEE Transactions on Automatic Control*, Vol. AC–12, pp. 290–293, 1967.

5

Classical design methods

5.1 Introduction

This chapter presents the classical design methods of discrete-time closed-loop control systems [1], [4], [5], [7], [9], [11]. These methods are separated into indirect and direct techniques.

1. Indirect techniques

Using these techniques, a discrete-time controller $G_c(z)$ is determined indirectly as follows. Initially, the continuous-time controller $G_c(s)$ is designed in the s-domain, using well-known classical techniques (e.g. root-locus, Bode, Nyquist, etc.). In determining $G_c(s)$, it is assumed that a hold circuit is placed in series with the controller. Then, based on the continuous-time controller $G_c(s)$, the discrete-time controller $G_c(z)$ may be determined using one of the discretization techniques presented in Section 3.3.3. The indirect techniques are presented in Section 5.2 below.

2. Direct techniques

These techniques start by deriving a discrete-time mathematical model of the continuous-time system under control. Subsequently, the design is carried out in the z-domain, in which the discrete-time controller $G_c(z)$ is directly determined. The design in the z-domain may be done using either the root-locus (Section 5.3) or Bode and Nyquist diagrams (Section 5.4).

In this chapter, special attention is given to PID discrete-time controller design (Section 5.5). The PID controller consists of three terms: the proportional (P), the integral (I) and the derivative (D). Hence, PID controllers are dynamic controllers. They are, therefore, superior to static controllers, since they offer greater flexibility in satisfying the closed-loop system's specifications. For this reason PID controllers are most popular in practice.

Finally, in Section 5.6, a brief description of the steady-state errors appearing in discrete-time systems is presented.

5.2 Discrete-time controllers derived from continuous-time controllers

5.2.1 *Discrete-time controller design using indirect techniques*

The practising control engineer often has greater knowledge and experience in designing continuous-time rather than discrete-time controllers. Moreover, many practical systems already incorporate a continuous-time controller which we desire to replace with a discrete-time controller.

The remarks above are the basic motives for the implementation of indirect design techniques for discrete-time controllers mentioned in the introduction. Indirect techniques take advantage of the knowledge and the experience one has of continuous-time systems. Furthermore, in cases where a continuous-time controller is already incorporated in the system under control, it facilitates the design of a discrete-time controller.

Consider the continuous-time closed-loop control system shown in Figure 5.1 and the discrete-time closed-loop control system shown in Figure 5.2. The indirect design technique for the design of a discrete-time controller may be stated as follows. Let the specifications of the closed-loop systems shown in Figures 5.1 and 5.2 be the same. Assume that a continuous-time controller $G_c(s)$, satisfying the specifications of the closed-loop system shown in Figure 5.1, has already been determined. Then, the discrete-time controller $G_c(z)$ shown in Figure 5.2 may be calculated from the continuous-time controller of Figure 5.1, using the discretization techniques

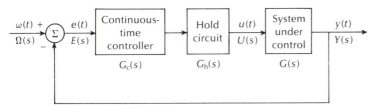

Figure 5.1 Continuous-time closed-loop control system with a hold circuit.

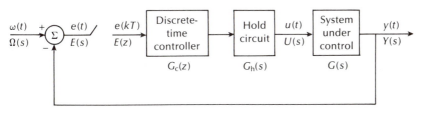

Figure 5.2 Discrete-time closed-loop system.

presented in Section 3.3.3. A popular technique for determining $G_c(z)$ from $G_c(s)$ is based on the invariant step response method given by (3.47), i.e. by the relation

$$G_c(z) = \mathscr{L}[G_h(s)G_c(s)] \tag{5.1}$$

Other popular techniques are pole–zero matching and the Tustin transformation method.

In applying indirect design techniques, a difficulty arises due to $G_h(s)$. We will study the simplest case, where the hold circuit is a zero-order circuit. Thus

$$G_h(s) = \frac{1 - e^{-Ts}}{s} \tag{5.2}$$

Even in this case, the term e^{-Ts} makes it difficult to determine the controller $G_c(s)$ of the closed-loop system of Figure 5.1. To facilitate the determination of $G_c(s)$, the term e^{-Ts} is approximated by the Padé form as follows:

$$e^{-Ts} = \frac{e^{-Ts/2}}{e^{Ts/2}} = \frac{1 - Ts/2 + (Ts)^2/8 - \ldots}{1 + Ts/2 + (Ts)^2/8 + \ldots} \simeq \frac{1 - Ts/2}{1 + Ts/2} \tag{5.3}$$

Substituting relation (5.3) in (5.2), we obtain

$$G_h(s) = \frac{1 - e^{-Ts}}{s} \simeq \frac{1}{s}\left(1 - \frac{1 - Ts/2}{1 + Ts/2}\right) = \frac{T}{Ts/2 + 1} \tag{5.4}$$

This approximate expression for $G_h(s)$ is quite satisfactory and for this reason it is very often used in practice. Clearly, this approximate expression involves the rational function (5.3) rather than the delay term e^{-Ts}, a fact which, as we mentioned above, facilitates determination of $G_c(s)$.

5.2.2 Specifications of the transient response of continuous-time systems

In this section, a brief review of the specifications of the transient response of continuous-time systems is given. These specifications are useful for the material that follows and, as is usually done, refer to the step response of a second-order system.

1. Overshoot

One of the basic characteristics of the transient response of a system is the overshoot v, which depends mainly on the damping ratio ζ. In the case of a second-order system, without zeros, i.e. for a system with transfer function of the form

$$H(s) = \frac{\omega_0}{s^2 + 2\zeta\omega_0 s + \omega_0^2}$$

it is approximately true that [3], [6], [8], [10]

$$\text{percentage overshoot} = v\% = 100 \exp\left(\frac{-\zeta\pi}{\sqrt{1-\zeta^2}}\right) \approx 100\left(1 - \frac{\zeta}{0.6}\right)$$

Therefore, for a desired percentage overshoot, the damping ratio would be

$$\zeta \geq (0.6)\left(1 - \frac{v\%}{100}\right)$$

2. Rise time

Another property which is of interest is the rise time t_r, which is defined as the time required for the response of the system to rise from 0.1 to 0.9 of its final value. For all values of ζ around 0.5, the rise time is approximately given by

$$t_r \approx 1.8/\omega_0$$

where ω_0 is the natural frequency of the system. Hence, to satisfy the above relation for the rise time, the natural frequency ω_0 should satisfy the condition

$$\omega_0 \geq 1.8/t_r$$

3. Settling time

Finally, another significant characteristic of the response in the time domain is the settling time t_s, which is defined as the time required for the response to reach and remain close (i.e. within a small error) to the final value. The settling time t_s is given by the relation

$$t_s = \beta/\zeta\omega_0$$

where β is a constant. In the case of an error tolerance of about 1%, the constant β takes on the value 4.6, whereas in the case of an error tolerance of about 2%, the constant β takes on the value 4. Hence, if we desire that the settling time should be smaller than a specified value, and for an error tolerance of about 1%, then

$$\zeta\omega_0 \geq 4.6/t_s$$

■ **REMARK 5.2.1** Theorem 2.3.1 requires that the sampling frequency f must be at least twice the highest frequency of the frequency spectrum of the continuous-time input signal. In practice, for a wide class of systems, the selection of the sampling period $T = 1/f$ is made using the following approximate method. Let q be the smallest time constant of the system. Then, T is chosen so that $T \in [0.1q, 0.5q]$.

■ **EXAMPLE 5.2.1** Consider the position control servomechanism shown in Figure 5.3(a). This system is designed to control the angle of the load automatically.

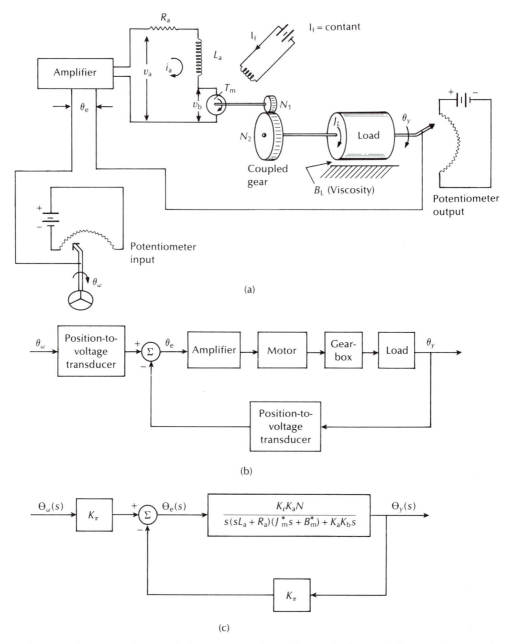

Figure 5.3 The position control servomechanism: (a) overall picture of the position control system; (b) closed-loop control system; (c) block diagram of the system.

Specifically, the output angle position θ_y is required to follow every change in the input (command) angle position θ_ω, as accurately as possible. This may be accomplished as follows. The input θ_ω is driven manually, rotating the slider of the input potentiometer by θ_ω degrees. The input angle θ_ω, as well as the output angle θ_y, are converted to electrical signals whose difference is the error θ_e. The error θ_e is amplified through a suitable amplifier whose output is used to excite the armature-controlled motor. The motor rotates the load through gears. Clearly, if $\theta_e = 0$, then $\theta_\omega = \theta_y$, meaning that the system has come to a standstill. In the case where $\theta_e \neq 0$, the motor will be rotated clockwise or counterclockwise depending on the sign of the error θ_e, until it reaches a standstill, i.e. until $\theta_e = 0$. Figure 5.3(b) shows a schematic diagram of the servomechanism.

The algebraic and differential equations describing the various subsystems are as follows:

1. The error detector:

$$\theta_e(t) = K_\pi[\theta_\omega(t) - \theta_y(t)] \tag{5.5}$$

   where K_π is the constant of the device.
2. The amplifier:

$$V_a(t) = K_\varepsilon \theta_e(t) \tag{5.6}$$

   where K_ε is the amplifying constant.
3. The motor: this is an armature-controlled motor with a fixed field current whose dynamic equations are approximated by

$$L_a \frac{di_a}{dt} + R_a i_a + v_b = v_a \quad \text{where} \quad v_b(t) = K_b \omega_m(t) \tag{5.7}$$

$$J_m^* \frac{d\omega_m}{dt} + B_m^* \omega_m = T_m \quad \text{where} \quad T_m(t) = K_a i_a(t) \tag{5.8}$$

   where $N = N_1/N_2$, $J_m^* = J_m + N^2 J_L$ and $B_m^* = B_m + N^2 B_L$. The constants J_m^* and B_m^* are determined under the assumption that the gearbox is ideal, i.e. that there are no losses. J_L is the moment of inertia and B_L is the viscosity coefficient of the load. Likewise, J_m and B_m, and J_m^* and B_m^*, are the moments of inertia and the viscosity coefficients of the motor and of the subsystem consisting of the motor, gears and load, respectively.
4. The output:

$$\frac{d\theta_m}{dt} = \omega_m \quad \text{and} \quad \theta_y = N\theta_m \tag{5.9}$$

The transfer function of the open-loop system is therefore

$$G(s) = \frac{\Theta_y(s)}{\Theta_e(s)} = \frac{K_\varepsilon K_a N}{s(sL_a + R_a)(J_m^* s + B_m^*) + K_a K_b s} \tag{5.10}$$

The transfer function of the closed-loop system (see Figure 5.3c) is consequently

$$H(s) = \frac{\Theta_y(s)}{\Theta_\omega(s)} = K_\pi \left(\frac{G(s)}{1 + K_\pi G(s)} \right)$$

$$= \frac{K_\pi K_f K_a N}{s(sL_a + R_a)(sJ_m^* + B_m^*) + K_a K_b s + K_\pi K_f K_a N} \tag{5.11}$$

Assuming that $L_a \approx 0$ the transfer function of the open-loop system reduces to

$$G(s) = \frac{K_f K_a N}{s(R_a J_m^* s + R_a B_m^* + K_a K_b)} = \frac{K}{s(As + B)} \tag{5.12}$$

where

$$K = \frac{K_f K_a N}{R_a}, \quad A = J_m^* \quad \text{and} \quad B = B_m^* + \frac{K_a K_b}{R_a} \tag{5.13}$$

Thus $G(s)$ is then a second-order system.

For simplicity, let $L_a \approx 0$, $K_\pi = 1$, $K = A = 1$ and $B = 2$. Then, the transfer function of the motor–gear–load system is $G(s) = 1/s(s+2)$. To this servomechanism, a controller $G_c(s)$ is introduced in series with a hold circuit $G_h(s)$ as shown in Figure 5.4. Of course, if the discrete-time controller $G_c(z)$ is introduced instead of the continuous-time controller $G_c(s)$, then the closed-loop system would be as shown in Figure 5.5. It is noted that in Figure 5.5 no approximation has been introduced to the hold circuit compared with Figure 5.4.

The specifications of the problem are as follows. For the closed-loop system shown in Figure 5.4, the step response should have maximum overshoot $v \leqslant 10\%$, the damping factor ζ should be around 0.6, the natural frequency should be $\omega_0 = 4$ rad/s and settling time t_s approximately 1.7 s. For the closed-loop system shown in Figure 5.5 the step response should have the same specifications as that of Figure 5.4.

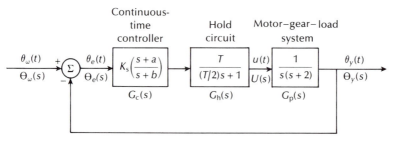

Figure 5.4 Continuous-time closed-loop control system of the position control servomechanism using a hold circuit.

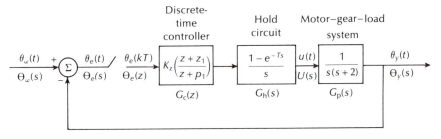

Figure 5.5 Discrete-time closed-loop control system of the position control servomechanism.

It is pointed out that, given ζ and ω_0, the parameters v and t_s may be estimated from the following two relations (see Section 5.2.2):

$$\text{percentage overshoot } v = 100 \exp\left(\frac{-\zeta\pi}{\sqrt{1-\zeta^2}}\right) \approx 10 \tag{5.14}$$

$$\text{settling time (up to 2\% of the final value)} = t_s \approx \frac{4}{\zeta\omega_0} \approx 1.7 \tag{5.15}$$

The design requirements are: determine $G_c(s)$ and $G_c(z)$ of the closed-loop systems shown in Figures 5.4 and 5.5, respectively. Compare the step responses of the two closed-loop systems.

■ **SOLUTION** To determine the approximate expression for $G_h(s)$ given by (5.4) we work as follows: the time constants of $G_p(s)$ are 1 and 0.5. Hence, the smallest time constant is $q=0.5$. Therefore, the sampling time T should lie between 0.05 and 0.25, in accordance with Remark 5.2.1. Thus let $T=0.15$, in which case the approximate transfer function of the hold circuit is $G_h(s)=2/(5+20)$.

The controller is of the form [3], [6], [8], [10]

$$G_c(s) = K_s\left(\frac{s+a}{s+b}\right)$$

We choose $a=2$ so that the term $(s+2)$ in the numerator will cancel the -2 pole of the system. We choose b and K_s so that the transfer function of the closed-loop system

$$H(s) = \frac{2K_s(s+2)}{(s+b)(s+20)s(s+2) + 2K_s(s+2)}$$

$$= \frac{2K_s}{s^3 + (20+b)s^2 + 20bs + 2K_s}$$

has denominator of the form

$$(s^2 + 2\zeta\omega_0 s + \omega_0^2)(s + x)$$

for some large positive x (so that the large negative pole at $-x$ does not influence the system behaviour). So we have

$$20 + b = 2\zeta\omega_0 + x$$
$$20b = \omega_0^2 + 2\zeta\omega_0 x$$
$$2K_s = \omega_0^2 x$$

where K_s, b, x are unknowns and ζ, ω_0 are given. Solving the above system of equations we obtain $K_s = 168.4$, $b = 5.85$, $x = 21$; thus x is indeed large. Finally, the controller transfer function is

$$G_c(s) = 168.4\left(\frac{s+2}{s+5.85}\right) \tag{5.16}$$

The transfer function $H(s)$ of the closed-loop system of Figure 5.4 is

$$H(s) = \frac{\Theta_y(s)}{\Theta_\omega(s)} = \frac{G_c(s)G_h(s)G_p(s)}{1 + G_c(s)G_h(s)G_p(s)}$$

$$= \frac{336.8}{s^3 + 25.85s^2 + 117s + 336.8} \tag{5.17}$$

To find $G_c(z)$ of the closed-loop system of Figure 5.5, it is sufficient to discretize $G_c(s)$ given in (5.16). To this end we will use the method of pole–zero matching presented in Section 3.3.3 (relations (3.48) to (3.51)). According to this method, $G_c(z)$ has the form

$$G_c(z) = K_z\left(\frac{z+z_1}{z+p_1}\right) \tag{5.18}$$

where the pole $s = -5.85$ of $G_c(s)$ is mapped into the pole $z = -p_1$ of $G_c(z)$ and the zero $s = -2$ of $G_c(s)$ is mapped into the zero $z = -z_1$ of $G_c(z)$. That is, we have that

$$p_1 = -e^{-bT} = -e^{-5.85(0.1)} = -0.5571$$
$$z_1 = -e^{-aT} = -e^{-2(0.1)} = -0.8187$$

The constant K_z of (5.18) is calculated so that the zero-frequency amplification constants of $G_c(z)$ and $G_c(s)G_h(s)$ are the same, i.e. so that the following holds (see relation (3.51)):

$$G_c(z = 1) = K_z\left(\frac{1 - 0.8187}{1 - 0.5571}\right) = G_c(s = 0)G_h(s = 0) = 168.4\left(\frac{0+2}{0+5.85}\right)\left(\frac{2}{0+20}\right)$$

Thus $K_z = 14.06$. It is noted that in the relation above the value of $G_h(s)$ for $s = 0$ was taken into consideration in order to obtain the total zero-frequency amplification for the continuous-time controller. Hence

$$G_c(z) = 14.06 \left(\frac{z - 0.8187}{z - 0.5571} \right) \tag{5.19}$$

In what follows, the responses of the closed-loop systems of Figures 5.4 and 5.5 are compared. To this end, the discrete-time transfer function of $\hat{G}(s) = G_h(s)G_p(s)$ is determined for $T = 0.1$. We have

$$\hat{G}(z) = \mathscr{Z}[G_h(s)G_p(s)] = \mathscr{Z} \left[\left(\frac{1 - e^{-0.1s}}{s} \right) \left(\frac{1}{s(s + 2)} \right) \right]$$

$$= (1 - z^{-1}) \mathscr{Z} \left[\frac{1}{s^2(s + 2)} \right] = (1 - z^{-1}) \mathscr{Z} \left[\frac{0.5}{s^2} - \frac{0.25}{s} + \frac{0.25}{s + 2} \right]$$

$$= \frac{z - 1}{z} \left(\frac{0.1}{(z - 1)^2} - \frac{0.25z}{z - 1} + \frac{0.25z}{z - e^{-0.2}} \right)$$

$$= \frac{0.0047(z + 0.9355)}{(z - 1)(z - 0.8187)} \tag{5.20}$$

The transfer function of the closed-loop system of Figure 5.5 will be

$$H(z) = \frac{G_c(z)\hat{G}(z)}{1 + G_c(z)\hat{G}(z)} = \frac{0.066z^{-1}(1 + 0.9355z^{-1})}{1 - 1.49z^{-1} + 0.619z^{-2}} \tag{5.21}$$

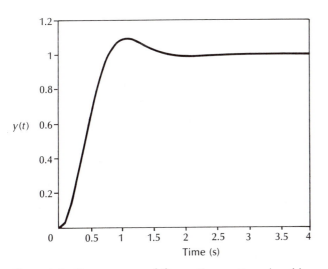

Figure 5.6 Step response of the continuous-time closed-loop system of Figure 5.4.

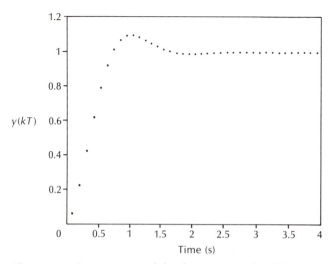

Figure 5.7 Step response of the discrete-time closed-loop system of Figure 5.5.

Figures 5.6 and 5.7 show the responses of the closed-loop systems of Figures 5.4 and 5.5, respectively, where it can be seen that the two step responses are almost the same.

■ **EXAMPLE 5.2.2** Consider the block diagram of an automatic orientation control system for a telescope on earth designed to track a bright star constantly (Figure 5.8). The closed-loop control system of the telescope should satisfy the following specifications:

(a) Overshoot $\leqslant 10\%$, a fact which implies that the damping factor ζ should be around 0.6 for a second-order system.
(b) Settling time $t_s \leqslant 5$ s, where a natural frequency of $\omega_0 \cong 1.33$ rad/s is required.

Determine the controller $G_c(s)$ and subsequently the equivalent discrete-time controller $G_c(z)$ so that the closed-loop system satisfies the specifications above.

■ **SOLUTION** To determine the approximate expression for $G_h(s)$ given by (5.4) we work as follows: the time constants of $G_p(s)$ are 1 and 10. Hence, the smallest time constant is $q = 1$. Therefore, the sampling time T should lie between 0.1 and 0.5, in accordance with Remark 5.2.1. Let us choose $T = 0.2s$. Then $G_h(s) = 2/(s + 10)$. Subsequently, using the approach presented in Example 5.2.1, an appropriate continuous-time controller, designed for the closed-loop system satisfying the specifications, is derived and has the following form [10]:

$$G_c(s) = K_s\left(\frac{s + a}{s + b}\right) = 9.08\left(\frac{10s + 1}{s + 1.81}\right) \tag{5.22}$$

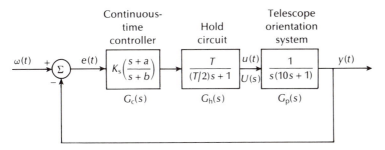

Figure 5.8 Continuous-time closed-loop control system of the telescope of Example 5.2.2.

To determine the respective discrete-time controller $G_c(z)$ we will apply the pole–zero matching method, in which case the discrete-time controller $G_c(z)$ has the following form:

$$G_c(z) = K_z \left(\frac{z + z_1}{z + p_1} \right) \tag{5.23}$$

where the pole $s = -1.81$ of $G_c(s)$ is mapped to the pole $z = -p_1$ of $G_c(z)$ and the zero $s = -0.1$ of $G_c(s)$ is mapped to the zero $z = -z_1$ of $G_c(z)$. That is, we have that

$$p_1 = -e^{-bT} = -e^{(-1.81)(0.2)} = -e^{-0.36} = -0.7$$
$$z_1 = -e^{-aT} = -e^{(-0.1)(0.2)} = -e^{-0.02} = -0.98$$

The constant K_z is calculated so that the zero-frequency amplification constants of $G_c(z)$ and $G_c(s)G_h(s)$ are the same, i.e. so that the following holds true (see relation (3.51)):

$$G_c(z=1) = K_z \left(\frac{1-0.98}{1-0.7} \right) = G_c(s=0)G_h(s=0) = 9.08 \left(\frac{0+1}{0+1.81} \right) \left(\frac{2}{0+10} \right)$$

Thus, $K_z \cong 15$. Hence

$$G_c(z) = 15 \left(\frac{z-0.98}{z-0.7} \right) \tag{5.24}$$

This relation may easily be converted to a difference equation suitable for realization using a digital computer controller. From relation (5.24) we obtain

$$G_c(z) = \frac{U(z)}{E(z)} = 15 \left(\frac{1-0.98z^{-1}}{1-0.7z^{-1}} \right)$$

which can be written as

$$U(z)(1-0.7z^{-1}) = 15E(z)(1-0.98z^{-1})$$

which may be expressed in the time domain in the form of a difference equation as follows:

$$u(k) = 0.7u(k-1) + 15[e(k) - 0.98e(k-1)]$$

This difference equation may be directly realized on a digital computer under the condition that during every sampling period the value of the control signal $u(k)$ (which is the controller output), as well as the value of the error signal $e(k)$ at each previous sampling instant, are stored in memory. More details on controller realization techniques are given in Chapter 10.

Having completed the design of the discrete-time controller, examination of the specifications of the closed-loop system of the telescope may now be carried out. We have three alternatives. The first would be to realize the controller on a computer, connect it to the telescope system and observe the efficiency of the overall system and, thus, find out if the desired specifications are actually satisfied. A second alternative would be to analyze the whole closed-loop system in the z-plane and theoretically determine the influence on the specifications of the transient response as a result of the discretization of the continuous-time controller. The third alternative would be to simulate the overall system (controller and telescope) on a computer in order to observe the response of the closed-loop system.

We will consider the second alternative. To this end, the discrete-time transfer function $\hat{G}(z)$ of $\hat{G}(s) = G_h(s)G_p(s)$ is calculated for $T = 0.2$. We have

$$\hat{G}(z) = \mathcal{L}[G_h(s)G_p(s)] = \mathcal{L}\left[\left(\frac{1 - e^{-0.2s}}{s}\right)\left(\frac{1}{s(10s+1)}\right)\right]$$

$$= (1 - z^{-1})\mathcal{L}\left[\frac{0.1}{s^2(s+0.1)}\right] = (1 - z^{-1})\mathcal{L}\left[\frac{1}{s^2} - \frac{1}{0.1s} + \frac{1}{0.1}\frac{1}{s+0.1}\right]$$

$$= \frac{z-1}{z}\left(\frac{0.2z}{(z-1)^2} - \frac{z}{0.1(z-1)} + \frac{1}{0.1}\frac{z}{z-e^{-0.02}}\right)$$

$$= 0.002\left(\frac{z+0.993}{(z-1)(z-0.98)}\right) \tag{5.25}$$

The poles of the discrete-time closed-loop system are the roots of the characteristic equation

$$1 + G_c(z)\hat{G}(z) = 0$$

whose roots are

$$z = 0.835 \pm j0.181$$

These two poles may be transformed to the s-plane using the transformation which maps the z-plane onto the s-plane and has the form

$$s = \frac{1}{T}\ln(z)$$

From the relation above it follows that the equivalent poles of the system on the s-plane are

$$s = -0.787 \pm j1.067$$

Hence, the specifications of the discrete-time closed-loop system are altered as follows:

$\zeta = 0.59$ (instead of $\zeta = 0.6$)

$\omega_0 = 1.326$ (instead of 1.333)

$t_s = 5$ s (same)

$v = 10\%$ (same)

Selecting a smaller sampling frequency makes the dynamic response of the system worse, since a greater hold time of the order of $T/2$ is introduced in the loop owing to the hold circuit and consequently increased delay. To see more clearly how this can happen, let us examine the case where the value $T = 1$ s, i.e. T lies outside the specified range $[0.1, \ 0.5]$. Repeating the above calculations for $T = 1$ s, we have

$$G_c(z) = 9.97\left(\frac{z - 0.905}{z - 0.002}\right) \tag{5.26}$$

and

$$\hat{G}(z) = 0.048\left(\frac{z + 0.967}{(z - 1)(z - 0.905)}\right) \tag{5.27}$$

Combining equations (5.26) and (5.27) and repeating the procedure for $T = 0.2$ s, we find that the poles of the closed system for $T = 1$ s are now

$$z = 0.2615 \pm j0.63$$
$$s = -0.383 \pm j1.177$$

Therefore, the specifications for the transient response change as follows:

$\zeta = 0.309$ (instead of the desired value $\zeta = 0.6$)

$\omega_0 = 1.238$ (instead of 1.333)

$t_s = 10.45$ s (instead of 5 s)

$v = 36\%$ (instead of 10%) (5.28)

It is clear from this example that the indirect design method for the discrete-time closed-loop system does not produce acceptable results in the case where the sampling period has not been chosen in accordance with Remark 5.2.1.

5.3 Controller design via the root-locus method

The root-locus method is a direct method for determining $G_c(z)$ and is applied as follows. Consider the closed-loop system shown in Figure 5.9. The transfer function $H(z)$ of the closed-loop system is

$$H(z) = \frac{G(z)}{1 + G(z)F(z)} \tag{5.29}$$

The characteristic equation of the closed-loop system is

$$1 + G(z)F(z) = 0 \tag{5.30}$$

For linear time-invariant systems, the open-loop transfer function $G(z)F(z)$ has the form

$$G(z)F(z) = K \frac{\prod_{i=1}^{m}(z + z_i)}{\prod_{i=1}^{n}(z + p_i)} \tag{5.31}$$

Substituting (5.31) in (5.30) yields the algebraic equation

$$\prod_{i=1}^{n}(z + p_i) + K \prod_{i=1}^{m}(z + z_i) = 0 \tag{5.32}$$

■ **DEFINITION 5.3.1** The root loci of the closed-loop system of Figure 5.9 are the loci of (5.32) in the z-domain as the parameter K varies from $-\infty$ to $+\infty$. Since the poles $-p_i$ and the zeros $-z_i$ are, in general, functions of the sampling period T, it follows that for each T there corresponds a root locus of (5.32), thus yielding a family of root loci for various values of T.

The construction of the root locus of (5.32) is straightforward. It is based on a set of rules which yield an approximate graphical representation of the root locus in the z-plane. These rules are presented in a variety of introductory control textbooks, and for this reason are not presented here.

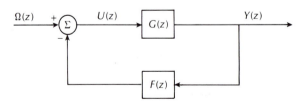

Figure 5.9 Discrete-time closed-loop system.

The following examples illustrate the construction of the root locus as the parameter K varies from 0 to $+\infty$. They also illustrate the influence on the root locus of the parameter T. Clearly, this influence of the sampling time T is a feature which appears in the case of discrete-time systems but not in continuous-time systems.

■ **EXAMPLE 5.3.1**   Consider the closed-loop system shown in Figure 5.10. Construct the root locus of the system for $K>0$ and for several values of the sampling period T. Note that for $a=2$, the system under control is the position control system presented in Example 5.2.1.

■ **SOLUTION**   Let $\hat{G}(s) = G_h(s)G_p(s)$. Then

$$\hat{G}(s) = \left(\frac{1 - e^{-sT}}{s}\right)\left(\frac{1}{s(s+a)}\right) = \frac{1}{s^2(s+a)}(1 - e^{-sT})$$

$$= \frac{1}{a^2}\left(\frac{a}{s^2} - \frac{1}{s} + \frac{1}{s+a}\right)(1 - e^{-sT})$$

Hence

$$\hat{g}(t) = \mathcal{L}^{-1}[\hat{G}(s)] = f(t) - f(t-T)\beta(t-T)$$

where $\beta(t-T)$ is the unit step function (2.2) and

$$f(t) = \mathcal{L}^{-1}\left[\frac{1}{a^2}\left(\frac{a}{s^2} - \frac{1}{s} + \frac{1}{s+a}\right)\right] = \frac{1}{a^2}(at - 1 + e^{-at})$$

Therefore

$$\hat{g}(kT) = f(kT) - f(kT-T)\beta(kT-T)$$

Applying relation (2.19b) we have

$$\hat{G}(z) = \mathcal{Z}[\hat{g}(kT)] = F(z) - z^{-1}F(z) = (1 - z^{-1})F(z)$$

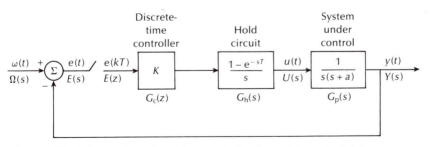

Figure 5.10   Discrete-time closed-loop control system of Example 5.3.1.

where

$$F(z) = \mathcal{Z}[f(kT)] = \frac{1}{a^2} \mathcal{Z}[akT - \beta(kT) + e^{-akT}]$$

$$= \frac{1}{a^2} \left(\frac{aTz}{(z-1)^2} - \frac{z}{z-1} + \frac{z}{z-e^{-aT}} \right)$$

Therefore

$$\hat{G}(z) = \mathcal{Z}[\hat{G}(s)] = \mathcal{Z}[G_h(s)G_p(s)] = \mathcal{Z}[\hat{g}(kT)] = (1 - z^{-1})F(z)$$

$$= \frac{1}{a^2} (1 - z^{-1}) \left(\frac{aTz}{(z-1)^2} - \frac{z}{z-1} + \frac{z}{z-e^{-aT}} \right)$$

$$= \frac{(aT + e^{-aT} - 1)}{a^2} \left(\frac{z + [(1 - aT e^{-aT} - e^{-aT})/(-1 + aT + e^{-aT})]}{(z-1)(z-e^{-aT})} \right)$$

or

$$\hat{G}(z) = K_0 \left(\frac{(z + z_1)}{(z-1)(z-p_1)} \right)$$

where

$$K_0 = \frac{1}{a^2} (aT + e^{-aT} - 1), \quad p_1 = e^{-aT} \quad \text{and} \quad z_1 = \frac{1 - Ta\, e^{-aT} - e^{-aT}}{-1 + aT + e^{-aT}}$$

Hence, the open-loop transfer function $G(z)F(z)$ for the present example is

$$G(z)F(z) = K\hat{G}(Z) = K \left(\frac{K_0(z + z_1)}{(z-1)(z-p_1)} \right)$$

Clearly, the constant K_0, the pole p_1 and the root $-z_1$ are functions of the sampling period T. For this reason, as T changes, so will the root locus of $G(z)F(z)$. Figure 5.11 presents the root loci for three different values T_1, T_2 and T_3 of T, where $T_1 < T_2 < T_3$.

Next, the special case where $a = 1$ and $T = 1$, 2 and 4 s will be studied. For $a = 1$ and $T = 1$, the open-loop transfer function $G(z)F(z)$ becomes

$$G(z)F(z) = K \left(\frac{0.368(z + 0.718)}{(z-1)(z-0.368)} \right)$$

For $a = 1$ and $T = 2$, the open-loop transfer function $G(z)F(z)$ becomes

$$G(z)F(z) = K \left(\frac{1.135(z + 0.523)}{(z-1)(z-0.135)} \right)$$

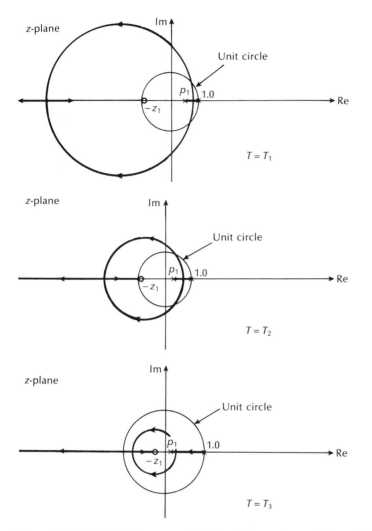

Figure 5.11 Root-loci diagrams of the closed-loop system of Figure 5.10, where $T_1 < T_2 < T_3$.

Finally, for $a = 1$ and $T = 4$, the open-loop transfer function $G(z)F(z)$ becomes

$$G(z)F(z) = K\left(\frac{3.018(z + 0.3)}{(z - 1)(z - 0.018)}\right)$$

Figure 5.12 presents the root loci for $a = 1$ and for the three cases of $T = 1, 2$ and 4 s. The influence of the parameter T can be observed here in greater detail than in Figure 5.11. Figure 5.12 shows that, for a fixed value of K, an increase in the

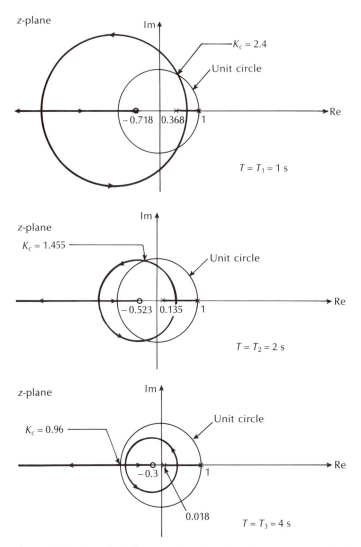

Figure 5.12 Root-loci diagrams for the closed-loop system of Figure 5.10 for $a = 1$ and $T = 1$, 2 and 4 s.

sampling time T would result in a less stable closed-loop system. On the contrary, a decrease in T results in a more stable system. As a matter of fact, the more the sampling period T tends to zero $(T \rightarrow 0)$, the more the behaviour of the closed-loop system approaches that of the continuous-time system (here, the continuous-time closed-loop system is stable for all positive values of K). Also, as the value of T increases, the critical value K_c decreases and vice versa. By critical value K_c we mean the particular value of K where the system becomes unstable.

■ **EXAMPLE 5.3.2** This example refers to the automatic control of a nuclear reactor (Figure 5.13a). The major control objective here is to maintain the output power within specified limits. This can be achieved as follows. The nuclear reaction releases energy in the form of heat. This energy is used for the production of steam. The steam is subsequently used to drive a turbine and, in turn, the turbine drives a generator which finally produces electrical power.

The input ω is a reference signal which corresponds to the desired output power. The two signals ω and y are compared and their difference $e = \omega - y$ is fed into the control unit. This device consists of special bars which, when they move towards the point where the nuclear reaction takes place, increase the output power y, and when they move away, decrease the output power y. When $y > \omega$, in this case the error e is negative, the bars move away from the point of the nuclear reaction and the output y is decreased. When $y < \omega$, in this case the error e is positive and the

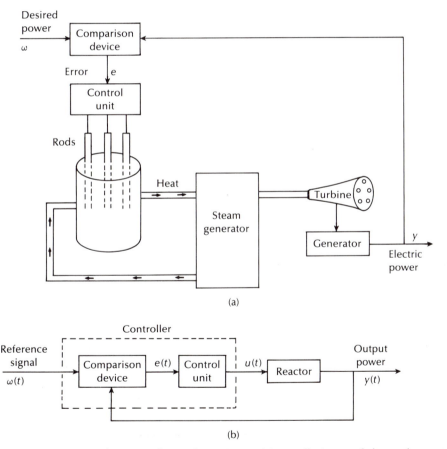

Figure 5.13 Control system of a nuclear reactor: (a) overall picture of the nuclear reactor; (b) block diagram of the system.

bars move towards the point of the nuclear reaction resulting in an increase of the output y.

A simplified block diagram of the control system of the reactor is shown in Figure 5.14, where the reactor is described approximately by a first-order system. A significant improvement in the behaviour of the closed-loop system is achieved if we apply a discrete-time controller, not of the proportional type (i.e. $G_c(z) = K$), but of the integral type, i.e.

$$G_c(z) = K\left(\frac{1}{1 - z^{-1}}\right) = K\left(\frac{z}{z - 1}\right)$$

(More on integral controllers, as well as on more complex controllers such as the PID controller, is presented in Section 5.5.)

Construct the root locus of the closed-loop system of Figure 5.14 for $T = 0.5$, 1 and 2 s.

■ **SOLUTION**   Let $\hat{G}(s) = G_h(s)G_a(s)$. Then

$$\mathscr{L}[\hat{G}(s)] = \mathscr{L}\left[\left(\frac{1 - e^{-Ts}}{s}\right)\left(\frac{1}{s + 1}\right)\right] = (1 - z^{-1})\mathscr{L}\left[\frac{1}{s(s + 1)}\right]$$

$$= (1 - z^{-1})\left(\frac{z}{z - 1} - \frac{z}{z - e^{-T}}\right) = \frac{1 - e^{-T}}{z - e^{-T}}$$

The open-loop transfer function $G(z)F(z)$ is

$$G(z)F(z) = G_c(z)\hat{G}(z) = K\left(\frac{z}{z - 1}\right)\left(\frac{1 - e^{-T}}{z - e^{-T}}\right)$$

For $T = 0.5$ s, the open-loop transfer function $G(z)F(z)$ becomes

$$G(z)F(z) = K\left(\frac{0.394z}{(z - 1)(z - 0.607)}\right)$$

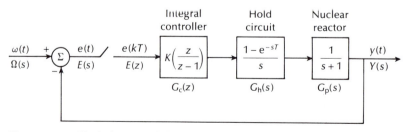

Figure 5.14   Block diagram of the closed-loop control system of a nuclear reactor, where the controller applied is an integral discrete-time controller.

For $T = 1$ s, the open-loop transfer function $G(z)F(z)$ becomes

$$G(z)F(z) = K\left(\frac{0.632z}{(z-1)(z-0.368)}\right)$$

Finally, for $T = 2$ s, the open-loop transfer function $G(z)F(z)$ becomes

$$G(z)F(z) = K\left(\frac{0.865z}{(z-1)(z-0.135)}\right)$$

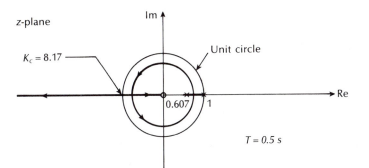

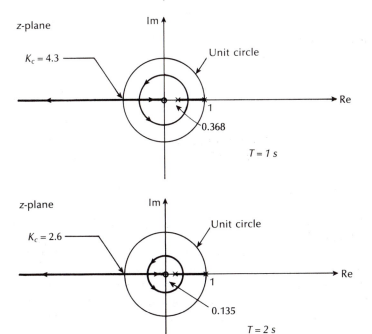

Figure 5.15 Root-loci diagrams for the closed-loop system of the nuclear reactor for $T = 0.5$, 1 and 2 s.

Figure 5.15 shows the root loci for the three cases of $T = 0.5$, 1 and 2 s. It can be seen that, for a fixed value of K, an increase in the sampling time T results in a less stable closed-loop system. In the opposite case, a decrease in the value of T results in a more stable system. In particular, the more the sampling period T approaches zero the more the behaviour of the closed-loop system approaches that of a continuous-time system. Also, as the value of T increases, the critical value K_c decreases and vice versa. By critical value K_c we mean the particular value of K where the system becomes unstable.

5.4 Controller design based on the frequency response

5.4.1 Introduction

The well-established frequency domain design controller techniques for continuous-time systems (i.e. Bode and Nyquist diagrams) can be extended to include the case of discrete-time systems. At first, one might think of carrying out this extension by using the mapping $z = e^{sT}$. Making use of this relation, the simple and easy to use logarithmic curves of the Bode diagrams for the continuous-time case cease to hold for discrete-time systems (that is why the extension via the relation $z = e^{sT}$ is not recommended). To maintain the simplicity of the logarithmic curves for discrete-time systems, we make use of the bilinear transformation

$$z = \frac{1 + Tw/2}{1 - Tw/2} \quad \text{or} \quad w = \frac{2}{T}\left(\frac{z-1}{z+1}\right) \tag{5.33}$$

The transformation of a function of s into a function of z based on the mapping $z = e^{sT}$, and subsequently the transformation of the resulting function of z into a function of w based on relation (5.33), are presented in Figure 5.16. The figure shows that the transformation of the left-half complex plane on the s-plane transforms into the unit circle in the z-plane via the relation $z = e^{sT}$, whereas the unit circle in the z-plane transforms into the left-half complex plane in the w-plane via the bilinear transformation (5.33).

At first sight, it seems that the frequency responses would be the same in both the s- and w-domains. This is actually true, the only difference being that the scales of the frequencies w and v are distorted, where v is the (hypothetical or abstract) frequency in the w-domain. This frequency distortion or warping may be observed if we set $w = jv$ and $z = e^{j\omega T}$ in equation (5.33) yielding

$$w|_{w=jv} = jv = \frac{2}{T}\left(\frac{z-1}{z+1}\right)\bigg|_{z=e^{j\omega T}} = \frac{2}{T}\left(\frac{e^{j\omega T}-1}{e^{j\omega T}+1}\right) = j\frac{2}{T}\tan\left(\frac{\omega T}{2}\right)$$

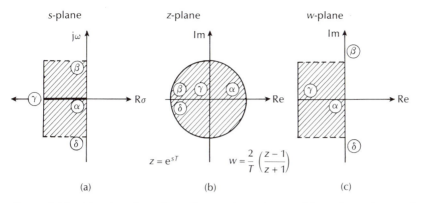

Figure 5.16 Mappings from the s-plane to the z-plane and from the z-plane to the w-plane.

Therefore

$$v = \frac{2}{T} \tan\left(\frac{\omega T}{2}\right) \tag{5.34}$$

Since

$$\tan\left(\frac{\omega T}{2}\right) = \frac{\omega T}{2} - \frac{(\omega T)^3}{8} + \dots \tag{5.35}$$

it follows that for small values of ωT we have that $\tan(\omega T/2) \simeq \omega T/2$. Substituting this result in equation (5.34) we have

$$v \simeq \omega \quad \text{for small } \omega T \tag{5.36}$$

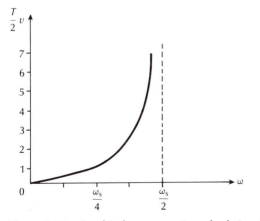

Figure 5.17 Graphical representation of relation (5.34).

Therefore, the frequencies ω and v are linearly related only if the product ωT is small. For larger ωT, equation (5.36) no longer holds true. Figure 5.17 shows the graphical representation of equation (5.34). The frequency range $-\omega_s/2 \leqslant \omega \leqslant \omega_s/2$ in the s-domain corresponds to the frequency range $-\infty \leqslant v \leqslant \infty$ in the w-domain, where ω_s is defined by the relation $(\omega_s/2)(T/2) = \pi/2$.

5.4.2   Bode diagrams

Using the foregoing results, one may readily design discrete-time controllers using Bode diagrams. To this end, consider the closed-loop system shown in Figure 5.18. Then, the five basic steps for the design of $G_c(z)$ are the following:

1. Determine $\hat{G}(z)$ from the relation $\hat{G}(z) = \mathscr{L}[\hat{G}(s)] = \mathscr{L}[G_h(s)G_p(s)]$
2. Determine $\hat{G}(w)$ using the bilinear transformation (5.33), yielding

$$\hat{G}(w) = \hat{G}(z)\big|_{z = (1 + Tw/2)/(1 - Tw/2)} \tag{5.37}$$

3. Set $w = jv$ in $\hat{G}(w)$ and draw the Bode diagrams of $\hat{G}(jv)$.
4. Determine the poles and zeros of $G_c(w)$ using similar techniques to those applied for continuous-time systems [3], [6], [8], [10].
5. Determine $G_c(z)$ from $G_c(w)$ using the bilinear transformation (5.33), yielding

$$G_c(z) = G_c(w)\big|_{w = (2/T)[(z - 1)/(z + 1)]} \tag{5.38}$$

Note that the specifications for the bandwidth are transformed from the s-domain to the w-domain using relation (5.34). Thus, if for example ω_b is the desired frequency bandwidth, then the design in the w-domain must be carried out for a frequency bandwidth v_b, where

$$v_b = \frac{2}{T} \tan\left(\frac{\omega_b T}{2}\right) \tag{5.39}$$

■■■ **EXAMPLE 5.4.1**   Consider the position servomechanism shown in Figure 5.5 of Example 5.2.1. Find a controller $G_c(z)$ such that the closed-loop system satisfies the following specifications: gain margin $K_g \geqslant 25$ db, phase margin $\phi_p \geqslant 70°$ and velocity error constant $K_v = 1\ \text{s}^{-1}$. The sampling period T is chosen to be 0.1 s.

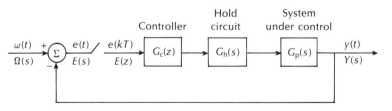

Figure 5.18   Discrete-time closed-loop control system.

■ **SOLUTION** Let $\hat{G}(s) = G_h(s)G_p(s)$. Then

$$\hat{G}(z) = \mathscr{L}[\hat{G}(s)] = \mathscr{L}\left[\frac{1 - e^{-Ts}}{s}\,\frac{1}{s(s+2)}\right] = (1 - z^{-1})\mathscr{L}\left[\frac{1}{s^2(s+2)}\right]$$

$$= 0.0047z^{-1}\left(\frac{1 + 0.935z^{-1}}{(1 - z^{-1})(1 - 0.819z^{-1})}\right) = (0.0047)\left(\frac{z + 0.935}{(z-1)(z-0.819)}\right)$$

For $T = 0.1$ s, the bilinear transformation (5.33) becomes

$$z = \frac{1 + Tw/2}{1 - Tw/2} = \frac{1 + 0.05w}{1 - 0.05w}$$

Substituting this transformation in $\hat{G}(z)$ we have

$$\hat{G}(w) = \frac{0.0047\{[(1 + 0.05w)/(1 - 0.05w)] + 0.935\}}{\{[(1 + 0.05w)/(1 - 0.05w)] - 1\}\{[(1 + 0.05w)/(1 - 0.05w)] - 0.8187\}}$$

$$= \frac{0.5(1 + 0.00167w)(1 - 0.05w)}{w(1 + 0.5w)}$$

The gain and phase Bode diagrams of $\hat{G}(j\upsilon) = \hat{G}(w = j\upsilon)$ are given in Figure 5.19. We select the following form for the regulator $G_c(w)$:

$$G_c(w) = K\left(\frac{1 + aw}{1 + bw}\right)$$

where a and b are constants. The open-loop transfer function is

$$G_c(w)\hat{G}(w) = K\left(\frac{1 + aw}{1 + bw}\right)\left(\frac{0.5(1 + 0.00167w)(1 - 0.05w)}{w(1 + 0.5w)}\right)$$

From the definition of the velocity error constant K_v we have

$$K_v = \lim_{w \to 0} [wG_c(w)\hat{G}(w)] = 0.5K = 1$$

and therefore $K = 2$. The parameters a and b can be determined by applying the respective techniques for continuous-time systems [3], [6], [8], [10], which yield $a = 0.8$ and $b = 0.5$. Hence

$$G_c(w) = 2\left(\frac{1 + 0.8w}{1 + 0.5w}\right)$$

The open-loop transfer function is

$$G_c(w)\hat{G}(w) = 2\left(\frac{1 + 0.8w}{1 + 0.5w}\right)\left(\frac{0.5(1 + 0.00167w)(1 - 0.05w)}{w(1 + 0.5w)}\right)$$

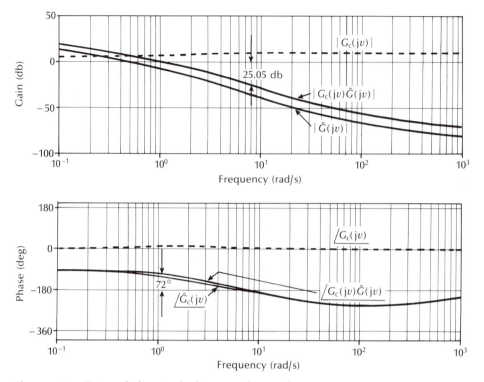

Figure 5.19 Gain and phase Bode diagrams of Example 5.4.1.

We note that $K_g \approx 25.05$ db, $\phi_p \approx 72°$ and $K_v = 1$ s^{-1}. Therefore the closed-loop design requirements are indeed satisfied. It remains to determine $G_c(z)$ from $G_c(w)$. To this end, we use the bilinear transformation (5.33) which, for $T = 0.1$ s, becomes

$$w = \frac{2}{T}\left(\frac{z-1}{z+1}\right) = \frac{2}{0.1}\left(\frac{z-1}{z+1}\right) = 20\left(\frac{z-1}{z+1}\right)$$

Thus, $G_c(z)$ has the form

$$G_c(z) = 2\left(\frac{1 + (0.8)(20)[(z-1)/(z+1)]}{1 + (0.5)(20)[(z-1)/(z+1)]}\right) = 3.09\left(\frac{z - 0.882}{z - 0.818}\right) = 3.09\left(\frac{1 - 0.882z^{-1}}{1 - 0.818z^{-1}}\right)$$

■ **EXAMPLE 5.4.2** This example refers to the automatic control of a space satellite orientation system, a simplified form of which is shown in Figure 5.20(a). This system is designed so that the satellite's telescope, which is mounted on the satellite, is constantly in perfect orientation with a star, despite disturbances such as collisions with meteorites, gravity forces, etc. The system operates as follows: the orientation of the telescope is achieved with the help of a reference star which is much brighter

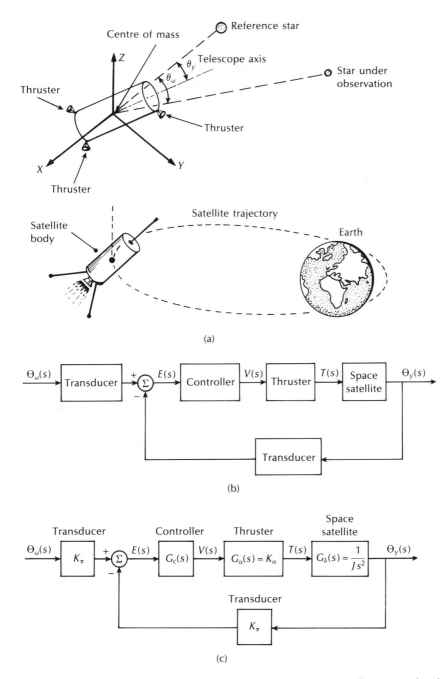

Figure 5.20 Space satellite orientation control system: (a) overall picture; (b) closed-loop control system; (c) block diagram of closed-loop system.

than the star under observation. It is initially assumed that the axis of the telescope is in line with the star under observation. At this position, the axis forms an angle of θ_w with the line which connects the telescope to the reference star. The angle θ_w is known in advance and is the system's excitation. If for any reason the angle θ_y, which is formed by the axis of the telescope with the reference star, is different from the desired angle θ_w, then the control system should align the Z-axis so that $\theta_y = \theta_w$. The block diagram of the closed-loop system is given in Figure 5.20(b), where the block designated 'thruster' is excited by the controller's output and produces torque forces about the X-axis. These forces will ultimately cause the correction in the angle θ_y, so that $\theta_y = \theta_w$.

The various blocks in Figure 5.20(c) are described, respectively, by the relations

$$E(s) = K_\pi[\Theta_w(s) - \Theta_y(s)] \tag{5.40a}$$

$$V(s) = G_c(s)E(s) \tag{5.40b}$$

$$T(s) = G_a(s)V(s) = K_a V(s) \tag{5.40c}$$

$$\Theta_y(s) = G_\delta(s)T(s) = \frac{1}{Js^2} T(s) \tag{5.40d}$$

where K_π is the conversion constant of the position angle to an electrical signal, $G_c(s)$ is the transfer function of the controller, K_a is the constant of the thruster in which the voltage $V(s)$ is converted to the torque $T(s)$, and $G_\delta(s) = 1/Js^2$ is the satellite's transfer function, where J is the moment of inertia of the satellite about the X-axis. It is noted that relation (5.40d) does not involve friction terms, since the satellite is assumed to be in space.

Find a discrete-time controller so that the closed-loop system has the following specifications: gain margin $K_g \geqslant 15$ db, phase margin $\phi_p \geqslant 45°$ and sampling time $T = 0.1$ s. For simplicity, assume that $K_\pi = K_a = J = 1$.

■ **SOLUTION** Consider the block diagram given in Figure 5.21. Let $\hat{G}(s) = G_h(s)G_\delta(s)$. Then

$$\hat{G}(z) = \mathscr{L}\left[\left(\frac{1 - e^{-Ts}}{s}\right)\left(\frac{1}{s^2}\right)\right] = (1 - z^{-1})\mathscr{L}\left[\frac{1}{s^3}\right] = (1 - z^{-1})\frac{T^2(1 + z^{-1})z^{-1}}{2(1 - z^{-1})^3}$$

Setting $T = 0.1$, we have

$$\hat{G}(z) = \frac{0.005(1 + z^{-1})z^{-1}}{(1 - z^{-1})^2} = \frac{0.005(z + 1)}{(z - 1)^2}$$

The bilinear transformation (5.33), for $T = 0.1$, becomes

$$z = \frac{1 + Tw/2}{1 - Tw/2} = \frac{1 + 0.05w}{1 - 0.05w}$$

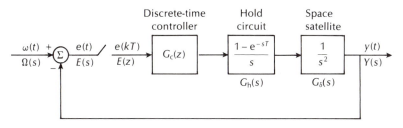

Figure 5.21 Block diagram of the discrete-time controller design problem for the automatic orientation of a satellite.

If the above bilinear transformation is substituted in $\hat{G}(z)$, we obtain

$$\hat{G}(w) = \frac{0.005\{[(1+0.05w)/(1-0.05w)]+1\}}{\{[(1+0.05w)/(1-0.05w)]-1\}^2} = \frac{1-0.05w}{w^2}$$

The transfer function $\hat{G}(jv) = \hat{G}(w=jv)$ will be

$$\hat{G}(jv) = \frac{1-0.05jv}{(jv)^2}$$

The gain and phase Bode diagrams of $\hat{G}(jv)$ are shown in Figure 5.22. Based on these diagrams and using continuous-time system controller design techniques [3], [6], [8], [10], we arrive at the controller:

$$G_c(w) = 4\left(\frac{0.8w+1}{0.07w+1}\right)$$

The open-loop transfer function is

$$G_c(w)\hat{G}(w) = 4\left(\frac{0.8w+1}{0.07w+1}\right)\left(\frac{1-0.05w}{w^2}\right)$$

Checking the above results we find that $K_g \approx 15.06$ db and $\phi_p \approx 46.6°$. Hence, the design requirements of the closed-loop system are satisfied. Finally it is necessary to determine $G_c(z)$ from $G_c(w)$. To this end, we use the bilinear transformation (5.33), which for $T = 0.1$ becomes

$$w = \frac{2}{T}\left(\frac{z-1}{z+1}\right) = \frac{2}{0.1}\left(\frac{z-1}{z+1}\right) = 20\left(\frac{z-1}{z+1}\right)$$

Hence, the controller $G_c(z)$ will be

$$G_c(z) = 4\left(\frac{(0.8)(20)[(z-1)/(z+1)]+1}{(0.07)(20)[(z-1)/(z+1)]+1}\right) = 28.33\left(\frac{z-0.882}{z-0.166}\right)$$

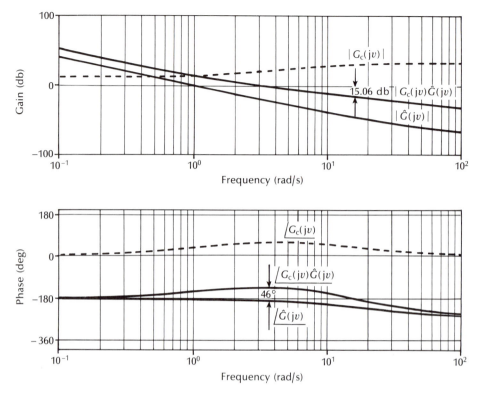

Figure 5.22 Gain and phase Bode diagrams of Example 5.4.2.

5.4.3 *Nyquist diagrams*

Consider a closed-loop system with open-loop transfer function $\tilde{G}(z) = \mathfrak{L}[G(s)F(s)]$. Since the z- and s-domains are related via the relation $z = e^{sT}$, it follows that the Nyquist diagram of $\tilde{G}(z)$ is the diagram of $\tilde{G}(e^{sT})$, as s traces the Nyquist path. In the z-domain, the Nyquist path is given by the relation

$$z = e^{sT}\big|_{s=j\omega} = e^{j\omega T} \tag{5.41}$$

and, therefore, the Nyquist path in the z-domain is the unit circle. Hence, to apply the Nyquist stability criterion for discrete-time systems, we draw the diagram of $\tilde{G}(e^{j\omega T})$ having the cyclic frequency ω as a parameter.

The following theorem holds.

■ **THEOREM 5.4.1** Assume that the transfer function $\mathfrak{L}[G(s)F(s)]$ does not have any poles outside the unit circle. Then, the closed-loop system is stable if the Nyquist diagram of $\mathfrak{L}[G(s)F(s)]$, for $z = e^{j\omega T}$, does not encircle the critical point $(-1, j0)$.

Theorem 5.4.1 is the equivalent of the Nyquist theorem for continuous-time systems [3], [6], [8], [10].

Clearly, the study of stability of discrete-time closed-loop systems, as well as the design of diskrete-time controllers, can be accomplished on the basis of Nyquist diagrams by extending the known techniques of continuous-time systems as done for the Bode diagrams in the previous section. This extension is straightforward and is not presented here (see for example [4]).

5.5 The PID controller

In discrete-time systems, as in the case of continuous-time systems, the PID controller is widely used in practice [2]. This section is devoted to the study of PID controllers. We will first study the proportional (P), the integral (I) and the derivative (D) actions and subsequently the composite PID controller. Furthermore, the classical PID controller design methods, suggested by Ziegler and Nichols [12], will be presented.

5.5.1 *Proportional controller*

For continuous-time systems, the proportional controller is described by the relation

$$u(t) = K_p e(t) \tag{5.42a}$$

and therefore

$$G_c(s) = K_p \tag{5.42b}$$

For discrete-time systems, the equivalent proportional controller is described by the relation

$$u(k) = K_p e(k) \tag{5.43a}$$

and therefore

$$G_c(z) = K_p \tag{5.43b}$$

5.5.2 *Integral controller*

For continuous-time systems, the integral controller is described by the integral equation

$$u(t) = \frac{K_p}{T_i} \int_{t_0}^{t} e(t) \, dt \tag{5.44a}$$

and therefore

$$G_c(s) = \frac{K_p}{T_i s} \tag{5.44b}$$

where the constant T_i is called the integration time constant or reset. In the case of discrete-time systems, the integral equation (5.44a) can be approximated by the difference equation

$$\frac{u(k) - u(k-1)}{T} = \frac{K_p}{T_i} e(k)$$

in which case

$$u(k) = u(k-1) + \frac{K_p T}{T_i} e(k) \tag{5.45a}$$

whose discrete transfer function is

$$G_c(z) = \frac{K_p T}{T_i(1-z)^{-1}} = \frac{K_p Tz}{T_i(z-1)} \tag{5.45b}$$

5.5.3 *Derivative controller*

For continuous-time systems, the derivative controller is described by the differential equation

$$u(k) = K_p T_d \frac{de(t)}{dt} \tag{5.46a}$$

and therefore

$$G_c(z) = K_p T_d s \tag{5.46b}$$

where the constant T_d is called the derivative or rate time constant. In the case of discrete-time systems, the differential equation (5.46a) can be approximated by the difference equation

$$u(k) = K_p T_d \left(\frac{e(k) - e(k-1)}{T} \right) \tag{5.47a}$$

whence

$$G_c(z) = K_p T_d \left(\frac{1 - z^{-1}}{T} \right) = \frac{K_p T_d}{T} \left(\frac{z-1}{z} \right) \tag{5.47b}$$

5.5.4 *The three-term PID controller*

Combining all the previous three control actions, we have the classical PID controller which finds extensive application in industrial control. For the continuous-time case, the controller in its basic form is described by the integro-differential equation

$$u(t) = K_p \left(e(t) + \frac{1}{T_i} \int_{t_0}^{t} e(t)\, dt + T_d \frac{de(t)}{dt} \right) \tag{5.48a}$$

whence

$$G_c(s) = K_p \left(1 + \frac{1}{T_i s} + T_d s \right) \tag{5.48b}$$

Figure 5.23 presents a block diagram of the three-term controller $G_c(s)$.

In the case of discrete-time systems, the PID controller can be described in its simplest form by the difference equation

$$u(k) = K_p \left(e(k) + \frac{T}{T_i} \sum_{i=0}^{k-1} e(i) + \frac{T_d}{T} [e(k) - e(k-1)] \right) \tag{5.49a}$$

where the middle term is the solution of equation (5.45a). Hence

$$G_c(z) = K_p \left[1 + \frac{T}{T_i} \left(\frac{z}{z-1} \right) + \frac{T_d}{T} \left(\frac{z-1}{z} \right) \right] \tag{5.49b}$$

After some algebraic manipulations, $G_c(z)$ may be written as

$$G_c(z) = K \left(\frac{z^2 - az + b}{z(z-1)} \right) \tag{5.50}$$

where

$$K = K_p \left(\frac{TT_i + T_d T_i + T^2}{T_i T} \right)$$

$$a = \frac{T_i T - T_d T_i}{TT_i + T_d T_i + T^2}$$

$$b = \frac{T_d T_i}{TT_i + T_d T_i + T^2}$$

Figure 5.24 presents the block diagram of the three-term controller $G_c(z)$.

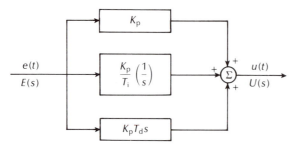

Figure 5.23   The block diagram of the continuous-time PID controller.

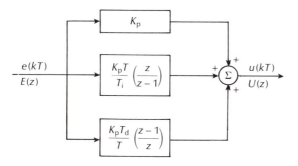

Figure 5.24   The block diagram of the discrete-time PID controller.

5.5.5 *Design of PID controllers using the Ziegler–Nichols methods*

The PID controller has the capability of simultaneously tuning three parameters, namely K_p, T_i and T_d. This allows a PID controller to satisfy the design requirements in many practical cases, a fact which makes the PID controller very useful in practice. The appropriate values of the parameters K_p, T_i and T_d may be chosen using the techniques presented thus far in the present chapter. However, this is usually a formidable task, even in cases where the design engineer has great experience of the subject. To facilitate the determination of the appropriate values of K_p, T_i and T_d, for cases where a mathematical model for the system under control is not available, Ziegler and Nichols [12] have suggested the following two simple and practically useful methods.

1. The transient response method

The system under control is excited with the unit step function. The shape of the transient response of the open-loop system has the general form shown in Figure 5.25. We introduce the parameters θ and t_d, where $\theta = \text{slope} = K/t_r = $ reaction time and $t_d = $ delay time. To achieve a damping factor of about $\zeta = 0.2$, the values of the parameters K_p, T_i and T_d are chosen according to Table 5.1.

Table 5.1 The values of the parameters K_p, T_i and T_d using the Ziegler–Nichols transient response method.

Controller		K_p	T_i	T_d
Proportional	P	$1/(t_d\theta)$		
Proportional–integral	PI	$0.9/(t_d\theta)$	$3t_d$	
Proportional–integral–derivative	PID	$1.2/(t_d\theta)$	$2t_d$	$0.5t_d$

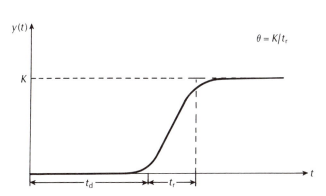

Figure 5.25 Step response of the system under control.

2. The stability limit method

We start tuning the system by controlling it only with a proportional controller. The parameter K_p is slowly increased until a persistent oscillation is reached. At this point, we note the value of the parameter K_p, denoted as $\tilde{K}_p$, as well as the respective oscillation period, denoted as $\tilde{T}$. Then, K_p, T_i and T_d are chosen according to Table 5.2.

Table 5.2 The values of the parameters K_p, T_i and T_d using the Ziegler–Nichols stability limit method.

Controller		K_p	T_i	T_d
Proportional	P	$0.5\,\tilde{K}_p$		
Proportional–integral	PI	$0.45\,\tilde{K}_p$	$\tilde{T}/1.2$	
Proportional integral–derivative	PID	$0.6\,\tilde{K}_p$	$\tilde{T}/2$	$\tilde{T}/8$

The above two simple rules hold for continuous-time systems and can also be applied to discrete-time systems as long as the sampling is fast, i.e. 20 times the highest bandwidth frequency, as is normally the case in practice. If sampling is not that fast, then the discrete-time PID controller may not produce satisfactory results.

5.6 Steady-state errors

Consider the discrete-time closed-loop system shown in Figure 5.26. Assume that the system under control is stable (a fact which will allow us to apply the final value theorem given by equation (2.27)). Define

$$\hat{G}(z) = \mathscr{L}[G_h(s)G(s)] = \mathscr{L}\left[\left(\frac{1 - e^{-Ts}}{s}\right)G(s)\right] = (1 - z^{-1})\mathscr{L}\left[\frac{G(s)}{s}\right]$$

Also define

$$\tilde{G}(z) = \mathscr{L}[G_h(s)G(s)F(s)] = (1 - z^{-1})\mathscr{L}\left[\frac{G(s)F(s)}{s}\right]$$

Then, the closed-loop transfer function $H(z)$ is

$$H(z) = \frac{Y(z)}{\Omega(z)} = \frac{\hat{G}(z)}{1 + \tilde{G}(z)} \tag{5.51}$$

The error $E(z)$ is given by

$$E(z) = \Omega(z) - B(z) = \Omega(z) - \tilde{G}(z)E(z)$$

and hence

$$E(z) = \left(\frac{1}{1 + \tilde{G}(z)}\right)\Omega(z) \tag{5.52}$$

The steady state error of $e(t)$, denoted by e_{ss}, is defined as

$$e_{ss} = \lim_{k \to \infty} e(kT) = \lim_{z \to 1} (1 - z^{-1})E(z) \tag{5.53}$$

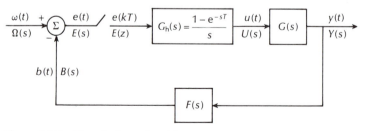

Figure 5.26 Discrete-time closed-loop system.

Relation (5.53) is known as the final value theorem which is defined by relation (2.27). If equation (5.52) is substituted in (5.53) then

$$e_{ss} = \lim_{z \to 1} \left[(1 - z^{-1}) \left(\frac{1}{1 + \tilde{G}(z)} \right) \Omega(z) \right] \qquad (5.54)$$

Next, we will consider three particular excitations $\omega(t)$, namely the step function, the ramp function and the acceleration function.

1. Step function

In this case, $\omega(t) = 1$ or $\omega(kT) = \beta(kT)$ and

$$\Omega(z) = \mathscr{Z}[\omega(kT)] = \mathscr{Z}[\beta(kT)] = \frac{1}{1 - z^{-1}}$$

Substituting the value of $\Omega(z)$ in equation (5.54) yields

$$e_{ss} = \lim_{z \to 1} \left[(1 - z^{-1}) \left(\frac{1}{1 + \tilde{G}(z)} \right) \left(\frac{1}{1 - z^{-1}} \right) \right] = \lim_{z \to 1} \left[\frac{1}{1 + \tilde{G}(z)} \right]$$

$$= \frac{1}{1 + K_p}, \quad K_p = \lim_{z \to 1} \tilde{G}(z) \qquad (5.55)$$

where K_p is called the **position error constant**.

2. Ramp function

In this case $\omega(t) = t$ or $\omega(kT) = r(kT)$ which is defined by the relation (2.4). Hence

$$\Omega(z) = \mathscr{Z}[\omega(kT)] = \mathscr{Z}[r(kT)] = \frac{Tz^{-1}}{(1 - z^{-1})^2}$$

Substituting the value of $\Omega(z)$ in equation (5.54) yields

$$e_{ss} = \lim_{z \to 1} \left[(1 - z^{-1}) \left(\frac{1}{1 + \tilde{G}(z)} \right) \left(\frac{Tz^{-1}}{(1 - z^{-1})^2} \right) \right] = \lim_{z \to 1} \left[\frac{T}{(1 - z^{-1}) \tilde{G}(z)} \right]$$

$$= \frac{1}{K_v}, \quad K_v = \lim_{z \to 1} \left[\frac{(1 - z^{-1}) \tilde{G}(z)}{T} \right] \qquad (5.56)$$

where K_v is called the **velocity error constant**.

3. Acceleration function

In this case $\omega(t) = \frac{1}{2} t^2$ or $\omega(kT) = \frac{1}{2}(kT)^2$ and

$$\Omega(z) = \mathscr{Z}[\omega(kT)] = \mathscr{Z}[\tfrac{1}{2}(kT)^2] = \frac{T^2(1 + z^{-1})z^{-1}}{2(1 - z^{-1})^3}$$

Substituting the value of $\Omega(z)$ in equation (5.54) yields

$$e_{ss} = \lim_{z \to 1}\left[(1 - z^{-1})\left(\frac{1}{1 + \tilde{G}(z)}\right)\left(\frac{T^2(1 + z^{-1})z^{-1}}{2(1 - z^{-1})^3}\right)\right] = \lim_{z \to 1}\left[\frac{T^2}{(1 - z^{-1})^2\tilde{G}(z)}\right]$$

$$= \frac{1}{K_a}, \quad K_a = \lim_{z \to 1}\left[\frac{(1 - z^{-1})^2\tilde{G}(z)}{T^2}\right] \tag{5.57}$$

where K_a is called the **acceleration error constant**.

■ **REMARK 5.6.1**   As in the case of continuous-time systems [3], [10], a discrete-time system is called a **type j system** when its transfer function $\tilde{G}(z)$ has the form

$$\tilde{G}(z) = \frac{1}{(z - 1)^j}\left[\frac{a(z)}{b(z)}\right] \tag{5.58}$$

where $a(z)$ and $b(z)$ are polynomials in z which do not involve the term $(z - 1)$.

■ **EXAMPLE 5.6.1**   Consider the position control servomechanism system of Example 5.2.1. Determine the constants K_p, K_v and K_a of the continuous- and discrete-time closed-loop systems.

■ **SOLUTION**   For the continuous-time case (Figure 5.4), we have

$$\tilde{G}(s) = G_c(s)G_h(s)G_p(s) = 168.2\left(\frac{s + 2}{s + 5.85}\right)\left(\frac{2}{s + 20}\right)\left(\frac{1}{s(s + 2)}\right)$$

Hence

$$K_p = \lim_{s \to 0} \tilde{G}(s) = \lim_{s \to 0} 168.2\left(\frac{s + 2}{s + 5.85}\right)\left(\frac{2}{s + 20}\right)\left(\frac{1}{s(s + 2)}\right) = \infty$$

$$K_v = \lim_{s \to 0} sG(s) = \lim_{s \to 0} s\left[168.2\left(\frac{s + 2}{s + 5.85}\right)\left(\frac{2}{s + 20}\right)\left(\frac{1}{s(s + 2)}\right)\right] = 2.875$$

$$K_a = \lim_{s \to 0} s^2\tilde{G}(s) = \lim_{s \to 0} s^2\left[168.2\left(\frac{s + 2}{s + 5.85}\right)\left(\frac{2}{s + 20}\right)\left(\frac{1}{s(s + 2)}\right)\right] = 0$$

For the discrete-time case (Figure 5.5), we have

$$\tilde{G}(z) = \mathscr{Z}[\tilde{G}(s)] = 14.06\left(\frac{z - 0.8187}{z - 0.5571}\right)\left(\frac{0.0047(z + 0.9355)}{(z - 1)(z - 0.8187)}\right)$$

$$= 0.066\left(\frac{z + 0.9355}{(z - 1)(z - 0.5571)}\right)$$

Hence

$$K_p = \lim_{z \to 1} \tilde{G}(z) = \lim_{z \to 1} \left[0.066 \left(\frac{z + 0.9355}{(z-1)(z-0.5571)} \right) \right] = \infty$$

$$K_v = \lim_{z \to 1} \left[\frac{(1-z^{-1})}{T} \tilde{G}(z) \right]$$

$$= \lim_{z \to 1} \left[\frac{(1-z^{-1})}{0.1} \right] \left[0.066 \left(\frac{z + 0.9355}{(z-1)(z-0.5571)} \right) \right] = 2.88$$

$$K_a = \lim_{z \to 1} \left[\frac{(1-z^{-1})^2}{T^2} \tilde{G}(z) \right]$$

$$= \lim_{z \to 1} \left[\frac{(1-z^{-1})^2}{(0.1)^2} \right] \left[0.066 \left(\frac{z + 0.9355}{(z-1)(z-0.5571)} \right) \right] = 0$$

5.7 Problems

1. Consider a continuous-time second-order system with poles on the left-half complex plane given by

$$s_{1,2} = -\zeta\omega \pm j\sqrt{1-\zeta^2}\,\omega = -\sigma \pm j\omega_d$$

where $\sigma = \zeta\omega$ and $\omega_d = \sqrt{1-\zeta^2}\,\omega$. The parameter ζ is the damping ratio with $0 \leq \zeta \leq 1$ and ω is the natural frequency of the system.

(a) Determine the root locus in the z-plane under the transformation $z = e^{sT}$ for the following:

1. The points on the s-plane which have a constant real part, i.e. $\sigma = \zeta\omega = $ constant.
2. The points on the s-plane of constant ζ.
3. The points on the s-plane of constant ω.

(b) The desired performance characteristics for the continuous-time system are specified in terms of the time domain quantities: percentage overshoot $v\%$, rise time t_r and settling time t_s corresponding to a $\pm 1\%$ tolerance. Translate these quantities to z-plane performance requirements for the equivalent ($z = e^{sT}$) discrete-time system.

2. Solve Example 5.2.1 with the following specifications:

(a) Maximum overshoot $\leq 10\%$ and natural frequency $\omega_0 = 2$ rad/s.
(b) Maximum overshoot $\leq 20\%$ and natural frequency $\omega_0 = 6$ rad/s.
(c) Maximum overshoot $\leq 10\%$ and natural frequency $\omega_0 = 6$ rad/s.

3. Solve Example 5.2.2 with the following specifications:

(a) Overshoot ≤ 10% and settling time $t_s \leq 2$ s with tolerance ±1%.
(b) Overshoot ≤ 20% and settling time to $t_s \leq 4$ s with tolerance ±2%.

4. Solve Example 5.4.1 with the following specifications: $K_g \geq 20$ db, $\phi_p \geq 50°$ and velocity error constant $K_v = 2$ s^{-1}.

5. Solve Example 5.4.2 with the following specifications: $K_g \geq 18$ db and $\phi_p \geq 60°$.

6. Consider the ball and beam system depicted in Figure 5.27. The beam is free to rotate in the plane of the page about an axis perpendicular to the page while the ball rolls in a groove along the beam. The control problem is that of maintaining the ball at a desired position by applying an input torque to the beam.

(a) A linear model for the system is $G(s) = 1/s^2$ as shown with a PD controller in Figure 5.28. Obtain an equivalent discrete-time system. The sampling period is $T = 0.1$ s.
(b) Design a discrete-time controller using the pole–zero matching method. Draw the unit step responses for the continuous- and the discrete-time systems.

7. Plastic extrusion is an industrial process. The extruders consist of a large barrel which is divided into several temperature zones with a hopper at one end and a die at the other. The polymer is fed into the barrel from the hopper and is pushed forward by a powerful screw. Simultaneously, it is heated while passing through the various temperature zones set to gradually increasing temperatures. The heat produced by the heaters in the barrel, together with the heat released from the friction between the raw polymer and the surfaces of the barrel and the screw, eventually causes the

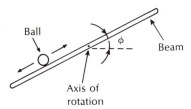

Figure 5.27 Ball and beam system.

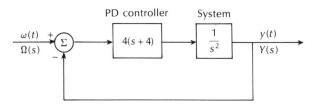

Figure 5.28 Continuous-time control system.

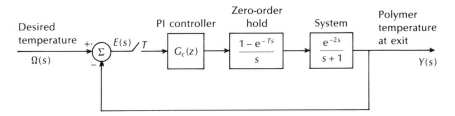

Figure 5.29 Temperature control system for plastic extrusion.

polymer to melt. The polymer is then pushed out of the die. The discrete-time system for the temperature control is shown in Figure 5.29.

The transfer function relating the angular velocity of the screw and the output temperature is $G(s) = e^{-2s}/(s+1)$, i.e. the system is of the first order, incorporating a delay of 2 s. The sampling period $T = 1$ s. Design a PI controller so that the dominant closed-loop poles have a damping ratio $\zeta = 0.5$ and the number of the output samples in a full cycle of the damped sinusoidal response is 10. Find the unit step response of the discrete-time system. Determine the velocity error coefficient K_v and the steady state error of the output due to a ramp input.

8. A photovoltaic system is mounted on a space station in order to develop the required power. The photovoltaic panels should follow the sun as accurately as possible in order to maximize energy production. The system uses a d.c. motor and the transfer function of the panel mount and the motor is

$$G(s) = \frac{1}{s(s+1)}$$

An optical sensor is available to track the sun's position accurately and forms a unity feedback (see Figure 5.30). The sampling period $T = 0.2$ s. Find a discrete-time controller such that the dominant closed-loop poles have damping ratio $\zeta = 0.5$ and there are eight output samples in a complete cycle of the damped sinusoidal response. Use the root-locus method in the z-plane to determine the transfer function of the required controller. Find the unit step response of the system and the velocity error coefficient K_v.

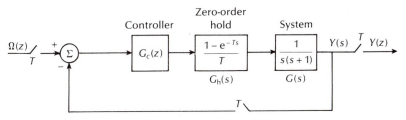

Figure 5.30 Discrete-time system for the positioning of the photovoltaic panels.

9. In this problem, the automatic control of a wheelchair will be studied. The automatic wheelchair is specially designed for handicapped people with a disability from the neck down. It consists of a control system which the handicapped person may operate by using his or her head, thus determining the direction as well as the speed of the chair. The direction is determined by a sensor situated on the head of the handicapped person in intervals of 90°, so that the person may choose one of four possible directions (motions): forward, backward, left and right. The speed is regulated by another sensor, whose output is proportional to the movement of the head. Clearly, in the present example, the person is part of the overall controller.

For simplicity, we assume that the wheelchair, as well as the sensory device on the head, are described by first-order transfer functions as shown in Figure 5.31(a). We also assume that the time delay, which is anticipated to appear in the visual feedback path, is negligible. More specifically, we assume that $K_1 = 1$, $K_2 = 10$, $\alpha = 1$, $\beta = 2$ and $F(s) = 1$. Suppose that we want to introduce a discrete-time controller to the system. Then, the closed-loop system would be as in Figure 5.31(b). Find $G_c(z)$ in order for the closed-loop system to have a gain margin $K_g \geq 12$ db, a phase margin $\phi_p \geq 50°$ and an error constant $K_v = 4$ s^{-1}. The sampling period is chosen to be 0.1 s.

10. Construct the root locus of Example 5.3.1 for the following values of the parameters a and T: $a = 4$ and $T = 0.1, 1, 5$ and 10.

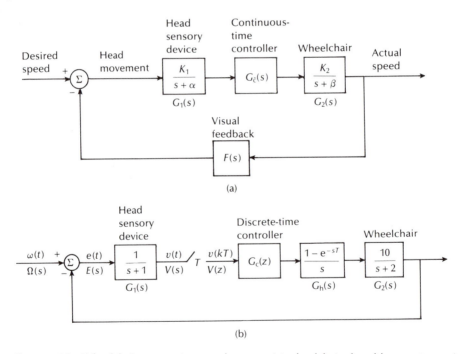

(a)

(b)

Figure 5.31 Wheelchair automatic control system: (a) wheelchair closed-loop system using a continuous-time controller; (b) wheelchair closed-loop system using a discrete-time controller.

11. Let the characteristic equation of a discrete-time system be

$$z^2 + a_1 Kz + a_0 K = 0$$

Draw the root locus and determine the range of values of K which yields a stable system (the variable K takes both positive and negative values) for the following values of a_1 and a_0:

(a) $a_1 = -0.5$ and $a_0 = 0.1$
(b) $a_1 = -0.2$ and $a_0 = 0.1$.

12. Consider the continuous-time system shown in Figure 5.32. Find the velocity error constant and the steady-state error due to a unit ramp input.

Discretize this system using a zero-order hold for the plant and an equivalent discrete-time controller for the integrator, obtained with the backward difference method. Show that the velocity error constant of the equivalent discrete-time system is the same as that obtained earlier. Determine an equivalent discrete-time controller using the pole–zero matching method with the requirement that the velocity error constant is the same.

13. Consider a continuous-time open-loop system with transfer function $G(s) = a/s(s + a)$. Close the loop with unity feedback and find the position and the velocity error constants. Use a zero-order hold and unity feedback to obtain a discrete-time equivalent system. Determine the new position and velocity error constants and compare with the continuous-time case.

14. Consider a discrete-time system with unity feedback (see Figure 5.33). The closed-loop transfer function is

$$\frac{Y(z)}{\Omega(z)} = \frac{K(z + b_0)}{z^2 + a_1 z + a_0}$$

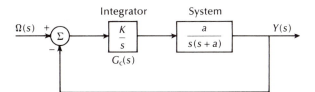

Figure 5.32 Continuous-time system with integrator.

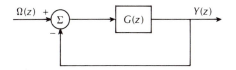

Figure 5.33 Discrete-time control system.

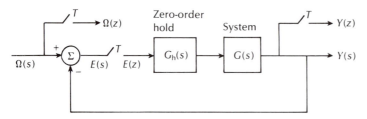

Figure 5.34 Discretized control system.

Let an integrator, i.e. a simple pole at $z = 1$, be included in the forward path (the combined system is of type 1). Determine the steady state error of the system excited by a ramp sequence.

15. Consider the system of Figure 5.34 with unity feedback where

$$G(s) = \frac{K}{s(s + a)}$$

(a) Determine the transfer function $Y(z)/\Omega(z)$ in terms of K, a and T (sampling period).

(b) Determine the root locus and the maximum value of K for a stable response with $T = 0.1$ s, 0.5 s and 1 s and $a = 2$.

(c) Find the steady state error characteristics for a unit step sequence and a ramp sequence for those values of K and T that yield a stable system response for $a = 1$ and $a = 2$.

5.8 **References**

Books

[1] J. Ackermann, *Sampled Data Control Systems*, Springer Verlag, New York, 1985.

[2] K.J. Åström and T. Hagglund, *Automatic Tuning of PID Controllers*, ISA, Sweden, 1988.

[3] G.F. Franklin, J.D. Powell and A. Emami-Naeini, *Feedback Control of Dynamic Systems*, Addison-Wesley, Reading, Massachusetts, 1986.

[4] G.F. Franklin, J.D. Powell and M.L. Workman, *Digital Control of Dynamic Systems*, Addison-Wesley, London, 1990 (Second Edition).

[5] C.H. Houpis and G.B. Lamont, *Digital Control Systems*, McGraw-Hill, New York, 1985.

[6] B.C. Kuo, *Automatic Control Systems*, Prentice Hall, Englewood Cliffs, New Jersey, 1975 (Third Edition).

[7] B.C. Kuo, *Digital Control Systems*, Holt-Saunders, Tokyo, 1980.

[8] K. Ogata, *Modern Control Engineering*, Prentice Hall, Englewood Cliffs, New Jersey, 1970.

[9] K. Ogata, *Discrete-Time Control Systems*, Prentice Hall, Englewood Cliffs, New Jersey, 1987.

[10] P.N. Paraskevopoulos, *Introduction to Automatic Control Systems*, in preparation.

[11] C.L. Phillips and H.T. Nagle Jr, *Digital Control System Analysis and Design*, Prentice Hall, Englewood Cliffs, New Jersey, 1984.

Papers

[12] J.G. Ziegler and N.B. Nichols, 'Optimum Settings for Automatic Controllers', *Transactions of the ASME*, Vol. 64, pp. 759–768, 1942.

6

State-space design methods

6.1 Introduction

In Chapter 5 we presented **classical** control design methods for discrete-time systems, where the system under control is described by its transfer function, while the controller sought has the form of a rational function. Both Chapters 6 and 7 refer to **modern** control design methods for discrete-time systems, where the system under control is described in state space, while the controller sought is a linear function of the state vector. Chapter 6 deals with the very popular feedback design problems of pole placement, deadbeat control and state observers, while Chapter 7 deals with the problem of optimal control.

Modern control methods are developed in the time domain, while classical methods are developed in the frequency domain. Modern methods cover the general case of multi-input multi-output systems, whereas classical methods refer mostly to single-input single-output systems. Finally, modern control methods are mainly analytical, whereas classical control methods are mainly graphical.

The design methods presented in this chapter, as compared with those presented in Chapter 7, are relatively easier to apply, since they require simple mathematics. For this and other reasons, all results in the present chapter (pole placement, deadbeat control and state vector observers) are becoming increasingly useful in practical applications.

6.2 The pole-placement design method

Consider the linear time-invariant discrete-time system

$$\mathbf{x}(k+1) = \mathbf{A}\mathbf{x}(k) + \mathbf{B}\mathbf{u}(k) \tag{6.1}$$

where we assume that all states are accessible and known. To this system we apply a linear state feedback control law of the form

$$\mathbf{u}(k) = -\mathbf{F}\mathbf{x}(k) \tag{6.2}$$

Then, the closed-loop system (see Figure 6.1) is given by the homogeneous equation

$$\mathbf{x}(k+1) = (\mathbf{A} - \mathbf{BF})\mathbf{x}(k) \tag{6.3}$$

Clearly, the controller matrix $\mathbf{F}$ may be chosen so as to improve the performance of the closed-loop system (6.3). One method to improve the performance of (6.3) is that of pole placement, which consists of finding a particular matrix $\mathbf{F}$ such that the poles of the closed-loop system (6.3) take on desirable preassigned values. Using this method, the behaviour of the open-loop system may be improved significantly. The method can stabilize an unstable system, increase or decrease the rate of response, widen or narrow the system bandwidth, increase or decrease the steady state error, etc. For these reasons, improving the system performance through the pole-placement method is widely used in practice.

The pole-placement or eigenvalue assignment problem can be defined as follows: let $\lambda_1, \lambda_2, \ldots, \lambda_n$ be the eigenvalues of the matrix $\mathbf{A}$ of the open-loop system (6.1) and $\hat{\lambda}_1, \hat{\lambda}_2, \ldots, \hat{\lambda}_n$ be the desired eigenvalues of the matrix $\mathbf{A} - \mathbf{BF}$ of the closed-loop system (6.3), where all complex eigenvalues exist in complex conjugate pairs. Also, let $p(z)$ and $\hat{p}(z)$ be the respective characteristic polynomials, i.e. let

$$p(z) = \prod_{i=1}^{n}(z - \lambda_i) = |z\mathbf{I} - \mathbf{A}| = z^n + a_1 z^{n-1} + \ldots + a_{n-1}z + a_n \tag{6.4}$$

$$\hat{p}(z) = \prod_{i=1}^{n}(z - \hat{\lambda}_i) = |z\mathbf{I} - \mathbf{A} - \mathbf{BF}| = z^n + \hat{a}_1 z^{n-1} + \ldots + \hat{a}_{n-1}z + \hat{a}_n \tag{6.5}$$

Then, we have to find a matrix $\mathbf{F}$ so that equation (6.5) is satisfied.

The pole-placement problem has attracted considerable attention for many years. The first significant results were established by Wonham in the late 1960s and are given by the following theorem [14].

■ **THEOREM 6.2.1** There exists a state feedback matrix $\mathbf{F}$ which assigns to the matrix $\mathbf{A} - \mathbf{BF}$ of the closed-loop system any arbitrary eigenvalues $\hat{\lambda}_1, \hat{\lambda}_2, \ldots, \hat{\lambda}_n$, if and only if the state vector of the open-loop system (6.1) is controllable, i.e. if and only if

$$\text{rank } \mathbf{S} = n \quad \text{where } \mathbf{S} = [\mathbf{B} \,\vdots\, \mathbf{AB} \,\vdots\, \mathbf{A}^2\mathbf{B} \,\vdots\, \cdots \,\vdots\, \mathbf{A}^{n-1}\mathbf{B}] \tag{6.6}$$

where all complex eigenvalues of the set $\{\hat{\lambda}_1, .., \hat{\lambda}_n\}$ appear in conjugate pairs.

According to this theorem, in cases where the open-loop system (6.1) is not controllable, at least one eigenvalue of the matrix $\mathbf{A}$ remains invariant under the state feedback law (6.2). In such cases, in order to assign all eigenvalues, one must apply dynamic controllers in which the feedback law (6.2) will involve derivative and integral terms in the state vector (a special category of dynamic controllers are the PID controllers presented in Chapter 5). Dynamic controllers have the disadvantage that they increase the order of the system.

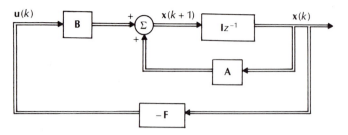

Figure 6.1   Closed-loop system with a linear state feedback law.

Subsequently, we will deal with the problem of determining a feedback matrix **F**, assuming that all eigenvalues of matrix **A** can be relocated if the system $(\mathbf{A}, \mathbf{B})$ is controllable. For simplicity, we will first study the case of single-input systems, in which the matrix **B** reduces to a column vector **b** and the matrix **F** reduces to a row vector $\mathbf{f}^T$. Equation (6.5) then becomes

$$\hat{p}(z) = \prod_{i=1}^{n} (z - \hat{\lambda}_i) = \left| z\mathbf{I} - \mathbf{A} + \mathbf{b}\mathbf{f}^T \right| = z^n + \hat{a}_1 z^{n-1} + \ldots + \hat{a}_{n-1}z + \hat{a}_n \tag{6.7}$$

The solution of (6.7) for **f** is unique. Several methods have been proposed for determining **f**. One of the most popular is due to Bass and Gura [4], which gives the following simple solution:

$$\mathbf{f} = [\mathbf{W}^T\mathbf{S}^T]^{-1}(\hat{\mathbf{a}} - \mathbf{a}) \tag{6.8}$$

where **S** is the controllability matrix defined in equation (6.6) and

$$\mathbf{W} = \begin{bmatrix} 1 & a_1 & \cdots & a_{n-1} \\ 0 & 1 & \cdots & a_{n-2} \\ \vdots & \vdots & & \vdots \\ 0 & 0 & \cdots & 1 \end{bmatrix}, \quad \hat{\mathbf{a}} = \begin{bmatrix} \hat{a}_1 \\ \hat{a}_2 \\ \vdots \\ \hat{a}_n \end{bmatrix} \quad \text{and} \quad \mathbf{a} = \begin{bmatrix} a_1 \\ a_2 \\ \vdots \\ a_n \end{bmatrix} \tag{6.9}$$

In the special case where the system under control is described in its phase-variable canonical form, i.e. **A** and **b** have the special forms $\mathbf{A}^*$ and $\mathbf{b}^*$, where

$$\mathbf{A}^* = \begin{bmatrix} 0 & 1 & 0 & \cdots & 0 \\ 0 & 0 & 1 & \cdots & 0 \\ 0 & 0 & 0 & \cdots & 0 \\ \vdots & \vdots & \vdots & & \vdots \\ 0 & 0 & 0 & \cdots & 1 \\ -a_n & -a_{n-1} & -a_{n-2} & \cdots & -a_1 \end{bmatrix} \quad \text{and} \quad \mathbf{b}^* = \begin{bmatrix} 0 \\ 0 \\ 0 \\ \vdots \\ 0 \\ 1 \end{bmatrix} \tag{6.10}$$

then it can be easily shown that the matrix $\mathbf{S}^* = [\mathbf{b}^* \vdots \mathbf{A}^*\mathbf{b}^* \vdots \mathbf{A}^{*2}\mathbf{b}^* \vdots \ldots \vdots \mathbf{A}^{*n-1}\mathbf{b}^*]$ is such that the product $\mathbf{W}^T\mathbf{S}^{*T}$ has the simple form

$$\mathbf{W}^T\mathbf{S}^{*T} = \tilde{\mathbf{I}} = \begin{bmatrix} 0 & 0 & \cdots & 0 & 1 \\ 0 & 0 & \cdots & 1 & 0 \\ \vdots & \vdots & & \vdots & \vdots \\ 1 & 0 & \cdots & 0 & 0 \end{bmatrix} \tag{6.11}$$

In this case, the vector $\mathbf{f}^*$ in expression (6.8) reduces to $\mathbf{f}^* = \tilde{\mathbf{I}}(\hat{\mathbf{a}} - \mathbf{a})$, i.e. to the following form [13]:

$$\mathbf{f}^* = \tilde{\mathbf{I}}(\hat{\mathbf{a}} - \mathbf{a}) = \begin{bmatrix} \hat{a}_n - a_n \\ \hat{a}_{n-1} - a_{n-1} \\ \vdots \\ \hat{a}_1 - a_1 \end{bmatrix} \tag{6.12}$$

where use is made of the property $(\tilde{\mathbf{I}})^{-1} = \tilde{\mathbf{I}}$. It is evident that expression (6.12) is extremely simple to apply, provided that the matrix $\mathbf{A}$ and the vector $\mathbf{b}$ of the system under control are in the phase-variable canonical form (6.10).

Another approach for computing $\mathbf{f}$ has been proposed by Ackermann, leading to the following expression [6]:

$$\mathbf{f}^T = \mathbf{e}^T\mathbf{S}^{-1}\hat{\mathbf{p}}(\mathbf{A}) \tag{6.13}$$

where the matrix $\mathbf{S}$ is given in (6.6), $\mathbf{e}^T = (0, 0, \ldots, 0, 1)$ and $\hat{\mathbf{p}}(\mathbf{A})$ is given by equation (6.5), in which the variable z is substituted by the matrix $\mathbf{A}$, i.e.

$$\hat{\mathbf{p}}(\mathbf{A}) = \mathbf{A}^n + \hat{a}_1\mathbf{A}^{n-1} + \cdots + \hat{a}_{n-1}\mathbf{A} + \hat{a}_n\mathbf{I} \tag{6.14}$$

The proof of equation (6.13) is presented in Example 6.2.3 below.

In the general case of multi-input systems, the determination of the matrix $\mathbf{F}$ is somewhat complicated [13]. A simple approach to the problem is to assume that $\mathbf{F}$ has the following outer product form:

$$\mathbf{F} = \mathbf{q}\mathbf{p}^T \tag{6.15}$$

where $\mathbf{q}$ and $\mathbf{p}$ are n-dimensional vectors. The matrix $\mathbf{A} - \mathbf{B}\mathbf{F}$ then becomes

$$\mathbf{A} - \mathbf{B}\mathbf{F} = \mathbf{A} - \mathbf{B}\mathbf{q}\mathbf{p}^T = \mathbf{A} - \boldsymbol{\beta}\mathbf{p}^T \quad \text{where } \boldsymbol{\beta} = \mathbf{B}\mathbf{q} \tag{6.16}$$

Therefore, assuming that $\mathbf{F}$ has the form (6.15), then the multi-input system case is reduced to the single-input case studied previously. In other words, the solution for the vector $\mathbf{p}$ is (6.8) or (6.13) and differs only in that the matrix $\mathbf{S}$ is now the matrix $\tilde{\mathbf{S}}$, having the form

$$\tilde{\mathbf{S}} = [\boldsymbol{\beta} \,\vdots\, \mathbf{A}\boldsymbol{\beta} \,\vdots\, \mathbf{A}^2\boldsymbol{\beta} \,\vdots\, \cdots \,\vdots\, \mathbf{A}^{n-1}\boldsymbol{\beta}] \tag{6.17}$$

The vector $\boldsymbol{\beta} = \mathbf{B}\mathbf{q}$ involves arbitrary parameters, which are the elements of the arbitrary vector $\mathbf{q}$. These arbitrary parameters can have any value, provided that rank $\tilde{\mathbf{S}} = n$. In cases where this condition cannot be satisfied, other approaches for determining $\mathbf{F}$ may be found in the literature [13].

■■■ **EXAMPLE 6.2.1**   Assume a system in the form (6.1), where

$$\mathbf{A} = \begin{bmatrix} 0 & 1 \\ -1 & 0 \end{bmatrix} \quad \text{and} \quad \mathbf{b} = \begin{bmatrix} 0 \\ 1 \end{bmatrix}$$

Find a vector $\mathbf{f}$ such that the closed-loop system has eigenvalues $\hat{\lambda}_1 = -1$ and $\hat{\lambda}_2 = 0.5$.

■■■ **SOLUTION**   We have

$$p(z) = |z\mathbf{I} - \mathbf{A}| = z^2 + 1 \quad \text{and} \quad \hat{p}(z) = (z - \hat{\lambda}_1)(z - \hat{\lambda}_2) = z^2 + 0.5z - 0.5$$

Method 1

Since the system is in phase-variable canonical form, the vector $\mathbf{f}$ can readily be determined by equation (6.12), as follows:

$$\mathbf{f} = \mathbf{f}^* = \begin{bmatrix} \hat{a}_2 - a_2 \\ \hat{a}_1 - a_1 \end{bmatrix} = \begin{bmatrix} -0.5 - 1 \\ 0.5 - 0 \end{bmatrix} = \begin{bmatrix} -1.5 \\ 0.5 \end{bmatrix}$$

Method 2

Here we use equation (6.8). Equations (6.9) and (6.6) give

$$\mathbf{W} = \begin{bmatrix} 1 & a_1 \\ 0 & 1 \end{bmatrix} = \begin{bmatrix} 1 & 0 \\ 0 & 1 \end{bmatrix} \quad \text{and} \quad \mathbf{S} = [\mathbf{b} \mathbin{\vdots} \mathbf{Ab}] = \begin{bmatrix} 0 & 1 \\ 1 & 0 \end{bmatrix}$$

Therefore

$$\mathbf{W}^T\mathbf{S}^T = \begin{bmatrix} 1 & 0 \\ 0 & 1 \end{bmatrix}\begin{bmatrix} 0 & 1 \\ 1 & 0 \end{bmatrix} = \begin{bmatrix} 0 & 1 \\ 1 & 0 \end{bmatrix} \quad \text{and} \quad (\mathbf{W}^T\mathbf{S}^T)^{-1} = \begin{bmatrix} 0 & 1 \\ 1 & 0 \end{bmatrix}$$

Hence

$$\mathbf{f} = (\mathbf{W}^T\mathbf{S}^T)^{-1}(\hat{\mathbf{a}} - \mathbf{a}) = \begin{bmatrix} 0 & 1 \\ 1 & 0 \end{bmatrix}\left(\begin{bmatrix} 0.5 \\ -0.5 \end{bmatrix} - \begin{bmatrix} 0 \\ 1 \end{bmatrix}\right) = \begin{bmatrix} 0 & 1 \\ 1 & 0 \end{bmatrix}\begin{bmatrix} 0.5 \\ -1.5 \end{bmatrix} = \begin{bmatrix} -1.5 \\ 0.5 \end{bmatrix}$$

Method 3

Here we apply equation (6.13). We have

$$\hat{\mathbf{p}}(\mathbf{A}) = \mathbf{A}^2 + \hat{a}_1\mathbf{A} + \hat{a}_2\mathbf{I} = \mathbf{A}^2 + \tfrac{1}{2}\mathbf{A} - \tfrac{1}{2}\mathbf{I}$$

$$= \begin{bmatrix} 0 & 1 \\ -1 & 0 \end{bmatrix}^2 + \tfrac{1}{2}\begin{bmatrix} 0 & 1 \\ -1 & 0 \end{bmatrix} - \tfrac{1}{2}\begin{bmatrix} 1 & 0 \\ 0 & 1 \end{bmatrix}$$

$$= \begin{bmatrix} -1 & 0 \\ 0 & -1 \end{bmatrix} + \begin{bmatrix} 0 & 1/2 \\ -1/2 & 0 \end{bmatrix} + \begin{bmatrix} -1/2 & 0 \\ 0 & -1/2 \end{bmatrix} = \begin{bmatrix} -3/2 & 1/2 \\ -1/2 & -3/2 \end{bmatrix}$$

$$\mathbf{S}^{-1} = [\mathbf{b} \mid \mathbf{Ab}]^{-1} = \begin{bmatrix} 0 & 1 \\ 1 & 0 \end{bmatrix}$$

Therefore

$$\mathbf{f}^{\mathsf{T}} = \mathbf{e}^{\mathsf{T}}\mathbf{S}^{-1}\hat{\mathbf{p}}(\mathbf{A}) = [0 \quad 1] \begin{bmatrix} 0 & 1 \\ 1 & 0 \end{bmatrix} \begin{bmatrix} -3/2 & 1/2 \\ -1/2 & -3/2 \end{bmatrix} = [-3/2 \quad 1/2]$$

It can be seen that the resulting three controller vectors derived by the three methods are identical. This is due to the fact that for single-input systems, $\mathbf{f}$ is unique.

■ **EXAMPLE 6.2.2** Assume a system in the form (6.1), where

$$\mathbf{A} = \begin{bmatrix} 0 & 1 & 0 \\ 0 & 0 & 1 \\ 1 & 0 & 0 \end{bmatrix} \quad \text{and} \quad \mathbf{b} = \begin{bmatrix} 0 \\ 0 \\ 1 \end{bmatrix}$$

Find a vector $\mathbf{f}$ such that the closed-loop system has eigenvalues $\hat{\lambda}_1 = -1/2$, $\hat{\lambda}_2 = 0$ and $\hat{\lambda}_3 = 0$.

■ **SOLUTION** We have

$$p(z) = |z\mathbf{I} - \mathbf{A}| = z^3 - 1 \quad \text{and} \quad \hat{p}(z) = (z - \hat{\lambda}_1)(z - \hat{\lambda}_2)(z - \hat{\lambda}_3) = z^3 + \tfrac{1}{2}z^2$$

Method 1

Since the system is in phase-variable canonical form, the vector $\mathbf{f}$ can readily be determined by equation (6.12), as follows:

$$\mathbf{f} = \mathbf{f}^* = \begin{bmatrix} \hat{a}_3 - a_3 \\ \hat{a}_2 - a_2 \\ \hat{a}_1 - a_1 \end{bmatrix} = \begin{bmatrix} 0 + 1 \\ 0 - 0 \\ 1/2 - 0 \end{bmatrix} = \begin{bmatrix} 1 \\ 0 \\ 1/2 \end{bmatrix}$$

Method 2

Here, we make use of equation (6.8). Equations (6.9) and (6.6) give

$$\mathbf{W} = \begin{bmatrix} 1 & a_1 & a_2 \\ 0 & 1 & a_1 \\ 0 & 0 & 1 \end{bmatrix} = \begin{bmatrix} 1 & 0 & 0 \\ 0 & 1 & 0 \\ 0 & 0 & 1 \end{bmatrix} \quad \text{and} \quad \mathbf{S} = [\mathbf{b} \vdots \mathbf{Ab} \vdots \mathbf{A}^2\mathbf{b}] = \begin{bmatrix} 0 & 0 & 1 \\ 0 & 1 & 0 \\ 1 & 0 & 0 \end{bmatrix}$$

Therefore

$$\mathbf{W}^T\mathbf{S}^T = \begin{bmatrix} 1 & 0 & 0 \\ 0 & 1 & 0 \\ 0 & 0 & 1 \end{bmatrix}\begin{bmatrix} 0 & 0 & 1 \\ 0 & 1 & 0 \\ 1 & 0 & 0 \end{bmatrix} = \begin{bmatrix} 0 & 0 & 1 \\ 0 & 1 & 0 \\ 1 & 0 & 0 \end{bmatrix} \quad \text{and} \quad (\mathbf{W}^T\mathbf{S}^T)^{-1} = \begin{bmatrix} 0 & 0 & 1 \\ 0 & 1 & 0 \\ 1 & 0 & 0 \end{bmatrix}$$

Hence

$$\mathbf{f} = (\mathbf{W}^T\mathbf{S}^T)^{-1}(\hat{\mathbf{a}} - \mathbf{a}) = \begin{bmatrix} 0 & 0 & 1 \\ 0 & 1 & 0 \\ 1 & 0 & 0 \end{bmatrix}\left\{\begin{bmatrix} 1/2 \\ 0 \\ 0 \end{bmatrix} - \begin{bmatrix} 0 \\ 0 \\ -1 \end{bmatrix}\right\} = \begin{bmatrix} 0 & 0 & 1 \\ 0 & 1 & 0 \\ 1 & 0 & 0 \end{bmatrix}\begin{bmatrix} 1/2 \\ 0 \\ 1 \end{bmatrix} = \begin{bmatrix} 1 \\ 0 \\ 1/2 \end{bmatrix}$$

Method 3

Here, we make use of equation (6.13). We have

$$\hat{\mathbf{p}}(\mathbf{A}) = \mathbf{A}^3 + \hat{a}_1\mathbf{A}^2 + \hat{a}_2\mathbf{A} + \hat{a}_3\mathbf{I} = \mathbf{A}^3 + \tfrac{1}{2}\mathbf{A}^2$$

$$= \begin{bmatrix} 0 & 1 & 0 \\ 0 & 0 & 1 \\ 1 & 0 & 0 \end{bmatrix}^3 + \tfrac{1}{2}\begin{bmatrix} 0 & 1 & 0 \\ 0 & 0 & 1 \\ 1 & 0 & 0 \end{bmatrix}^2$$

$$= \begin{bmatrix} 1 & 0 & 0 \\ 0 & 1 & 0 \\ 0 & 0 & 1 \end{bmatrix} + \begin{bmatrix} 0 & 0 & 1/2 \\ 1/2 & 0 & 0 \\ 0 & 1/2 & 0 \end{bmatrix} = \begin{bmatrix} 1 & 0 & 1/2 \\ 1/2 & 1 & 0 \\ 0 & 1/2 & 1 \end{bmatrix}$$

$$\mathbf{S}^{-1} = [\mathbf{b} \vdots \mathbf{Ab} \vdots \mathbf{A}^2\mathbf{b}]^{-1} = \begin{bmatrix} 0 & 0 & 1 \\ 0 & 1 & 0 \\ 1 & 0 & 0 \end{bmatrix}$$

Therefore

$$\mathbf{f}^T = \mathbf{e}^T\mathbf{S}^{-1}\hat{\mathbf{p}}(\mathbf{A}) = [0 \quad 0 \quad 1]\begin{bmatrix} 0 & 0 & 1 \\ 0 & 1 & 0 \\ 1 & 0 & 0 \end{bmatrix}\begin{bmatrix} 1 & 0 & 1/2 \\ 1/2 & 1 & 0 \\ 0 & 1/2 & 1 \end{bmatrix} = [1 \quad 0 \quad 1/2]$$

The resulting three controller vectors derived by the three methods are identical. This is due to the fact that for single-input systems, $\mathbf{f}$ is unique.

■ **EXAMPLE 6.2.3** Consider the system (6.1). Let the transformation matrix **T**, where

$$\mathbf{T} = \mathbf{W}^{-1}\mathbf{S}^{-1} \tag{6.18}$$

transform the given system (6.1) into the system

$$\mathbf{z}(k+1) = \tilde{\mathbf{A}}\mathbf{z}(k) + \tilde{\mathbf{b}}u(k) \tag{6.19a}$$

where $\tilde{\mathbf{A}}$ and $\tilde{\mathbf{b}}$ are in phase-variable canonical form, as follows:

$$\tilde{\mathbf{A}} = \begin{bmatrix} -a_1 & -a_2 & \cdots & -a_{n-1} & -a_n \\ 1 & 0 & \cdots & 0 & 0 \\ \vdots & \vdots & & \vdots & \vdots \\ 0 & 0 & \cdots & 1 & 0 \end{bmatrix} \quad \text{and} \quad \tilde{\mathbf{b}} = \begin{bmatrix} 1 \\ 0 \\ \vdots \\ 0 \end{bmatrix} \tag{6.19b}$$

For the two systems $(\mathbf{A}, \mathbf{b})$ and $(\tilde{\mathbf{A}}, \tilde{\mathbf{b}})$, the following relations hold:

$$\mathbf{z}(k) = \mathbf{T}\mathbf{x}(k), \quad \tilde{\mathbf{A}} = \mathbf{T}\mathbf{A}\mathbf{T}^{-1}, \quad \tilde{\mathbf{b}} = \mathbf{T}\mathbf{b}, \quad \tilde{\mathbf{S}} = \mathbf{T}\mathbf{S} \quad \text{and} \quad \tilde{\mathbf{S}} = \mathbf{W}^{-1} \tag{6.20}$$

Derive equation (6.13).

■ **SOLUTION** Let $\tilde{\mathbf{f}}^{\mathrm{T}}$ be the unknown vector sought for the transformed system (6.19). It is clear that

$$\tilde{\mathbf{f}}^{\mathrm{T}} = (\hat{a}_1 - a_1, \hat{a}_2 - a_2, \ldots, \hat{a}_n - a_n) \tag{6.21}$$

The Cayley–Hamilton theorem states that every square matrix satisfies its own characteristic equation (see Appendix A). Since two similar matrices, as for example $\mathbf{A}$ and $\tilde{\mathbf{A}}$, have the same characteristic polynomial, it follows that the Cayley–Hamilton theorem for the matrix $\tilde{\mathbf{A}}$ is

$$\tilde{\mathbf{A}}^n + a_1\tilde{\mathbf{A}}^{n-1} + \cdots + a_{n-1}\tilde{\mathbf{A}} + a_n\mathbf{I} = \mathbf{0}$$

Therefore, the matrix $\tilde{\mathbf{A}}^n$ is

$$\tilde{\mathbf{A}}^n = -a_1\tilde{\mathbf{A}}^{n-1} - \cdots - a_{n-1}\tilde{\mathbf{A}} - a_n\mathbf{I} \tag{6.22}$$

Introducing equation (6.22) into the following equation

$$\hat{\mathbf{p}}(\tilde{\mathbf{A}}) = \tilde{\mathbf{A}}^n + \hat{a}_1\tilde{\mathbf{A}}^{n-1} + \cdots + \hat{a}_{n-1}\tilde{\mathbf{A}} + \hat{a}_n\mathbf{I} \tag{6.23}$$

we have

$$\hat{\mathbf{p}}(\tilde{\mathbf{A}}) = (\hat{a}_1 - a_1)\tilde{\mathbf{A}}^{n-1} + \cdots + (\hat{a}_{n-1} - a_{n-1})\tilde{\mathbf{A}} + (\hat{a}_n - a_n)\mathbf{I} \tag{6.24}$$

Examining the matrix $\tilde{\mathbf{A}}^k$, $k = 0, 1, \ldots, n-1$, we notice that all elements of its last row are zero, except for the $(n-k)$th element, which is equal to unity. Therefore, the vector $\tilde{\mathbf{f}}^{\mathrm{T}}$, given by equation (6.21), can be written as

$$\tilde{\mathbf{f}}^{\mathrm{T}} = [0 \ 0 \ \cdots \ 0 \ 1]\hat{\mathbf{p}}(\tilde{\mathbf{A}}) \tag{6.25}$$

If we introduce equation $\tilde{\mathbf{A}} = \mathbf{T}\mathbf{A}\mathbf{T}^{-1}$ into equation (6.25), then the vector $\mathbf{f}^{\mathrm{T}}$ of the initial system (6.1) will be

$$\mathbf{f}^{\mathrm{T}} = \tilde{\mathbf{f}}^{\mathrm{T}}\mathbf{T} = [0 \ 0 \ \cdots \ 0 \ 1]\hat{\mathbf{p}}(\mathbf{T}\mathbf{A}\mathbf{T}^{-1})\mathbf{T} = [0 \ 0 \ \cdots \ 0 \ 1]\mathbf{T}\hat{\mathbf{p}}(\mathbf{A})$$

Since $[0 \ 0 \ \cdots \ 0 \ 1]\tilde{\mathbf{S}} = [0 \ 0 \ \cdots \ 0 \ 1]$, and making use of the equation (6.18), we finally arrive at the expression

$$\mathbf{f}^{\mathrm{T}} = \mathbf{e}^{\mathrm{T}}\mathbf{W}^{-1}\mathbf{S}^{-1}\hat{\mathbf{p}}(\mathbf{A}) = \mathbf{e}^{\mathrm{T}}\tilde{\mathbf{S}}\mathbf{S}^{-1}\hat{\mathbf{p}}(\mathbf{A}) = \mathbf{e}^{\mathrm{T}}\mathbf{S}^{-1}\hat{\mathbf{p}}(\mathbf{A})$$

which is identical to expression (6.13).

■ **EXAMPLE 6.2.4** Consider the system

$$\mathbf{x}(k+1) = \mathbf{A}\mathbf{x}(k) + \mathbf{b}u(k)$$

where $\mathbf{x}(k) \in \mathbb{R}^3$ and $u(k) \in \mathbb{R}$. Show that the controllable subspace of the system remains invariant under the feedback law $u(k) = \mathbf{f}^{\mathrm{T}}\mathbf{x}(k) + v(k)$.

■ **SOLUTION** The controllable subspace $\mathscr{E}$ of the open-loop system is

$$\mathscr{E} = \mathrm{Im}[\mathbf{b} \,\vdots\, \mathbf{Ab} \,\vdots\, \mathbf{A}^2\mathbf{b}]$$

The closed-loop system is

$$\mathbf{x}(k+1) = (\mathbf{A} + \mathbf{bf}^{\mathrm{T}})\mathbf{x}(k) + \mathbf{b}u(k)$$

The controllable subspace $\mathscr{E}_\kappa$ of the closed-loop system is

$$\mathscr{E}_\kappa = \mathrm{Im}[\mathbf{b} \,\vdots\, (\mathbf{A} + \mathbf{bf}^{\mathrm{T}})\mathbf{b} \,\vdots\, (\mathbf{A} + \mathbf{bf}^{\mathrm{T}})^2\mathbf{b}]$$

Let $\mathbf{x}$ be a vector in $\mathscr{E}$. Then, there are z_1, z_2 and z_3 such that

$$\mathbf{x} = \mathbf{b}z_1 + \mathbf{Ab}z_2 + \mathbf{A}^2\mathbf{b}z_3$$

If we let

$$\tilde{w}_1 = z_1 - \mathbf{f}^{\mathrm{T}}\mathbf{b}z_2 - \mathbf{f}^{\mathrm{T}}\mathbf{Ab}z_3$$
$$\tilde{w}_2 = z_2 - \mathbf{f}^{\mathrm{T}}\mathbf{b}z_3$$
$$\tilde{w}_3 = z_3$$

we observe that $\mathbf{x} \in \mathscr{E}_\kappa$. Let $\mathbf{x}_\kappa$ be a vector in $\mathscr{E}_\kappa$. Then, there are w_1, w_2 and w_3 such that

$$\mathbf{x}_\kappa = \mathbf{b}w_1 + (\mathbf{A} + \mathbf{bf}^{\mathrm{T}})\mathbf{b}w_2 + (\mathbf{A} + \mathbf{bf}^{\mathrm{T}})^2\mathbf{b}w_3$$
$$= \mathbf{b}(w_1 + \mathbf{f}^{\mathrm{T}}\mathbf{b}w_2 + \mathbf{f}^{\mathrm{T}}\mathbf{Ab}w_3 + \mathbf{f}^{\mathrm{T}}\mathbf{bf}^{\mathrm{T}}\mathbf{b}w_3) + \mathbf{Ab}(w_2 + \mathbf{f}^{\mathrm{T}}\mathbf{b}w_3) + \mathbf{A}^2\mathbf{b}w_3$$

If we let

$$\tilde{z}_1 = w_1 + \mathbf{f}^{\mathrm{T}}\mathbf{b}w_2 + \mathbf{f}^{\mathrm{T}}\mathbf{Ab}w_3 + \mathbf{f}^{\mathrm{T}}\mathbf{bf}^{\mathrm{T}}\mathbf{b}w_3$$
$$\tilde{z}_2 = w_2 + \mathbf{f}^{\mathrm{T}}\mathbf{b}w_3$$
$$\tilde{z}_3 = w_3$$

we note that $\mathbf{x}_\kappa \in \mathscr{E}$. Since every $\mathbf{x} \in \mathscr{E}$ also belongs to $\mathscr{E}_\kappa$ and every $\mathbf{x}_\kappa \in \mathscr{E}_\kappa$ also belongs to $\mathscr{E}$, it follows that $\mathscr{E} = \mathscr{E}_\kappa$.

■ **EXAMPLE 6.2.5**   Consider the continuous-time system

$$\dot{\mathbf{x}}(t) = \bar{\mathbf{A}}\mathbf{x}(t) + \bar{\mathbf{b}}u(t) + \bar{\mathbf{d}}\zeta(t) \quad y(t) = \bar{\mathbf{c}}^{\mathsf{T}}\mathbf{x}(t)$$

where

$$\bar{\mathbf{A}} = \begin{bmatrix} 1 & -1 & 0 \\ 0 & 0 & 0 \\ 0 & 0 & 0 \end{bmatrix}, \quad \bar{\mathbf{b}} = \begin{bmatrix} 1 \\ 1 \\ 1 \end{bmatrix}, \quad \bar{\mathbf{c}} = \begin{bmatrix} 1 \\ 0 \\ 0 \end{bmatrix} \quad \text{and} \quad \bar{\mathbf{d}} = \begin{bmatrix} 0 \\ 0 \\ 1 \end{bmatrix}$$

and where $\mathbf{x}(t)$, $u(t)$, $y(t)$ and $\zeta(t)$ are the state, the input, the output and the (unknown) disturbance, respectively.

(a) Find the equivalent discrete-time mathematical state-space model.
(b) Examine the controllability and the observability of both the continuous- and the discrete-time systems, in relation to the sampling period T. Discuss the results.
(c) Let the sampling period $T = 1$ s. A state feedback of the form $u(k) = -\mathbf{f}^{\mathsf{T}}\mathbf{x}(k) = -f_1 x_1 - f_2 x_2 - f_3 x_3$ is applied to the discrete-time system. Determine a vector $\mathbf{f}$ such that the closed-loop system is stable, while simultaneously eliminating (rejecting) the influence of the disturbance in the output. The stability requirement for the closed-loop system is that of Definition 4.2.1.

■ **SOLUTION**
(a) The equivalent discrete-time mathematical model will be

$$\mathbf{x}[(k+1)T] = \mathbf{A}\mathbf{x}(kT) + \mathbf{b}u(kT) + \mathbf{d}\zeta(kT), \quad y(kT) = \mathbf{c}^{\mathsf{T}}\mathbf{x}(kT)$$

where $\mathbf{c} = \bar{\mathbf{c}}$ and $\mathbf{A}$, $\mathbf{b}$ and $\mathbf{d}$ are given by (see Section 3.3.4)

$$\mathbf{A} = e^{\bar{\mathbf{A}}T} = [\mathscr{L}^{-1}[s\mathbf{I} - \bar{\mathbf{A}}]^{-1}]_{t=T} = \begin{bmatrix} e^t & 1-e^t & 0 \\ 0 & 1 & 0 \\ 0 & 0 & 1 \end{bmatrix}_{t=T} = \begin{bmatrix} e^T & 1-e^T & 0 \\ 0 & 1 & 0 \\ 0 & 0 & 1 \end{bmatrix}$$

$$\mathbf{b} = \left[\int_0^T e^{\bar{\mathbf{A}}\lambda}\, d\lambda\right]\bar{\mathbf{b}} = \left[\int_0^T \begin{bmatrix} e^\lambda & 1-e^\lambda & 0 \\ 0 & 1 & 0 \\ 0 & 0 & 1 \end{bmatrix} d\lambda\right]\bar{\mathbf{b}} = \begin{bmatrix} T \\ T \\ T \end{bmatrix}$$

$$\mathbf{d} = \left[\int_0^T e^{\bar{\mathbf{A}}\lambda}\, d\lambda\right]\bar{\mathbf{d}} = \left[\int_0^T \begin{bmatrix} e^\lambda & 1-e^\lambda & 0 \\ 0 & 1 & 0 \\ 0 & 0 & 1 \end{bmatrix} d\lambda\right]\bar{\mathbf{d}} = \begin{bmatrix} 0 \\ 0 \\ T \end{bmatrix}$$

(b) The continuous-time system controllability matrix $\bar{\mathbf{S}}$ is

$$\bar{\mathbf{S}} = [\bar{\mathbf{b}} \mathrel{\vdots} \bar{\mathbf{A}}\bar{\mathbf{b}} \mathrel{\vdots} \bar{\mathbf{A}}^2\bar{\mathbf{b}}] = \begin{bmatrix} 1 & 0 & 0 \\ 1 & 0 & 0 \\ 1 & 0 & 0 \end{bmatrix} \quad \text{and} \quad \text{rank } \bar{\mathbf{S}} = 1$$

The discrete-time system controllability matrix $\mathbf{S}$ is

$$\mathbf{S} = [\mathbf{b} \mid \mathbf{Ab} \mid \mathbf{A}^2\mathbf{b}] = \begin{bmatrix} T & T & T \\ T & T & T \\ T & T & T \end{bmatrix} \quad \text{and} \quad \text{rank } \mathbf{S} = 1$$

The continuous-time system observability matrix $\bar{\mathbf{R}}$ is

$$\bar{\mathbf{R}} = \begin{bmatrix} \bar{\mathbf{c}}^T \\ \bar{\mathbf{c}}^T\bar{\mathbf{A}} \\ \bar{\mathbf{c}}^T\bar{\mathbf{A}}^2 \end{bmatrix} = \begin{bmatrix} 1 & 0 & 0 \\ 1 & -1 & 0 \\ 1 & -1 & 0 \end{bmatrix} \quad \text{and} \quad \text{rank } \bar{\mathbf{R}} = 2$$

The discrete-time system observability matrix $\mathbf{R}$ is

$$\mathbf{R} = \begin{bmatrix} \mathbf{c}^T \\ \mathbf{c}^T\mathbf{A} \\ \mathbf{c}^T\mathbf{A}^2 \end{bmatrix} = \begin{bmatrix} 1 & 0 & 0 \\ e^T & 1 - e^T & 0 \\ e^{2T} & 1 - 2e^{2T} & 0 \end{bmatrix} \quad \text{and} \quad \text{rank } \mathbf{R} = 2$$

We note that, for the present problem, the ranks of the controllability and observability matrices remain invariant after discretization.

(c) In order to meet the problem requirements, there must be $\mathbf{c}^T(\mathbf{A} - \mathbf{bf}^T)^k\mathbf{d} = 0$, $k = 0, 1, \ldots$, and the closed-loop system, with characteristic polynomial $|z\mathbf{I} - \mathbf{A} + \mathbf{bf}^T|$, must be stable. The first condition, according to the Cayley–Hamilton theorem, is reduced to the following three equations:

$$\mathbf{c}^T\mathbf{d} = 0$$
$$\mathbf{c}^T(\mathbf{A} - \mathbf{bf}^T)\mathbf{d} = 0$$
$$\mathbf{c}^T(\mathbf{A} - \mathbf{bf}^T)^2\mathbf{d} = 0$$

The first equation is always true. The second equation becomes $\mathbf{c}^T(\mathbf{A} - \mathbf{bf}^T)\mathbf{d} = f_3 = 0$. For $f_3 = 0$, it can be easily shown that the third equation is always satisfied. Since $f_3 = 0$, the characteristic polynomial $\hat{p}(z)$ of the closed-loop system becomes

$$\hat{p}(z) = (z - 1)\{z^2 - [e + (1 - f_1 - f_2)]z + e(1 - f_1 - f_2)\}$$

where we have substituted $T = 1$ s. Since $\hat{p}(z)$ has a root $z = 1$ on the unit circle, the remaining roots must be located inside the unit circle in order to meet the requirements of Definition 4.2.1. Letting $k = 1 - f_1 - f_2$ and applying the Jury criterion to the polynomial $\hat{p}(z) = z^2 - (e + k)z + ek$, we arrive at the inequality $k^2 < 1/e^2$. Choosing $f_1 = 0.99$ and $f_2 = 0$, i.e. $k = 0.01$, we note that the inequality is satisfied. Therefore, the choice of the feedback law

$$u = \mathbf{f}^T\mathbf{x} = [0.99 \ 0 \ 0]\mathbf{x} = 0.99x_1$$

meets the requirements of part (c).

6.3 Deadbeat control

The solution for $\mathbf{x}(k)$ of the closed-loop system (6.3) gives

$$\mathbf{x}(k) = [\mathbf{A} - \mathbf{BF}]^k \mathbf{x}(0) \tag{6.26}$$

where $\mathbf{x}(0)$ is the initial state vector. If the eigenvalues $\hat{\lambda}_1, \hat{\lambda}_2, ..., \hat{\lambda}_n$ of the matrix $\mathbf{A} - \mathbf{BF}$ lie inside the unit circle, then it is well known that the system is asymptotically stable, which means that $\mathbf{x}(k) \rightarrow \mathbf{0}$ as $k \rightarrow \infty$. Of course, this also applies to the particular case where all eigenvalues $\hat{\lambda}_i = 0$, $i = 1, 2, ..., n$. However, for this special case where $\hat{\lambda}_i = 0$, $i = 1, 2, ..., n$, an interesting fact arises: the vector $\mathbf{x}(k)$ becomes zero for $k \leq n$. In other words, the closed-loop system will arrive and remain at rest in at most n sampling periods. This type of control is called **deadbeat control** and has great theoretical and practical interest.

Clearly, in order to achieve deadbeat control, we must determine the matrix $\mathbf{F}$ such that all eigenvalues of the matrix $\mathbf{A} - \mathbf{BF}$ are zero, i.e. such that

$$\hat{p}(z) = |z\mathbf{I} - \mathbf{A} + \mathbf{BF}| = z^n \tag{6.27}$$

According to the results of the previous section, this is always possible, provided the system $(\mathbf{A}, \mathbf{B})$ is controllable.

In deadbeat control the settling time is less than or equal to nT. Yet, the sampling period T is expected to affect greatly the amplitude of the control signal $\mathbf{u}(kT)$. More specifically, the faster we desire $\mathbf{x}(kT)$ to settle to zero (i.e. small nT), the greater the control effort (i.e. the greater the amplitude of the elements of $\mathbf{u}(kT)$). For this reason, we select T so as to satisfy the Shannon sampling criterion (Theorem 2.3.1), while keeping the magnitude of $\mathbf{u}(kT)$ within acceptable limits. Otherwise, deadbeat control is not recommended for practical applications.

Deadbeat control is a phenomenon of discrete-time systems. There is no such equivalent phenomenon for continuous-time systems.

Instead of giving a formal general proof of deadbeat control we will prove deadbeat control for the particular case where the matrix $\mathbf{A}$ and the vector $\mathbf{b}$ are in phase-variable canonical form (6.10) and the vector $\mathbf{f}$ is given by equation (6.12). Since for deadbeat control it is necessary that $\hat{p}(z) = z^n$, then, according to (6.12), the vector $\mathbf{f}$ is $\mathbf{f}^T = [-a_n, -a_{n-1}, .., -a_1]$. For this particular $\mathbf{f}$, the matrix $\mathbf{A} - \mathbf{bf}^T$ takes on the following nilpotent form:

$$\mathbf{A} - \mathbf{bf}^T = \begin{bmatrix} 0 & 1 & 0 & \cdots & 0 \\ 0 & 0 & 1 & \cdots & 0 \\ \vdots & \vdots & \vdots & \ddots & \vdots \\ 0 & 0 & 0 & \cdots & 1 \\ 0 & 0 & 0 & \cdots & 0 \end{bmatrix}$$

It can easily be shown that $[\mathbf{A} - \mathbf{bf}^T]^n = \mathbf{0}$. Hence, $\mathbf{f}^T = [-a_n, -a_{n-1}, ..., -a_1]$ is a deadbeat vector since (6.26) becomes

$$\mathbf{x}(k) = [\mathbf{A} - \mathbf{bf}^T]^k \mathbf{x}(0) = \mathbf{0} \quad \text{for } k \geq n$$

■ **EXAMPLE 6.3.1**   Consider the double-integrator system

$$Y(s) = \frac{1}{s^2} U(s) \quad \text{or} \quad y^{(2)}(t) = u(t)$$

From the practical point of view, such a mathematical model describes systems like the satellite orientation system presented in Example 5.4.2 (Figure 5.20). A state-space description of this system is given by

$$\dot{\mathbf{x}}(t) = \bar{\mathbf{A}}\mathbf{x}(t) + \bar{\mathbf{b}}u(t) \quad y(t) = \bar{\mathbf{c}}^{\mathsf{T}}\mathbf{x}(t)$$

where

$$\mathbf{x} = \begin{bmatrix} x_1 \\ x_2 \end{bmatrix} = \begin{bmatrix} y \\ \dot{y} \end{bmatrix}, \quad \bar{\mathbf{A}} = \begin{bmatrix} 0 & 1 \\ 0 & 0 \end{bmatrix}, \quad \bar{\mathbf{b}} = \begin{bmatrix} 0 \\ 1 \end{bmatrix} \quad \text{and} \quad \bar{\mathbf{c}} = \begin{bmatrix} 1 \\ 0 \end{bmatrix}$$

The corresponding discrete-time model will be

$$\mathbf{x}[(k+1)T] = \mathbf{A}\mathbf{x}(kT) + \mathbf{b}u(kT), \quad y(kT) = \mathbf{c}^{\mathsf{T}}\mathbf{x}(kT)$$

where $\mathbf{c} = \bar{\mathbf{c}}$ and $\mathbf{A}$ and $\mathbf{b}$ are given by the equations (see Section 3.3.4)

$$\mathbf{A} = \mathrm{e}^{\bar{\mathbf{A}}T} = [\mathscr{L}^{-1}[s\mathbf{I} - \bar{\mathbf{A}}]^{-1}]_{t=T} = \begin{bmatrix} 1 & t \\ 0 & 1 \end{bmatrix}_{t=T} = \begin{bmatrix} 1 & T \\ 0 & 1 \end{bmatrix}$$

$$\mathbf{b} = \left[\int_0^T \mathrm{e}^{\bar{\mathbf{A}}\lambda} \, \mathrm{d}\lambda \right] \bar{\mathbf{b}} = \left[\int_0^T \begin{bmatrix} 1 & \lambda \\ 0 & 1 \end{bmatrix} \mathrm{d}\lambda \right] \bar{\mathbf{b}} = \begin{bmatrix} T^2/2 \\ T \end{bmatrix}$$

Determine a vector $\mathbf{f}$ for deadbeat control. Determine $u(0)$ and $u(T)$ for various values of T with initial conditions $\mathbf{x}^{\mathsf{T}}(0) = [1, 1]$.

■ **SOLUTION**   We will determine the vector $\mathbf{f}$ using equation (6.8). We have $p(z) = |z\mathbf{I} - \mathbf{A}| = (z - 1)^2 = z^2 - 2z + 1$ and $\hat{p}(z) = z^2$. Furthermore

$$\mathbf{W} = \begin{bmatrix} 1 & a_1 \\ 0 & 1 \end{bmatrix} = \begin{bmatrix} 1 & -2 \\ 0 & 1 \end{bmatrix} \quad \text{and} \quad \mathbf{S} = [\mathbf{b} \mid \mathbf{Ab}] = \begin{bmatrix} T^2/2 & 3T^2/2 \\ T & T \end{bmatrix}$$

$$\mathbf{W}^{\mathsf{T}}\mathbf{S}^{\mathsf{T}} = \begin{bmatrix} T^2/2 & T \\ T^2/2 & -T \end{bmatrix} \quad \text{and} \quad (\hat{\mathbf{a}} - \mathbf{a}) = \begin{bmatrix} 0 \\ 0 \end{bmatrix} - \begin{bmatrix} -2 \\ 1 \end{bmatrix} = \begin{bmatrix} 2 \\ -1 \end{bmatrix}$$

Finally, we have

$$\mathbf{f} = [\mathbf{W}^{\mathsf{T}}\mathbf{S}^{\mathsf{T}}]^{-1}(\hat{\mathbf{a}} - \mathbf{a}) = \begin{bmatrix} 1/T^2 & 1/T^2 \\ 1/2T & -1/2T \end{bmatrix} \begin{bmatrix} 2 \\ -1 \end{bmatrix} = \begin{bmatrix} 1/T^2 \\ 3/2T \end{bmatrix}$$

The closed-loop system matrix $\mathbf{A} - \mathbf{bf}^T$ will be

$$\mathbf{A} - \mathbf{bf}^T = \begin{bmatrix} 1 & T \\ 0 & 1 \end{bmatrix} - \begin{bmatrix} T^2/2 \\ T \end{bmatrix} [1/T^2 \quad 3/2T] = \begin{bmatrix} 1/2 & T/4 \\ -1/T & -1/2 \end{bmatrix}$$

The closed-loop system characteristic polynomial $\hat{p}(z)$ is

$$\hat{p}(z) = |z\mathbf{I} - \mathbf{A} + \mathbf{bf}^T| = \begin{vmatrix} z - 1/2 & -T/4 \\ 1/T & z + 1/2 \end{vmatrix} = z^2$$

We also have $(\mathbf{A} - \mathbf{bf}^T)^q = \mathbf{0}$, for $q > 1$. Therefore $\mathbf{x}(kT) = \mathbf{0}$ for $k > 1$, which means that the system will come to a standstill at the second sampling instant. Let us now investigate the control signal $u(kT)$. We have

$$u(kT) = -\mathbf{f}^T \mathbf{x}(kT) = -[1/T^2 \quad 3/2T]\mathbf{x}(kT)$$

For $k = 0$, $u(kT)$ becomes

$$u(0) = -[1/T^2 \quad 3/2T]\mathbf{x}(0) = -\left(\frac{1}{T^2} x_1(0) + \frac{3}{2T} x_2(0) \right)$$

For $k = 1$, $u(kT)$ becomes

$$u(T) = \mathbf{f}^T(\mathbf{A} - \mathbf{bf}^T)\mathbf{x}(0) = \frac{1}{T^2} x_1(0) + \frac{1}{2T} x_2(0)$$

For $k = 2$, $u(kT)$ becomes

$$u(2T) = \mathbf{f}^T(\mathbf{A} - \mathbf{bf}^T)^2 \mathbf{x}(0) = 0$$

For $\mathbf{x}^T(0) = [x_1(0), x_2(0)] = [1, 1]$ we have

$$u(0) = -\left(\frac{1}{T^2} + \frac{3}{2T} \right) \quad \text{and} \quad u(T) = \frac{1}{T^2} + \frac{1}{2T}$$

Table 6.1 summarizes the values of $u(0)$ and $u(T)$, while Figure 6.2 presents the trajectories of $y(kT)$ and $u(kT)$ for $T = 2$, 1 and 0.5 s, when $x^T(0) = [1, 1]$. From the

Table 6.1 Deadbeat control signals of the double-integrator system of Example 6.3.1 for different sampling periods.

T	2	1	0.5
$u(0)$	-1	-2.5	-7
$u(T)$	0.5	1.5	5

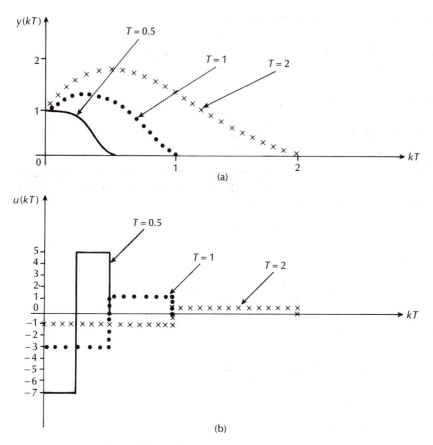

Figure 6.2 Waveforms of (a) $y(kT)$ for $T = 0.5$, 1 and 2 s and of (b) deadbeat control $u(kT)$ for $T = 0.5$, 1 and 2 s for the double-integrator system of Example 6.3.1.

foregoing results, one can draw the conclusion that the smaller the value of T, the bigger the value of $u(kT)$, and vice versa.

6.4 State observers

In designing a closed-loop system using modern control techniques, the control strategy applied is usually a feedback loop involving feedback of the system state vector **x**. Examples of such strategies are the state feedback law **u** = −**Fx** for pole placement and deadbeat control presented in the previous sections. This means that for this type of feedback law to be applicable, the entire state vector **x** must be available (measurable).

om the output equation $y(kT) = \mathbf{c}^T\mathbf{x}(kT) = [1,0]\mathbf{x}(kT) = x_1(kT)$ it follows that (kT) is directly measured since $x_1(kT) = y(kT)$. The state variable $x_2(kT)$ is otained by using the foregoing expression for $\mathbf{x}(kT)$ which gives

$$x_2(kT) = \frac{y(kT) - y(kT - T)}{T} + \frac{T}{2}u(kT - T)$$

The direct state vector estimate has the advantage of determining the state ariables in n steps, at most. The drawback of this method is its susceptibility to sturbances. For this reason it is important to have other alternative methods. A idely known alternative method is the state vector estimation (or reconstruction) sing an observer, first proposed by Luenberger [10–12], which we present in the ollowing section.

4.2   State vector reconstruction using a Luenberger observer

Consider the system (6.28). Assume that state vector $\mathbf{x}(k)$ is given approximately by the state vector $\hat{\mathbf{x}}(k)$ of the model

$$\hat{\mathbf{x}}(k+1) = \hat{\mathbf{A}}\hat{\mathbf{x}}(k) + \hat{\mathbf{B}}\mathbf{u}(k) + \mathbf{K}\mathbf{y}(k) \tag{6.33}$$

where $\hat{\mathbf{x}} \in \mathbb{R}^n$ and the matrices $\hat{\mathbf{A}}$, $\hat{\mathbf{B}}$ and $\mathbf{K}$ are unknown. System (6.33) is called the **tate observer** of system (6.28). In particular, the observer is a dynamic system aving two inputs, the input vector $\mathbf{u}(k)$ and the output vector $\mathbf{y}(k)$ of the initial ystem (6.28) (see Figure 6.3) and whose matrices $\hat{\mathbf{A}}$, $\hat{\mathbf{B}}$ and $\hat{\mathbf{K}}$ are such that $\hat{\mathbf{x}}(k)$ is as lose as possible to $\mathbf{x}(k)$. In cases where $\hat{\mathbf{x}}(k)$ and $\mathbf{x}(k)$ are of equal dimension, then he observer is referred to as a **full-order** observer. This case is studied in the present ection. When the dimension of $\hat{\mathbf{x}}(k)$ is smaller than that of $\mathbf{x}(k)$, then the observer is eferred to as a **reduced-order** observer which is studied in Section 6.4.3.

Define the state error

$$\mathbf{e}(k) = \mathbf{x}(k) - \hat{\mathbf{x}}(k) \tag{6.34}$$

The definition of the problem of designing the observer (6.33) is to determine appro-riate matrices $\hat{\mathbf{A}}$, $\hat{\mathbf{B}}$ and $\mathbf{K}$, such that the error $\mathbf{e}(k)$ tends to zero as fast as possible.

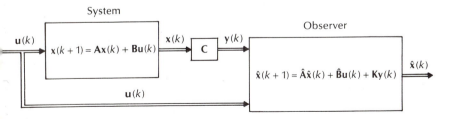

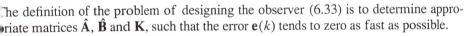

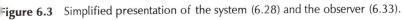

Figure 6.3   Simplified presentation of the system (6.28) and the observer (6.33).

In practice, however, it very often happens that not all state variables of a system are accessible to measurement. This obstacle can be circumvented if a mathematical model for the system is available, in which case it is possible to estimate (reconstruct) the state vector [4], [5], [9–12].

We will present two methods for estimating the state vector. The first method is the direct estimation of the state vector based on input and output measurements (Section 6.4.1), while the second is the estimation of the state vector using a dynamic system called a state observer (Sections 6.4.2–6.4.4). The system model is assumed known and of the following form:

$$\mathbf{x}(k+1) = \mathbf{A}\mathbf{x}(k) + \mathbf{B}\mathbf{u}(k), \quad \mathbf{x} \in \mathbb{R}^n \quad \text{and} \quad \mathbf{u} \in \mathbb{R}^m \tag{6.28a}$$

$$\mathbf{y}(k) = \mathbf{C}\mathbf{x}(k), \quad \mathbf{y} \in \mathbb{R}^p \tag{6.28b}$$

6.4.1   Direct state vector estimation

The direct state vector estimate will be obtained on the basis of the measured output sequences $\mathbf{y}(k)$, $\mathbf{y}(k-1)$, ... and input sequences $\mathbf{u}(k)$, $\mathbf{u}(k-1)$, ... To facilitate the presentation, we assume that the system (6.28) has only one output. Then, using (6.28), we obtain

$$y(k-n+1) = \mathbf{c}^T\mathbf{x}(k-n+1)$$

$$y(k-n+2) = \mathbf{c}^T\mathbf{A}\mathbf{x}(k-n+1) + \mathbf{c}^T\mathbf{B}\mathbf{u}(k-n+1)$$

$$\vdots$$

$$y(k) = \mathbf{c}^T\mathbf{A}^{n-1}\mathbf{x}(k-n+1) + \mathbf{c}^T\mathbf{A}^{n-2}\mathbf{B}\mathbf{u}(k-n+1) + \cdots + \mathbf{c}^T\mathbf{B}\mathbf{u}(k-1)$$

These equations can be written more compactly as

$$\begin{bmatrix} y(k-n+1) \\ y(k-n+2) \\ \vdots \\ y(k) \end{bmatrix} = \mathbf{R}\mathbf{x}(k-n+1) + \mathbf{\Omega}\begin{bmatrix} \mathbf{u}(k-n+1) \\ \mathbf{u}(k-n+2) \\ \vdots \\ \mathbf{u}(k-1) \end{bmatrix} \tag{6.29}$$

where

$$\mathbf{R} = \begin{bmatrix} \mathbf{c}^T \\ \mathbf{c}^T\mathbf{A} \\ \vdots \\ \mathbf{c}^T\mathbf{A}^{n-1} \end{bmatrix} \quad \text{and} \quad \mathbf{\Omega} = \begin{bmatrix} 0 & 0 & \dots & 0 \\ \mathbf{c}^T\mathbf{B} & 0 & \dots & 0 \\ \vdots & \vdots & & \vdots \\ \mathbf{c}^T\mathbf{A}^{n-2}\mathbf{B} & \mathbf{c}^T\mathbf{A}^{n-3}\mathbf{B} & \dots & \mathbf{c}^T\mathbf{B} \end{bmatrix} \tag{6.30}$$

where $\mathbf{R}$ is the well-known observability matrix of the system (6.28). If the system (6.28) is observable, then the matrix $\mathbf{R}$ is non-singular and the state vector

$\mathbf{x}(k-n+1)$ can consequently be estimated by using equation (6.29) as follows:

$$\mathbf{x}(k-n+1) = \mathbf{R}^{-1}\begin{bmatrix} y(k-n+1) \\ y(k-n+2) \\ \vdots \\ y(k) \end{bmatrix} - \mathbf{R}^{-1}\mathbf{\Omega}\begin{bmatrix} \mathbf{u}(k-n+1) \\ \mathbf{u}(k-n+2) \\ \vdots \\ \mathbf{u}(k-1) \end{bmatrix}$$

Repeated use of equation (6.28) leads to

$$\mathbf{x}(k-n+2) = \mathbf{A}\mathbf{x}(k-n+1) + \mathbf{B}\mathbf{u}(k-n+1)$$

$$= \mathbf{A}\left[\mathbf{R}^{-1}\begin{bmatrix} y(k-n+1) \\ y(k-n+2) \\ \vdots \\ y(k) \end{bmatrix} - \mathbf{R}^{-1}\mathbf{\Omega}\begin{bmatrix} \mathbf{u}(k-n+1) \\ \mathbf{u}(k-n+2) \\ \vdots \\ \mathbf{u}(k-1) \end{bmatrix}\right] + \mathbf{B}\mathbf{u}(k-n+1)$$

$$= \mathbf{A}\mathbf{R}^{-1}\begin{bmatrix} y(k-n+1) \\ y(k-n+2) \\ \vdots \\ y(k) \end{bmatrix} - \mathbf{A}\mathbf{R}^{-1}\mathbf{\Omega}\begin{bmatrix} \mathbf{u}(k-n+1) \\ \mathbf{u}(k-n+2) \\ \vdots \\ \mathbf{u}(k-1) \end{bmatrix} + \mathbf{B}\mathbf{u}(k-n+1)$$

Furthermore

$$\mathbf{x}(k-n+3) = \mathbf{A}\mathbf{x}(k-n+2) + \mathbf{B}\mathbf{u}(k-n+2)$$

$$= \mathbf{A}^2\mathbf{R}^{-1}\begin{bmatrix} y(k-n+1) \\ y(k-n+2) \\ \vdots \\ y(k) \end{bmatrix} - \mathbf{A}^2\mathbf{R}^{-1}\mathbf{\Omega}\begin{bmatrix} \mathbf{u}(k-n+1) \\ \mathbf{u}(k-n+2) \\ \vdots \\ \mathbf{u}(k-1) \end{bmatrix} + \mathbf{A}\mathbf{B}\mathbf{u}(k-n+1) + \mathbf{B}\mathbf{u}(k-n+2)$$

and so on. Finally, we arrive at the desired expression

$$\mathbf{x}(k) = \mathbf{A}^{n-1}\mathbf{R}^{-1}\begin{bmatrix} y(k-n+1) \\ y(k-n+2) \\ \vdots \\ y(k) \end{bmatrix} + \mathbf{\Psi}\begin{bmatrix} \mathbf{u}(k-n+1) \\ \mathbf{u}(k-n+2) \\ \vdots \\ \mathbf{u}(k-1) \end{bmatrix} \tag{6.31}$$

where

$$\mathbf{\Psi} = [\mathbf{A}^{n-2}\mathbf{B} \quad \mathbf{A}^{n-3}\mathbf{B} \quad \cdots \quad \mathbf{B}] - \mathbf{A}^{n-1}\mathbf{R}^{-1}\mathbf{\Omega} \tag{6.32}$$

From equation (6.31) it is clear that the state vector is given as a linear combination of $y(k)$, $y(k-1)$, ..., $y(k-n+1)$ and $\mathbf{u}(k)$, $\mathbf{u}(k-1)$, ..., $\mathbf{u}(k-n+1)$. Equation (6.31) shows that the estimation of $\mathbf{x}(k)$ will be achieved after n steps at most. For this reason, an observer of the form of equation (6.31) is called a **deadbeat observer**. The following theorem holds.

■ **THEOREM 6.4.1** The state vector of system (6.28) can be estimat input and output measurements, if and only if the system is observ

■ **EXAMPLE 6.4.1** Consider the double-integrator system of E discrete-time state-space representation of this system, derived i has the form

$$\mathbf{x}[(k+1)T] = \mathbf{A}(T)\mathbf{x}(kT) + \mathbf{b}(T)\mathbf{u}(kT), \quad y(kT) = \mathbf{c}^{\mathrm{T}}\mathbf{x}(kT)$$

where

$$\mathbf{A}(T) = \begin{bmatrix} 1 & T \\ 0 & 1 \end{bmatrix}, \quad \mathbf{b}(T) = \begin{bmatrix} T^2/2 \\ T \end{bmatrix} \quad \text{and} \quad \mathbf{c} = \begin{bmatrix} 1 \\ 0 \end{bmatrix}$$

Estimate the state variables in terms of $u(kT)$ and $y(kT)$.

■ **SOLUTION** The observability matrix $\mathbf{R}$ is

$$\mathbf{R} = \begin{bmatrix} \mathbf{c}^{\mathrm{T}} \\ \mathbf{c}^{\mathrm{T}}\mathbf{A} \end{bmatrix} = \begin{bmatrix} 1 & 0 \\ 1 & T \end{bmatrix} \quad \text{and} \quad \mathbf{R}^{-1} = \frac{1}{T}\begin{bmatrix} T & 0 \\ -1 & 1 \end{bmatrix} = \begin{bmatrix} 1 & 0 \\ -1/T & 1/T \end{bmatrix}$$

We also have

$$\mathbf{R}^{-1}\mathbf{\Omega} = \mathbf{R}^{-1}\begin{bmatrix} 0 \\ \mathbf{c}^{\mathrm{T}}\mathbf{b} \end{bmatrix} = \begin{bmatrix} 1 & 0 \\ -1/T & 1/T \end{bmatrix}\begin{bmatrix} 0 \\ T^2/2 \end{bmatrix} = \begin{bmatrix} 0 \\ T/2 \end{bmatrix}$$

and therefore

$$\mathbf{\Psi} = \mathbf{B} - \mathbf{A}\mathbf{R}^{-1}\mathbf{\Omega} = \begin{bmatrix} T^2/2 \\ T \end{bmatrix} - \begin{bmatrix} 1 & T \\ 0 & 1 \end{bmatrix}\begin{bmatrix} 0 \\ T/2 \end{bmatrix} = \begin{bmatrix} 0 \\ T/2 \end{bmatrix}$$

We further have

$$\mathbf{A}^{n-1}\mathbf{R}^{-1} = \mathbf{A}\mathbf{R}^{-1} = \begin{bmatrix} 1 & T \\ 0 & 1 \end{bmatrix}\begin{bmatrix} 1 & 0 \\ -1/T & 1/T \end{bmatrix} = \begin{bmatrix} 0 & 1 \\ -1/T & 1/T \end{bmatrix}$$

Hence, equation (6.31) finally becomes

$$\mathbf{x}(kT) = \begin{bmatrix} 0 & 1 \\ -1/T & 1/T \end{bmatrix}\begin{bmatrix} y(kT-T) \\ y(kT) \end{bmatrix} + \begin{bmatrix} 0 \\ T/2 \end{bmatrix}u(kT-T)$$

or

$$\mathbf{x}(kT) = \begin{bmatrix} 1 \\ 1/T \end{bmatrix}y(kT) + \begin{bmatrix} 0 \\ -1/T \end{bmatrix}y(kT-T) + \begin{bmatrix} 0 \\ T/2 \end{bmatrix}u(kT-T)$$

To solve the problem, we proceed as follows. Using equations (6.28) and (6.33), it can be shown that the error $\mathbf{e}(k)$ satisfies the difference equation

$$\mathbf{e}(k+1) = \mathbf{x}(k+1) - \hat{\mathbf{x}}(k+1) = \mathbf{A}\mathbf{x}(k) + \mathbf{B}\mathbf{u}(k) - \hat{\mathbf{A}}[\mathbf{x}(k) - \mathbf{e}(k)] - \hat{\mathbf{B}}\mathbf{u}(k) - \mathbf{K}\mathbf{C}\mathbf{x}(k)$$

or

$$\mathbf{e}(k+1) = \hat{\mathbf{A}}\mathbf{e}(k) + [\mathbf{A} - \mathbf{K}\mathbf{C} - \hat{\mathbf{A}}]\mathbf{x}(k) + [\mathbf{B} - \hat{\mathbf{B}}]\mathbf{u}(k)$$

For the error $\mathbf{e}(k)$ to tend to zero, independently of $\mathbf{x}(k)$ and $\mathbf{u}(k)$, the following three conditions must be satisfied simultaneously:

1. $\hat{\mathbf{A}} = \mathbf{A} - \mathbf{K}\mathbf{C}$
2. $\hat{\mathbf{B}} = \mathbf{B}$
3. matrix $\hat{\mathbf{A}}$ is stable

From the above we conclude that the error $\mathbf{e}(k)$ satisfies the difference equation

$$\mathbf{e}(k+1) = \hat{\mathbf{A}}\mathbf{e}(k) = [\mathbf{A} - \mathbf{K}\mathbf{C}]\mathbf{e}(k)$$

while the state observer (6.33) takes on the form

$$\hat{\mathbf{x}}(k+1) = [\mathbf{A} - \mathbf{K}\mathbf{C}]\hat{\mathbf{x}}(k) + \mathbf{B}\mathbf{u}(k) + \mathbf{K}\mathbf{y}(k) \qquad (6.35a)$$

or

$$\hat{\mathbf{x}}(k+1) = \mathbf{A}\hat{\mathbf{x}}(k) + \mathbf{B}\mathbf{u}(k) + \mathbf{K}[\mathbf{y}(k) - \mathbf{C}\hat{\mathbf{x}}(k)] \qquad (6.35b)$$

According to equation (6.35a), the observer can be considered as a system having the matrices $\mathbf{A}$, $\mathbf{B}$ and $\mathbf{C}$ of the original system together with an arbitrary matrix $\mathbf{K}$. The matrix $\mathbf{K}$ must be chosen so that the eigenvalues of the matrix $\hat{\mathbf{A}} = \mathbf{A} - \mathbf{K}\mathbf{C}$ effectively force the error $\mathbf{e}(k)$ to zero as fast as possible. According to equation (6.35b), the observer appears to be exactly the original system plus an additional term $\mathbf{K}[\mathbf{y}(k) - \mathbf{C}\hat{\mathbf{x}}(k)]$. The term $\mathbf{r}(k) = \mathbf{y}(k) - \hat{\mathbf{y}}(k) = \mathbf{y}(k) - \mathbf{C}\hat{\mathbf{x}}(k)$ can be considered as a corrective term, often called a **residual**. Of course, if $\hat{\mathbf{x}}(k) = \mathbf{x}(k)$, then $\mathbf{r}(k) = \mathbf{0}$. Therefore, a residual exists if the system output vector $\mathbf{y}(k)$ and the observer vector $\hat{\mathbf{y}}(k) = \mathbf{C}\hat{\mathbf{x}}(k)$ are different.

■ **REMARK 6.4.1**   To construct the observer it is necessary to construct the original system itself, plus the corrective term $\mathbf{K}[\mathbf{y}(k) - \mathbf{C}\hat{\mathbf{x}}(k)]$. One may then ask: why not build the model $\hat{\mathbf{x}}(k+1) = \mathbf{A}\hat{\mathbf{x}}(k) + \mathbf{B}\mathbf{u}(k)$ of the original system with initial condition $\hat{\mathbf{x}}(k_0)$, and on the basis of this model estimate the state vector $\hat{\mathbf{x}}(k)$? Such an approach presents certain serious drawbacks and is not used in practice. The most important drawback is the following. Since $\hat{\mathbf{x}}(k_0)$ is only an estimate of $\mathbf{x}(k_0)$, the initial condition $\mathbf{x}(k_0)$ of the system and the initial condition $\hat{\mathbf{x}}(k_0)$ of the model differ in most cases. As a result, $\hat{\mathbf{x}}(k)$ may not converge fast enough to $\mathbf{x}(k)$. In order to secure rapid convergence of $\hat{\mathbf{x}}(k)$ to $\mathbf{x}(k)$, we add the term $\mathbf{K}[\mathbf{y}(k) - \mathbf{C}\hat{\mathbf{x}}(k)]$ to the model $\hat{\mathbf{x}}(k+1) = \mathbf{A}\hat{\mathbf{x}}(k) + \mathbf{B}\mathbf{u}(k)$ resulting in an observer of the form (6.35b). Under the assumption that the system $(\mathbf{A}, \mathbf{C})$ is observable, the matrix $\mathbf{K}$ provides adequate design flexibility, as shown below, so that $\hat{\mathbf{x}}(k)$ converges to $\mathbf{x}(k)$ very fast.

The block diagram for the state observer (6.35) is presented in Figure 6.4.

The state observer design problem therefore reduces to one of determining an appropriate matrix $\mathbf{K}$, such that all eigenvalues of the matrix $\hat{\mathbf{A}} = \mathbf{A} - \mathbf{KC}$ lie inside the unit circle. A closer look at the problem reveals that it comes down to one of solving a pole-placement problem for the matrix $\hat{\mathbf{A}} = \mathbf{A} - \mathbf{KC}$. As a matter of fact, this problem is dual to the pole-placement problem discussed earlier in Section 6.2. In what follows, we will use the results of Section 6.2 to solve the observer design problem.

As already mentioned in Section 6.2, the necessary and sufficient condition for a matrix $\mathbf{F}$ to exist, and the matrix $\mathbf{A} - \mathbf{BF}$ may have any desired eigenvalues, is that the system $(\mathbf{A}, \mathbf{B})$ is controllable, i.e.

$$\text{rank } \mathbf{S} = n \quad \text{where } \mathbf{S} = [\mathbf{B} \mathbin{\vdots} \mathbf{AB} \mathbin{\vdots} \cdots \mathbin{\vdots} \mathbf{A}^{n-1}\mathbf{B}] \tag{6.36}$$

In the case of the observer, the necessary and sufficient condition for a matrix $\mathbf{K}$ to exist, so that the matrix $\hat{\mathbf{A}} = \mathbf{A} - \mathbf{KC}$ or, equivalently, the matrix $\hat{\mathbf{A}}^{\mathrm{T}} = \mathbf{A}^{\mathrm{T}} - \mathbf{C}^{\mathrm{T}}\mathbf{K}^{\mathrm{T}}$ has any desired eigenvalues, is that the system $(\mathbf{A}^{\mathrm{T}}, \mathbf{C}^{\mathrm{T}})$ is controllable or, equivalently, that the system $(\mathbf{A}, \mathbf{C})$ is observable, i.e.

$$\text{rank } \mathbf{R} = n \quad \text{where } \mathbf{R}^{\mathrm{T}} = [\mathbf{C}^{\mathrm{T}} \mathbin{\vdots} \mathbf{A}^{\mathrm{T}}\mathbf{C}^{\mathrm{T}} \mathbin{\vdots} \cdots \mathbin{\vdots} (\mathbf{A}^{\mathrm{T}})^{n-1}\mathbf{C}^{\mathrm{T}}] \tag{6.37}$$

For the system $(\mathbf{A}, \mathbf{B}, \mathbf{C})$, we say that the conditions (6.36) and (6.37) are dual.

We will first consider the single-output case for system (6.28). Then the matrix $\mathbf{C}$ reduces to a row vector $\mathbf{c}^{\mathrm{T}}$, thus reducing $\mathbf{R}$ to the $n \times n$ matrix

$$\mathbf{R}^{\mathrm{T}} = [\mathbf{c} \mathbin{\vdots} \mathbf{A}^{\mathrm{T}}\mathbf{c} \mathbin{\vdots} \cdots \mathbin{\vdots} (\mathbf{A}^{\mathrm{T}})^{n-1}\mathbf{c}]$$

The matrix $\hat{\mathbf{A}}$ becomes $\hat{\mathbf{A}} = \mathbf{A} - \mathbf{kc}^{\mathrm{T}}$, where $\mathbf{k}^{\mathrm{T}} = [k_1, k_2, \ldots, k_n]$. As in Section 6.2, we define

$$p(z) = |z\mathbf{I} - \mathbf{A}| = z^n + a_1 z^{n-1} + \ldots + a_n = \prod_{i=1}^{n}(z - \lambda_i)$$

$$\hat{p}(z) = |z\mathbf{I} - \hat{\mathbf{A}}| = z^n + \hat{a}_1 z^{n-1} + \ldots + \hat{a}_n = \prod_{i=1}^{n}(z - \hat{\lambda}_i)$$

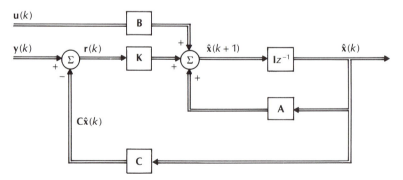

Figure 6.4 Block diagram of the observer (6.35).

where λ_i are the eigenvalues of system (6.28) and $\hat{\lambda}_i$ are the desired eigenvalues of the observer (6.35). Hence, the problem here is to find $\mathbf{k}$ so that the observer has the desired eigenvalues $\hat{\lambda}_1, \hat{\lambda}_2, \ldots, \hat{\lambda}_n$. The vector $\mathbf{k}$ sought is uniquely defined. In Section 6.2, two alternative methods were presented to solve for $\mathbf{k}$. Applying one of these methods yields the following solution:

$$\mathbf{k} = [\mathbf{W}^T \mathbf{R}]^{-1} (\hat{\mathbf{a}} - \mathbf{a}) \tag{6.38}$$

where

$$\mathbf{W} = \begin{bmatrix} 1 & a_1 & \cdots & a_{n-1} \\ 0 & 1 & \cdots & a_{n-2} \\ \vdots & \vdots & & \vdots \\ 0 & 0 & \cdots & 1 \end{bmatrix}, \quad \mathbf{R} = \begin{bmatrix} \mathbf{c}^T \\ \mathbf{c}^T \mathbf{A} \\ \vdots \\ \mathbf{c}^T \mathbf{A}^{n-1} \end{bmatrix}, \quad \hat{\mathbf{a}} = \begin{bmatrix} \hat{a}_1 \\ \hat{a}_2 \\ \vdots \\ \hat{a}_n \end{bmatrix} \quad \text{and} \quad \mathbf{a} = \begin{bmatrix} a_1 \\ a_2 \\ \vdots \\ a_n \end{bmatrix}$$

The solution (6.38) corresponds to the solution (6.8).

For the multi-output case, determining the matrix $\mathbf{K}$, as before, is usually a complicated task. A simple approach to the problem is to assume that $\mathbf{K}$ has the following outer product form:

$$\mathbf{K} = \mathbf{q}\mathbf{p}^T \tag{6.39}$$

where $\mathbf{q}$ and $\mathbf{p}$ are n-dimensional vectors. Then $\hat{\mathbf{A}} = \mathbf{A} - \mathbf{K}\mathbf{C} = \mathbf{A} - \mathbf{q}\mathbf{p}^T\mathbf{C} = \mathbf{A} - \mathbf{q}\boldsymbol{\gamma}^T$, where $\boldsymbol{\gamma}^T = \mathbf{p}^T\mathbf{C}$. Therefore, assuming $\mathbf{K}$ of the form (6.39), the multi-output case reduces to the single-output case studied previously. Hence the solution for $\mathbf{q}$ is given by (6.38), where the matrix $\mathbf{R}$ must be replaced by the matrix $\tilde{\mathbf{R}}$, where $\tilde{\mathbf{R}}^T = [\boldsymbol{\gamma} \mid \mathbf{A}^T\boldsymbol{\gamma} \mid \cdots \mid (\mathbf{A}^T)^{n-1}\boldsymbol{\gamma}]$. It is noted that the vector $\boldsymbol{\gamma} = \mathbf{C}^T\mathbf{p}$ involves arbitrary parameters which are the elements of the arbitrary vector $\mathbf{p}$. These arbitrary parameters may take any values as long as rank $\tilde{\mathbf{R}} = n$. If the condition rank $\tilde{\mathbf{R}} = n$ cannot be satisfied, other methods for determining $\mathbf{K}$ may be found in the literature.

■ **EXAMPLE 6.4.2** Consider the double-integrator system of Example 6.3.1, described in state space by the model (6.28), where

$$\mathbf{A} = \begin{bmatrix} 1 & T \\ 0 & 1 \end{bmatrix}, \quad \mathbf{b} = \begin{bmatrix} T^2/2 \\ T \end{bmatrix} \quad \text{and} \quad \mathbf{c} = \begin{bmatrix} 1 \\ 0 \end{bmatrix}$$

Determine a state observer such that the error $\mathbf{e}(k) = \mathbf{x}(k) - \hat{\mathbf{x}}(k)$ is deadbeat.

■ **SOLUTION** We have

$$\mathbf{R} = \begin{bmatrix} \mathbf{c}^T \\ \mathbf{c}^T\mathbf{A} \end{bmatrix} = \begin{bmatrix} 1 & 0 \\ 1 & T \end{bmatrix}$$

Since rank $\mathbf{R} = 2$, it follows that the system is observable. We also have

$$p(z) = |z\mathbf{I} - \mathbf{A}| = z^2 - 2z + 1$$

$$\hat{p}(z) = |z\mathbf{I} - \hat{\mathbf{A}}| = |z\mathbf{I} - \mathbf{A} - \mathbf{kc}^T| = z^2$$

$$\mathbf{W} = \begin{bmatrix} 1 & a_1 \\ 0 & 1 \end{bmatrix} = \begin{bmatrix} 1 & -2 \\ 0 & 1 \end{bmatrix}, \quad \hat{\mathbf{a}} = \begin{bmatrix} 0 \\ 0 \end{bmatrix} \quad \text{and} \quad \mathbf{a} = \begin{bmatrix} -2 \\ 1 \end{bmatrix}$$

$$\mathbf{W}^T\mathbf{R} = \begin{bmatrix} 1 & 0 \\ -2 & 1 \end{bmatrix}\begin{bmatrix} 1 & 0 \\ 1 & T \end{bmatrix} = \begin{bmatrix} 1 & 0 \\ -1 & T \end{bmatrix} \quad \text{and} \quad (\mathbf{W}^T\mathbf{R})^{-1} = \begin{bmatrix} 1 & 0 \\ 1/T & 1/T \end{bmatrix}$$

Hence

$$\mathbf{k} = [\mathbf{W}^T\mathbf{R}]^{-1}(\hat{\mathbf{a}} - \mathbf{a}) = \begin{bmatrix} 1 & 0 \\ 1/T & 1/T \end{bmatrix}\begin{bmatrix} 2 \\ -1 \end{bmatrix} = \begin{bmatrix} 2 \\ 1/T \end{bmatrix}$$

Checking the results, we have

$$\hat{\mathbf{A}} = \mathbf{A} - \mathbf{kc}^T = \begin{bmatrix} 1 & T \\ 0 & 1 \end{bmatrix} - \begin{bmatrix} 2 & 0 \\ 1/T & 0 \end{bmatrix} = \begin{bmatrix} -1 & T \\ -1/T & 1 \end{bmatrix}$$

Now, consider the error $\mathbf{e}(k) = \mathbf{x}(k) - \hat{\mathbf{x}}(k) = [\mathbf{A} - \mathbf{kc}^T]\mathbf{e}(k-1)$. For $k = 0$, the error $\mathbf{e}(k)$ becomes

$$\mathbf{e}(0) = \begin{bmatrix} e_1(0) \\ e_2(0) \end{bmatrix} = \mathbf{x}(0) - \hat{\mathbf{x}}(0) = \begin{bmatrix} x_1(0) \\ x_2(0) \end{bmatrix} - \begin{bmatrix} \hat{x}_1(0) \\ \hat{x}_2(0) \end{bmatrix} = \begin{bmatrix} \alpha \\ \beta \end{bmatrix}$$

For $k = 1$, the error $\mathbf{e}(k)$ becomes

$$\mathbf{e}(1) = [\mathbf{A} - \mathbf{kc}^T]\mathbf{e}(0) = \begin{bmatrix} -1 & T \\ -1/T & 1 \end{bmatrix}\begin{bmatrix} \alpha \\ \beta \end{bmatrix}$$

For $k = 2$, the error $\mathbf{e}(k)$ becomes

$$\mathbf{e}(2) = [\mathbf{A} - \mathbf{kc}^T]^2\mathbf{e}(0) = \begin{bmatrix} -1 & T \\ -1/T & 1 \end{bmatrix}^2\begin{bmatrix} \alpha \\ \beta \end{bmatrix} = \begin{bmatrix} 0 & 0 \\ 0 & 0 \end{bmatrix}\begin{bmatrix} \alpha \\ \beta \end{bmatrix} = \begin{bmatrix} 0 \\ 0 \end{bmatrix}$$

Hence, the error $\mathbf{e}(k)$ comes to zero in two steps. We have, therefore, a deadbeat observer.

6.4.3 Reduced-order observers

Suppose that the matrix $\mathbf{C}$ in the output equation $\mathbf{y}(k) = \mathbf{C}\mathbf{x}(k)$ is square and non-singular. Then $\mathbf{x}(k) = \mathbf{C}^{-1}\mathbf{y}(k)$, thus eliminating the need for an observer.

Now, assume that only one of the state variables is not accessible to measurement. Then, it is reasonable to expect that the required state observer will not be of order n, but of lower order. This is in fact true, and can be stated as a theorem.

■ **THEOREM 6.4.2** If system (6.28) is observable, then the smallest possible order of the state observer is $n - p$.

We will next present some useful results regarding the design of reduced-order observers. To this end, we assume that the vector $\mathbf{x}(k)$ and the matrices $\mathbf{A}$ and $\mathbf{B}$ may be decomposed as follows:

$$\mathbf{x}(k) = \begin{bmatrix} \mathbf{q}_1(k) \\ \mathbf{q}_2(k) \end{bmatrix}, \quad \mathbf{A} = \begin{bmatrix} \mathbf{A}_{11} & \mathbf{A}_{12} \\ \mathbf{A}_{21} & \mathbf{A}_{22} \end{bmatrix} \quad \text{and} \quad \mathbf{B} = \begin{bmatrix} \mathbf{B}_1 \\ \mathbf{B}_2 \end{bmatrix}$$

Thus

$$\mathbf{q}_1(k+1) = \mathbf{A}_{11}\mathbf{q}_1(k) + \mathbf{A}_{12}\mathbf{q}_2(k) + \mathbf{B}_1\mathbf{u}(k)$$
$$\mathbf{q}_2(k+1) = \mathbf{A}_{21}\mathbf{q}_1(k) + \mathbf{A}_{22}\mathbf{q}_2(k) + \mathbf{B}_2\mathbf{u}(k)$$

$$(6.40)$$

where $\mathbf{q}_1(k)$ is a vector whose elements are all the measurable state variables of $\mathbf{x}(k)$, i.e.

$$\mathbf{y}(k) = \mathbf{C}_1\mathbf{q}_1(k) \quad \text{with} \quad |\mathbf{C}_1| \neq 0 \tag{6.41}$$

In those cases where the system is not in the form (6.40) and (6.41), then it can easily be converted to this form by using an appropriate transformation matrix. The observer of the form (6.35b) for the system (6.40) and (6.41) will be

$$\hat{\mathbf{q}}_1(k+1) = \mathbf{A}_{11}\hat{\mathbf{q}}_1(k) + \mathbf{A}_{12}\hat{\mathbf{q}}_2(k) + \mathbf{B}_1\mathbf{u}(k) + \mathbf{K}_1[\mathbf{y}(k) - \mathbf{C}_1\hat{\mathbf{q}}_1(k)] \tag{6.42a}$$

$$\hat{\mathbf{q}}_2(k+1) = \mathbf{A}_{21}\hat{\mathbf{q}}_1(k) + \mathbf{A}_{22}\hat{\mathbf{q}}_2(k) + \mathbf{B}_2\mathbf{u}(k) + \mathbf{K}_2[\mathbf{y}(k) - \mathbf{C}_1\hat{\mathbf{q}}_1(k)] \tag{6.42b}$$

According to (6.41), we have

$$\mathbf{q}_1(k) = \hat{\mathbf{q}}_1(k) = \mathbf{C}_1^{-1}\mathbf{y}(k) \tag{6.43}$$

Therefore, there is no need for the observer (6.42a), while the observer (6.42b) becomes

$$\hat{\mathbf{q}}_2(k+1) = \mathbf{A}_{22}\hat{\mathbf{q}}_2(k) + \mathbf{B}_2\mathbf{u}(k) + \mathbf{A}_{21}\mathbf{C}_1^{-1}\mathbf{y}(k) \tag{6.44}$$

where use was made of equation (6.43). The observer (6.44) is a dynamic system of order equal to the number of state variables which are not accessible to measurement.

It is obvious that for the observer (6.44), the submatrix $\mathbf{A}_{22}$ plays an important role. If $\mathbf{A}_{22}$ has by luck satisfactory eigenvalues, then system (6.44) suffices for the estimation of $\mathbf{q}_2(k)$. On the other hand, if the eigenvalues of $\mathbf{A}_{22}$ are not satisfactory, then the following observer is proposed for estimating $\mathbf{q}_2(k)$:

$$\hat{\mathbf{q}}_2(k) = \mathbf{\Phi}\mathbf{y}(k) + \mathbf{v}(k) \tag{6.45}$$

where $\mathbf{v}(k)$ is a $(n - p)$ vector governed by the vector difference equation

$$\mathbf{v}(k + 1) = \mathbf{F}\mathbf{v}(k) + \mathbf{H}\mathbf{u}(k) + \mathbf{G}\mathbf{y}(k) \tag{6.46}$$

We define the error as before, i.e. let

$$\mathbf{e}(k) = \mathbf{x}(k) - \hat{\mathbf{x}}(k) = \begin{bmatrix} \mathbf{q}_1(k) - \hat{\mathbf{q}}_1(k) \\ \mathbf{q}_2(k) - \hat{\mathbf{q}}_2(k) \end{bmatrix} = \begin{bmatrix} \mathbf{e}_1(k) \\ \mathbf{e}_2(k) \end{bmatrix} = \begin{bmatrix} \mathbf{0} \\ \mathbf{e}_2(k) \end{bmatrix}$$

The difference equation for $\mathbf{e}_2(k)$ will be

$$\mathbf{e}_2(k + 1) = \mathbf{q}_2(k + 1) - \hat{\mathbf{q}}_2(k + 1)$$
$$= \mathbf{A}_{21}\mathbf{q}_1(k) + \mathbf{A}_{22}\mathbf{q}_2(k) + \mathbf{B}_2\mathbf{u}(k) - \mathbf{\Phi}\mathbf{y}(k + 1) - \mathbf{v}(k + 1)$$

After some algebraic manipulations and simplifications, we have

$$\mathbf{e}_2(k + 1) = \mathbf{F}\mathbf{e}_2(k) + (\mathbf{A}_{21} - \mathbf{\Phi}\mathbf{C}_1\mathbf{A}_{11} - \mathbf{G}\mathbf{C}_1 + \mathbf{F}\mathbf{\Phi}\mathbf{C}_1)\mathbf{q}_1(k)$$
$$+ (\mathbf{A}_{22} - \mathbf{\Phi}\mathbf{C}_1\mathbf{A}_{12} - \mathbf{F})\mathbf{q}_2(k) + (\mathbf{B}_2 - \mathbf{\Phi}\mathbf{C}_1\mathbf{B}_1 - \mathbf{H})\mathbf{u}(k) \tag{6.47}$$

In order for $\mathbf{e}_2(k)$ to be independent of $\mathbf{q}_1(k)$, $\mathbf{q}_2(k)$ and $\mathbf{u}(k)$, as well as to tend rapidly to zero, the following conditions must hold:

1. $\mathbf{G}\mathbf{C}_1 = \mathbf{A}_{21} - \mathbf{\Phi}\mathbf{C}_1\mathbf{A}_{11} + \mathbf{F}\mathbf{\Phi}\mathbf{C}_1$ or $\mathbf{G} = (\mathbf{A}_{21} - \mathbf{\Phi}\mathbf{C}_1\mathbf{A}_{11})\mathbf{C}_1^{-1} + \mathbf{F}\mathbf{\Phi}$ \hfill (6.48a)

2. $\mathbf{F} = \mathbf{A}_{22} - \mathbf{\Phi}\mathbf{C}_1\mathbf{A}_{12}$ \hfill (6.48b)

3. $\mathbf{H} = \mathbf{B}_2 - \mathbf{\Phi}\mathbf{C}_1\mathbf{B}_1$ \hfill (6.48c)

4. matrix $\mathbf{F}$ is stable \hfill (6.48d)

If the foregoing conditions are met, then (6.47) becomes

$$\mathbf{e}_2(k + 1) = \mathbf{F}\mathbf{e}_2(k)$$

The matrix $\mathbf{\Phi}$ may be chosen such that the matrix $\mathbf{F} = \mathbf{A}_{22} - \mathbf{\Phi}\mathbf{C}_1\mathbf{A}_{12}$ has any desired eigenvalues, as long as the system $(\mathbf{A}_{22}, \mathbf{C}_1\mathbf{A}_{12})$ is observable, i.e. as long as

$$\text{rank } \mathbf{R}_1 = n - p \quad \text{where } \mathbf{R}_1^T = [[\mathbf{C}_1\mathbf{A}_{12}]^T \mid \mathbf{A}_{22}^T[\mathbf{C}_1\mathbf{A}_{12}]^T \mid \cdots \mid [\mathbf{A}_{22}^T]^{n-p-1}[\mathbf{C}_1\mathbf{A}_{12}]^T]$$

The following useful theorem has been proved [7].

■ **THEOREM 6.4.3**   The pair $(\mathbf{A}_{22}, \mathbf{C}_1\mathbf{A}_{12})$ is observable, if and only if the pair $(\mathbf{A}, \mathbf{C})$ is observable.

The final form of the observer (6.46) is

$$\mathbf{v}(k+1) = \mathbf{Fv}(k) + \mathbf{Hu}(k) + [(\mathbf{A}_{21} - \boldsymbol{\Phi}\mathbf{C}_1\mathbf{A}_{11})\mathbf{C}_1^{-1} + \mathbf{F}\boldsymbol{\Phi}]\mathbf{y}(k)$$

or

$$\mathbf{v}(k+1) = \mathbf{F}\hat{\mathbf{q}}_2(k) + \mathbf{Hu}(k) + (\mathbf{A}_{21} - \boldsymbol{\Phi}\mathbf{C}_1\mathbf{A}_{11})\mathbf{C}_1^{-1}\mathbf{y}(k) \qquad (6.49)$$

The block diagram of the observer (6.49) is presented in Figure 6.5.

6.4.4 Closed-loop system design using state observers

Consider the system

$$\mathbf{x}(k+1) = \mathbf{Ax}(k) + \mathbf{Bu}(k), \quad \mathbf{y}(k) = \mathbf{Cx}(k) \qquad (6.50)$$

with the state observer

$$\hat{\mathbf{x}}(k+1) = \mathbf{A}\hat{\mathbf{x}}(k) + \mathbf{Bu}(k) + \mathbf{K}_2[\mathbf{y}(k) - \mathbf{C}\hat{\mathbf{x}}(k)] \qquad (6.51)$$

Apply the control law

$$\mathbf{u}(k) = -\mathbf{K}_1\hat{\mathbf{x}}(k) \qquad (6.52)$$

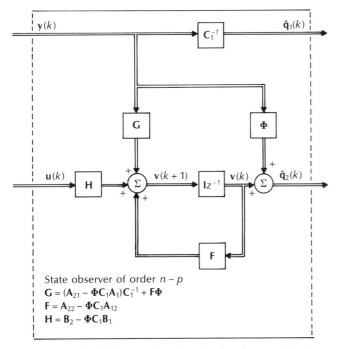

Figure 6.5 Block diagram of reduced-order observer (6.49).

Then, system (6.50) becomes

$$\mathbf{x}(k+1) = \mathbf{A}\mathbf{x}(k) - \mathbf{B}\mathbf{K}_1\hat{\mathbf{x}}(k) \tag{6.53}$$

and the observer (6.51) takes the form

$$\hat{\mathbf{x}}(k+1) = \mathbf{A}\hat{\mathbf{x}}(k) - \mathbf{B}\mathbf{K}_1\hat{\mathbf{x}}(k) + \mathbf{K}_2[\mathbf{C}\mathbf{x}(k) - \mathbf{C}\hat{\mathbf{x}}(k)] \tag{6.54}$$

If we use the definition $\mathbf{e}(k) = \mathbf{x}(k) - \hat{\mathbf{x}}(k)$, then equation (6.53) becomes

$$\mathbf{x}(k+1) = \mathbf{A}\mathbf{x}(k) - \mathbf{B}\mathbf{K}_1[\mathbf{x}(k) - \mathbf{e}(k)]$$

or

$$\mathbf{x}(k+1) = (\mathbf{A} - \mathbf{B}\mathbf{K}_1)\mathbf{x}(k) + \mathbf{B}\mathbf{K}_1\mathbf{e}(k) \tag{6.55}$$

Subtracting (6.54) from (6.53), we have

$$\mathbf{e}(k+1) = (\mathbf{A} - \mathbf{K}_2\mathbf{C})\mathbf{e}(k) \tag{6.56}$$

The foregoing results are very interesting, because they illustrate the fact that the matrix $\mathbf{K}_1$ of the closed-loop system (6.55) and the matrix $\mathbf{K}_2$ of the error equation (6.56) can be designed independently of each other. Indeed, if system $(\mathbf{A}, \mathbf{B})$ is controllable, then the matrix $\mathbf{K}_1$ of the state feedback law (6.52) can be chosen so that the poles of the closed-loop system (6.55) have any desired arbitrary values. The same applies to the error equation (6.56), where, if the system $(\mathbf{A}, \mathbf{C})$ is observable, then the matrix $\mathbf{K}_2$ of the observer (6.51) can be chosen so as to force the error to go rapidly to zero. This property, where the two design problems (the observer and the closed-loop system) can be handled independently, is called the **separation principle**. This principle is clearly a very important design feature, since it reduces a rather difficult design task to two 'separate' simpler design problems.

Figure 6.6 presents the closed-loop system (6.55) and the error equation (6.56), while Figure 6.7 gives the block diagram representation of the closed-loop system with state observer.

Finally, the transfer function $\mathbf{G}_c(z)$ of the compensator defined by the equation $\mathbf{U}(z) = \mathbf{G}_c(z)\mathbf{Y}(z)$ will be

$$\mathbf{G}_c(z) = -\mathbf{K}_1(z\mathbf{I} - \mathbf{A} + \mathbf{B}\mathbf{K}_1 + \mathbf{K}_2\mathbf{C})^{-1}\mathbf{K}_2 \tag{6.57}$$

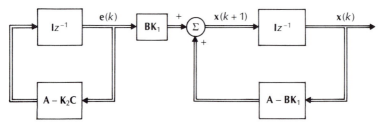

Figure 6.6 Representation of the state equation (6.55) and the error equation (6.56).

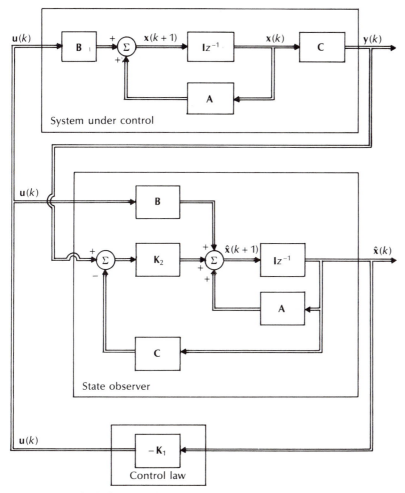

Figure 6.7   Block diagram of closed-loop system with state observer.

The results above cover the case of the full-order observer (order n). In the case of a reduced-order observer, e.g. of an observer of order $n - p$, similar results can be derived relatively easily.

■ **EXAMPLE 6.4.3**   Consider the system

$$\mathbf{x}(k + 1) = \mathbf{Ax}(k) + \mathbf{Bu}(k), \quad y(k) = \mathbf{c}^{\mathsf{T}}\mathbf{x}(k)$$

where

$$\mathbf{A} = \begin{bmatrix} 6 & 1 & 0 \\ -11 & 0 & 1 \\ 6 & 0 & 0 \end{bmatrix}, \quad \mathbf{B} = \begin{bmatrix} 1 & 0 \\ 0 & 1 \\ 1 & 0 \end{bmatrix} \quad \text{and} \quad \mathbf{c} = \begin{bmatrix} 1 \\ 0 \\ 0 \end{bmatrix}$$

Design:

(a) A full-order state observer, i.e. of order $n = 3$.
(b) A reduced-order state observer, i.e. of order $n - p = 3 - 1 = 2$.
(c) The closed-loop system for both cases.

■ **SOLUTION**

(a) We examine the system's observability. We have

$$\mathbf{R}^T = [\mathbf{c} \mathrel{\vdots} \mathbf{A}^T\mathbf{c} \mathrel{\vdots} (\mathbf{A}^T)^2\mathbf{c}] = \begin{bmatrix} 1 & 6 & 25 \\ 0 & 1 & 6 \\ 0 & 0 & 1 \end{bmatrix}$$

Since rank $\mathbf{R} = 3$, there exists a full-order state observer having the form

$$\hat{\mathbf{x}}(k+1) = [\mathbf{A} - \mathbf{k}\mathbf{c}^T]\hat{\mathbf{x}}(k) + \mathbf{B}u(k) + \mathbf{k}y(k)$$

The characteristic polynomial of the open-loop system is $p(z) = z^3 - 6z^2 + 11z - 6$. Suppose that the desired observer characteristic polynomial is chosen as $\hat{p}(z) = z^2(z - 0.5) = z^3 - 0.5z^2$. From equation (6.38) we have

$$\mathbf{W} = \begin{bmatrix} 1 & -6 & 11 \\ 0 & 1 & -6 \\ 0 & 0 & 1 \end{bmatrix}, \quad \hat{\mathbf{a}} = \begin{bmatrix} \hat{a}_1 \\ \hat{a}_2 \\ \hat{a}_3 \end{bmatrix} = \begin{bmatrix} 0.5 \\ 0 \\ 0 \end{bmatrix} \quad \text{and} \quad \mathbf{a} = \begin{bmatrix} a_1 \\ a_2 \\ a_3 \end{bmatrix} = \begin{bmatrix} -6 \\ 11 \\ -6 \end{bmatrix}$$

and hence

$$\mathbf{k} = \begin{bmatrix} k_1 \\ k_2 \\ k_3 \end{bmatrix} = [\mathbf{W}^T\mathbf{R}]^{-1}(\hat{\mathbf{a}} - \mathbf{a}) = \begin{bmatrix} 5.5 \\ -11 \\ 6 \end{bmatrix}$$

(b) The system of the present example is in the form (6.40), where

$$A_{11} = 6, \quad \mathbf{A}_{12} = [1 \quad 0], \quad \mathbf{A}_{21} = \begin{bmatrix} -11 \\ 6 \end{bmatrix} \quad \text{and} \quad \mathbf{A}_{22} = \begin{bmatrix} 0 & 1 \\ 0 & 0 \end{bmatrix}$$

$$\mathbf{B}_1 = [1 \quad 0], \quad \mathbf{B}_2 = \begin{bmatrix} 0 & 1 \\ 1 & 0 \end{bmatrix} \quad \text{and} \quad c_1 = 1$$

and where

$$\mathbf{q}_1(k) = x_1(k) \quad \text{and} \quad \mathbf{q}_2(k) = \begin{bmatrix} x_2(k) \\ x_3(k) \end{bmatrix}$$

Here, $\mathbf{q}_1(k) = x_1(k) = y(k)$. The proposed observer for the estimation of the vector $\mathbf{q}_2(k)$ is

$$\mathbf{q}_2(k) = \boldsymbol{\phi}y(k) + \mathbf{v}(k)$$

where $\mathbf{v}(k)$ is a two-dimensional vector described by the vector difference equation

$$\mathbf{v}(k+1) = \mathbf{F}\mathbf{v}(k) + \mathbf{H}u(k) + \mathbf{g}y(k)$$

and where

$$\mathbf{F} = \mathbf{A}_{22} - \boldsymbol{\phi}c_1\mathbf{A}_{12} = \begin{bmatrix} 0 & 1 \\ 0 & 0 \end{bmatrix} - \begin{bmatrix} \phi_1 \\ \phi_2 \end{bmatrix} [1 \quad 0] = \begin{bmatrix} -\phi_1 & 1 \\ -\phi_2 & 0 \end{bmatrix}$$

$$\mathbf{g} = (\mathbf{A}_{21} - \boldsymbol{\phi}c_1\mathbf{A}_{11})c_1^{-1} + \mathbf{F}\boldsymbol{\phi}$$

$$= \begin{bmatrix} -11 \\ 6 \end{bmatrix} - \begin{bmatrix} \phi_1 \\ \phi_2 \end{bmatrix} 6 + \begin{bmatrix} -\phi_1 & 1 \\ -\phi_2 & 0 \end{bmatrix} \begin{bmatrix} \phi_1 \\ \phi_2 \end{bmatrix} = \begin{bmatrix} -11 - 6\phi_1 - \phi_1^2 + \phi_2 \\ 6 - 6\phi_2 - \phi_1\phi_2 \end{bmatrix}$$

$$\mathbf{H} = \mathbf{B}_2 - \boldsymbol{\phi}c_1\mathbf{B}_1 = \begin{bmatrix} 0 & 1 \\ 1 & 0 \end{bmatrix} - \begin{bmatrix} \phi_1 \\ \phi_2 \end{bmatrix} [1 \quad 0] = \begin{bmatrix} -\phi_1 & 1 \\ 1 - \phi_2 & 0 \end{bmatrix}$$

Since

$$\text{rank } \mathbf{R}_1^T = \text{rank}[[c_1\mathbf{A}_{12}]^T \mid \mathbf{A}_{22}^T[c_1\mathbf{A}_{12}]^T] = \text{rank} \begin{bmatrix} 1 & 0 \\ 0 & 1 \end{bmatrix} = 2$$

we can find a vector $\boldsymbol{\phi}$ such that the matrix $\mathbf{F}$ has the desired eigenvalues. The characteristic polynomial of $\mathbf{A}_{22}$ is $p_2(z) = z^2$. Let $\hat{p}_2(z) = z^2 - 0.5z$ be the desired characteristic polynomial of matrix $\mathbf{F}$. From equation (6.38) we have

$$\mathbf{W} = \begin{bmatrix} 1 & 0 \\ 0 & 1 \end{bmatrix}, \quad \hat{\mathbf{a}} = \begin{bmatrix} -0.5 \\ 0 \end{bmatrix} \quad \text{and} \quad \mathbf{a} = \begin{bmatrix} 0 \\ 0 \end{bmatrix}$$

and therefore

$$\boldsymbol{\phi} = \begin{bmatrix} \phi_1 \\ \phi_2 \end{bmatrix} = [\mathbf{W}^T\mathbf{R}_1]^{-1}(\hat{\mathbf{a}} - \mathbf{a}) = \begin{bmatrix} -0.5 \\ 0 \end{bmatrix}$$

Introducing the value of $\boldsymbol{\phi}$ into $\mathbf{g}$, $\mathbf{F}$ and $\mathbf{H}$ yields

$$\mathbf{g} = \begin{bmatrix} -8.25 \\ 6 \end{bmatrix}, \quad \mathbf{F} = \begin{bmatrix} 0.5 & 1 \\ 0 & 0 \end{bmatrix} \quad \text{and} \quad \mathbf{H} = \begin{bmatrix} 0.5 & 1 \\ 1 & 0 \end{bmatrix}$$

(c) Let $p_c(z) = z^3$ be the desired characteristic polynomial of the closed-loop system. The system is controllable because rank $\mathbf{S} = \text{rank}[\mathbf{B} \mid \mathbf{AB} \mid \mathbf{A}^2\mathbf{B}] = 3$. Consequently, a feedback matrix $\mathbf{K}_1$ exists such that the closed-loop system poles are the roots of $p_c(z) = z^3$. According to the equations (6.15) and (6.16), the following matrix is determined:

$$\mathbf{K}_1 = \begin{bmatrix} 6 & 1 & 0 \\ 0 & 0 & 1 \end{bmatrix}$$

Checking, we have that $|z\mathbf{I} - (\mathbf{A} - \mathbf{BK}_1)| = z^3 = p_c(z)$. Of course, in the case of a full-order observer, $\mathbf{k}_2 = \mathbf{k}$, where $\mathbf{k}$ was determined in part (a) above. In the case of a reduced-order observer, $\mathbf{k}_2 = \boldsymbol{\phi}$, where $\boldsymbol{\phi}$ was determined in part (b) above.

■ **EXAMPLE 6.4.4**   Consider the system

$$\mathbf{x}(k+1) = \mathbf{A}\mathbf{x}(k) + \mathbf{b}u(k), \quad y(k) = \mathbf{c}^{\mathsf{T}}\mathbf{x}(k)$$

where

$$\mathbf{A} = \begin{bmatrix} 0 & 1 \\ 0 & -\gamma \end{bmatrix}, \quad \mathbf{b} = \begin{bmatrix} 0 \\ \beta \end{bmatrix} \quad \text{and} \quad \mathbf{c} = \begin{bmatrix} 1 \\ 0 \end{bmatrix}$$

In this example we suppose that only the state $x_1(k) = y(k)$ can be directly measured. Design:

(a) A full-order state observer, i.e. of order $n = 2$.
(b) A reduced-order state observer, i.e. of order $n - p = 2 - 1 = 1$. In other words, find an observer to estimate only the state $x_2(k)$, which we assume is not accessible to measurement. Note that x_1 is measurable since $x_1(k) = y(k)$.
(c) The closed-loop system for both cases.

■ **SOLUTION**
(a) We examine the system's observability. We have

$$\mathbf{R}^{\mathsf{T}} = [\mathbf{c} \vdots \mathbf{A}^{\mathsf{T}}\mathbf{c}] = \begin{bmatrix} 1 & 0 \\ 0 & 1 \end{bmatrix}$$

Since rank $\mathbf{R} = n = 2$, there exists a full-order state observer having the form

$$\hat{\mathbf{x}}(k+1) = [\mathbf{A} - \mathbf{k}\mathbf{c}^{\mathsf{T}}]\hat{\mathbf{x}}(k) + \mathbf{B}u(k) + \mathbf{k}y(k)$$

The characteristic polynomials $p(z)$ and $\hat{p}(z)$ of the open-loop system and of the observer are $p(z) = |z\mathbf{I} - \mathbf{A}| = z^2 + \gamma z$ and $\hat{p}(z) = |z\mathbf{I} - (\mathbf{A} - \mathbf{k}\mathbf{c}^{\mathsf{T}})| = z^2 + \hat{a}_1 z + \hat{a}_2$, respectively. From equation (6.38) we have

$$\mathbf{W} = \begin{bmatrix} 1 & \gamma \\ 0 & 1 \end{bmatrix}, \quad \hat{\mathbf{a}} = \begin{bmatrix} \hat{a}_1 \\ \hat{a}_2 \end{bmatrix} \quad \text{and} \quad \mathbf{a} = \begin{bmatrix} \gamma \\ 0 \end{bmatrix}$$

and thus

$$\mathbf{k} = \begin{bmatrix} k_1 \\ k_2 \end{bmatrix} = [\mathbf{W}^{\mathsf{T}}\mathbf{R}]^{-1}(\hat{\mathbf{a}} - \mathbf{a}) = \begin{bmatrix} \hat{a}_1 - \gamma \\ \hat{a}_2 - \gamma(\hat{a}_1 - \gamma) \end{bmatrix}$$

From a practical point of view, we choose $\hat{a}_1$ and $\hat{a}_2$ in $\hat{p}(z)$ so that the error $\mathbf{e}(k) = \mathbf{x}(k) - \hat{\mathbf{x}}(k)$ tends rapidly to zero. Of course, both roots of $\hat{p}(z)$ must be inside the unit circle.

(b) The system of the present example is in the form (6.40), where $A_{11} = 0$, $A_{12} = 1$, $A_{21} = 0$, $A_{22} = -\gamma$, $b_1 = 0$, $b_2 = \beta$ and $c_1 = 1$. Here $\mathbf{q}_1(k) = x_1(k)$ and $\mathbf{q}_2(k) = x_2(k)$. Moreover, $q_1(k) = x_1(k) = y(k)$. For the estimation of $x_2(k)$ the proposed observer is

$$q_2(k) = \hat{x}_2(k) = \phi y(k) + v(k)$$

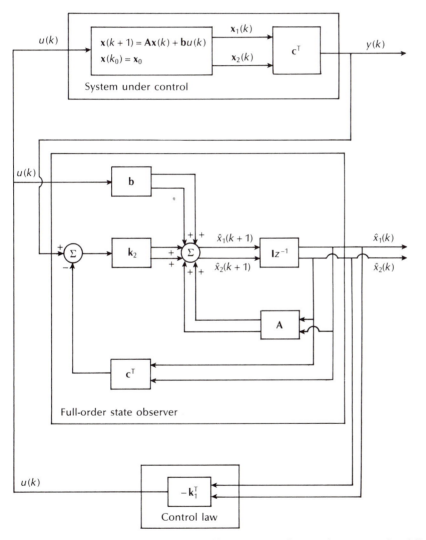

Figure 6.8 Block diagram of the closed-loop system of Example 6.4.4, with a full-order state observer.

where $v(k)$ is a scalar function governed by the difference equation

$$v(k+1) = fv(k) + hu(k) + gy(k)$$

and where

$$f = A_{22} - \phi c_1 A_{12} = -\gamma - \phi$$
$$g = (A_{21} - \phi c_1 A_{11})c_1^{-1} + f\phi = (-\gamma - \phi)\phi = -\gamma\phi - \phi^2$$
$$h = B_2 - \phi c_1 B_1 = \beta$$

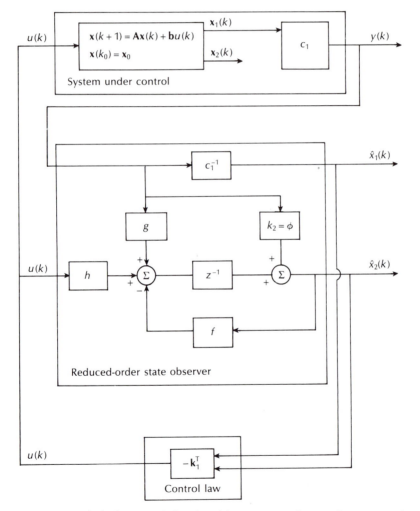

Figure 6.9 Block diagram of the closed-loop system of Example 6.4.4, with reduced-order state observer.

Since rank $R_1^T = \text{rank}[(c_1 A_{12})^T] = \text{rank } 1 = 1$ we can find a ϕ such that f has the desired eigenvalue. Let $-\rho$ be the desired eigenvalue of f. Then, $\phi = \rho - \gamma$. Introducing the value of ϕ into g and f we have

$$g = -\rho^2 + \rho\gamma \quad \text{and} \quad f = -\rho$$

(c) Let $p_c(z) = z^2 + \gamma_1 z + \gamma_2$ be the desired characteristic polynomial of the closed-loop system. The parameters γ_1 and γ_2 are arbitrary, but they will ultimately be specified in order to meet closed-loop system requirements. The system is controllable since rank $\mathbf{S} = \text{rank}[\mathbf{b} \vdots \mathbf{Ab}] = 2$. Therefore, we can choose a feedback vector $\mathbf{k}_1$ such that the closed-loop system poles are the roots of $p_c(z) = z^2 + \gamma_1 z + \gamma_2$. According to the results of Section 6.2, the following vector is determined:

$$\mathbf{k}_1^T = \begin{bmatrix} \dfrac{\gamma_2}{\beta} & -\dfrac{\gamma - \gamma_1}{\beta} \end{bmatrix}$$

Checking the results, we have $|z\mathbf{I} - (\mathbf{A} - \mathbf{b}\mathbf{k}_1^T)| = z^2 + \gamma_1 z + \gamma_2 = p_c(z)$. Of course, in the case of a full-order observer, $\mathbf{k}_2 = \mathbf{k}$, where $\mathbf{k}$ has been determined in part (a) above. In the case of a reduced-order observer, $\mathbf{k}_2 = \phi$, where ϕ has been determined in part (b) above. The block diagram representations of the closed-loop systems in both cases are given in Figures 6.8 and 6.9.

6.5 Problems

1. Consider the system in the form (6.1), where

$$\mathbf{A} = \begin{bmatrix} 0 & 1 & 0 \\ 0 & 0 & 1 \\ 1 & -3 & 3 \end{bmatrix} \quad \text{and} \quad \mathbf{b} = \begin{bmatrix} 0 \\ 0 \\ 1 \end{bmatrix}$$

Find the vector $\mathbf{f}$ of the control law $u(k) = \mathbf{f}^T \mathbf{x}(k)$, so that the closed-loop system has the triple eigenvalue $\hat{\lambda}_1 = \hat{\lambda}_2 = \hat{\lambda}_3 = 0$.

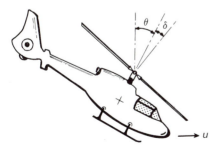

Figure 6.10 A simplified model of a helicopter.

2. In Example 6.2.2, find a vector **f** such that the closed-loop system has the triple eigenvalue $\hat{\lambda}_1 = \hat{\lambda}_2 = \hat{\lambda}_3 = -\frac{1}{4}$.

3. The lateral motion of a helicopter can be approximately described by the following third-order linear continuous-time state-space model (see Figure 6.10, p.187) [2]:

$$\dot{\mathbf{x}}(t) = \tilde{\mathbf{F}}\mathbf{x}(t) + \tilde{\mathbf{g}}\delta(t)$$

or

$$\begin{bmatrix} \dot{q}(t) \\ \dot{\theta}(t) \\ \dot{u}(t) \end{bmatrix} = \begin{bmatrix} -0.4 & 0 & -0.01 \\ 1 & 0 & 0 \\ -1.4 & 9.8 & -0.02 \end{bmatrix} \begin{bmatrix} q(t) \\ \theta(t) \\ u(t) \end{bmatrix} + \begin{bmatrix} 6.3 \\ 0 \\ 9.8 \end{bmatrix} \delta(t)$$

where

$q(t) =$ the angular pitch rate,
$\theta(t) =$ the pitch angle of the fuselage,
$u(t) =$ the horizontal velocity of the helicopter,
$\delta(t) =$ the rotor inclination angle.

(a) Discretize the system, with sampling period $T = 0.3$ s.
(b) Find the discrete-time state feedback law

$$\delta(kT) = \mathbf{f}^{\mathrm{T}}\mathbf{x}(kT)$$

so that the eigenvalues of the closed-loop system are $\hat{\lambda}_1 = 0.5$, $\hat{\lambda}_2 = -0.4$ and $\hat{\lambda}_3 = 0.3$.

4. The dynamical equations describing the yaw motion of a tanker are (see Figure 6.11) [8]

$$\dot{\mathbf{x}}(t) = \tilde{\mathbf{F}}\mathbf{x}(t) + \tilde{\mathbf{g}}u(t), \quad y(t) = \mathbf{c}^{\mathrm{T}}\mathbf{x}(t)$$

where

$$\mathbf{x}(t) = [v(t), r(t), \psi(t)]^{\mathrm{T}} \quad u(t) = \delta(t)$$

$$\tilde{\mathbf{F}} = \begin{bmatrix} \bar{f}_{11} & \bar{f}_{12} & 0 \\ \bar{f}_{21} & \bar{f}_{22} & 0 \\ 0 & 1 & 0 \end{bmatrix}, \quad \bar{\mathbf{g}} = \begin{bmatrix} \bar{g}_{11} \\ \bar{g}_{21} \\ 0 \end{bmatrix} \quad \text{and} \quad \mathbf{c}^{\mathrm{T}} = [1 \quad 0 \quad 0]$$

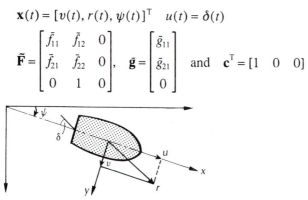

Figure 6.11 Tanker model and defining quantities.

and

$v(t)$ = the y-component of the tanker velocity,
$r(t)$ = the tanker velocity,
$\psi(t)$ = the axial inclination of the tanker relative to the given frame of reference,
$\delta(t)$ = the rudder orientation with respect to the axial direction.

The parameters may be taken as

$$\tilde{f}_{11} = -0.44, \quad \tilde{f}_{12} = -0.28, \quad \tilde{f}_{21} = -2.67 \quad \text{and} \quad \tilde{f}_{22} = -2.04$$
$$\tilde{g}_{11} = 0.07 \quad \text{and} \quad \tilde{g}_{21} = -0.53$$

Discretize the system and obtain the state feedback law

$$\delta(kT) = \mathbf{f}^\mathsf{T}\mathbf{x}(kT)$$

which yields the closed-loop poles $\hat{\lambda}_1 = -0.5$, $\hat{\lambda}_2 = 0.3$ and $\hat{\lambda}_3 = -0.4$. The sampling period has the value $T = 0.05$ s.

5. Consider the discrete-time system

$$\mathbf{x}(k+1) = \mathbf{A}\mathbf{x}(k) + \mathbf{B}u(k), \quad \mathbf{y}(k) = \mathbf{C}\mathbf{x}(k)$$

where $\mathbf{A} \in \mathbb{R}^{n \times n}$, $\mathbf{B} \in \mathbb{R}^{n \times 1}$ and $\mathbf{C} \in \mathbb{R}^{1 \times m}$. It is assumed that the state vector $\mathbf{x}(k)$ is controllable. It is also assumed that the rank of the following $(n+1) \times (n+1)$ matrix

$$\begin{bmatrix} \mathbf{A} & \mathbf{B} \\ -\mathbf{C} & \mathbf{0} \end{bmatrix}$$

is $n+1$. Prove that for the system

$$\mathbf{z}(k+1) = \hat{\mathbf{A}}\mathbf{z}(k) + \hat{\mathbf{B}}\mathbf{w}(k)$$

where

$$\hat{\mathbf{A}} = \begin{bmatrix} \mathbf{A} & \mathbf{B} \\ -\mathbf{C} & \mathbf{0} \end{bmatrix}, \quad \hat{\mathbf{B}} = \begin{bmatrix} \mathbf{B} \\ \mathbf{0} \end{bmatrix} \quad \text{and} \quad \mathbf{w}(k) = \mathbf{u}(k) - \mathbf{u}(\infty)$$

it is possible to select the eigenvalues arbitrarily using linear state feedback.

6. A d.c. motor which drives an inertial load at a specific position is described by the equations

$$\dot{\theta}(t) = \omega(t) \quad \text{and} \quad \dot{\omega}(t) = -\alpha\omega(t) + \beta u(t)$$

where $\theta(t)$ is the angular displacement of the load, $\omega(t)$ is the angular velocity of the load, $u(t)$ is the applied voltage, and α, β are constants which depend on the physical parameters of the motor and the load and are defined as

$$\alpha = -\frac{K^2}{JR} \quad \text{and} \quad \beta = \frac{K}{JR}$$

In these expressions, R is the resistance of the windings in the motor, K is a constant and J is the moment of inertia of the load. If θ_r is the desired angular displacement which is considered to be constant, let

$$e(t) = \theta(t) - \theta_r$$

to obtain the continuous-time state-space description for the motor:

$$\dot{\mathbf{x}} = \begin{bmatrix} \dot{e}(t) \\ \dot{\omega}(t) \end{bmatrix} = \begin{bmatrix} 0 & 0 \\ 0 & -\alpha \end{bmatrix} \begin{bmatrix} e(t) \\ \omega(t) \end{bmatrix} + \begin{bmatrix} 0 \\ \beta \end{bmatrix} u(t) = \overline{\mathbf{F}}\mathbf{x}(t) + \overline{\mathbf{g}}u(t)$$

Find a discrete-time description of the system. Obtain, if it exists, a state feedback law which relocates the system eigenvalues to correspond to the roots of the discrete-time characteristic polynomial

$$p(z) = z^2 + a_1 z + a_0$$

where a_1 and a_0 are some prespecified constants.

7. Consider the continuous-time model for lateral helicopter motion in problem 3. Use as the sampling period $T = 0.3$ s and find the discrete-time state feedback law for deadbeat control.

8. Consider the continuous-time model for the yaw motion of the tanker given in problem 4. If the sampling period $T = 0.05$ s, determine the discrete-time state feedback law for deadbeat control.

9. Solve Example 6.4.1 for a triple-integrator system.

10. Solve Example 6.4.2 for a triple-integrator system.

11. Consider the continuous-time model for the lateral helicopter motion in problem 3 and its discretization with sampling period $T = 0.3$ s. Let the angular pitch rate be the output of the system and find:
(a) a full-order state observer,
(b) a full-order deadbeat observer, and
(c) a reduced order observer.

12. Consider the continuous-time model for the yaw motion of the tanker given in problem 4 and its discretization with sampling period $T = 0.05$ s. If the axial inclination of the tanker is the output of the system, find:
(a) a full-order state observer, $n = 3$,
(b) a full-order deadbeat observer, and
(c) a reduced-order observer.

13. Consider the motion of a satellite (see Figure 6.12) about an axis perpendicular to the page. The equation of motion relative to an inertial frame of reference is [3]

$$J\ddot{\theta} = T_C + T_D$$

where J is the moment of inertia of the satellite about an axis through its centre of mass, T_C is the control torque applied by the thrusters, T_D is the resultant of the external disturbance torques and θ is the angular position measured relative to some inertial frame of reference.

Let

$$u = T_C/J, \quad w_D = T_D/J$$

Then

$$\ddot{\theta}(t) = u(t) + w_D$$

or, taking the Laplace transform,

$$\Theta(s) = \frac{1}{s^2}[U(s) + W_D(s)]$$

If the disturbances are negligible,

$$\frac{\Theta(s)}{U(s)} = G(s) = \frac{1}{s^2}$$

In discrete time, with u being applied through a zero-order hold,

$$G(z) = \frac{\Theta(z)}{U(z)} = \left(\frac{T^2}{2}\right)\frac{z+1}{(z-1)^2}$$

The sampling period $T = 0.05$ s. Find:

(a) a discrete-time state feedback law such that the closed-loop system eigenvalues are $\hat{\lambda}_1 = -0.4$ and $\hat{\lambda}_2 = 0.5$,

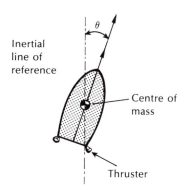

Figure 6.12 Satellite control system.

(b) a deadbeat controller,
(c) a full-order state observer,
(d) a full-order deadbeat observer,
(e) a reduced-order observer.

14. Consider a discrete-time system with the state-space description:

$$\mathbf{x}(k+1) = \begin{bmatrix} 0 & 1 & 0 \\ 0 & 0 & 1 \\ 1 & -3 & 3 \end{bmatrix} \mathbf{x}(k) + \begin{bmatrix} 0 \\ 0 \\ 1 \end{bmatrix} u(k)$$

$$y(k) = [1 \ 1 \ 0]\mathbf{x}(k)$$

Find:

(a) the state feedback law in the form $u(k) = -\mathbf{f}^T\mathbf{x}(k)$ so that the closed-loop system eigenvalues are $\hat{\lambda}_1 = \hat{\lambda}_2 = 0$ and $\hat{\lambda}_3 = 0.5$;
(b) the full-order state observer, so that the error in the state estimation becomes zero in at most three sampling periods.

15. Consider a discrete-time system with the state-space description:

$$\mathbf{x}(k+1) = \begin{bmatrix} 0 & 1 & 0 \\ 0 & 0 & 1 \\ -4 & 4 & -1 \end{bmatrix} \mathbf{x}(k) + \begin{bmatrix} 0 \\ 0 \\ 1 \end{bmatrix} u(k)$$

$$y(k) = [1 \ 0 \ 1]\mathbf{x}(k)$$

Find:

(a) the state feedback law in the form $u(k) = -\mathbf{f}^T\mathbf{x}(k)$, which yields the closed-loop system eigenvalues $\hat{\lambda}_1 = \hat{\lambda}_2 = 0.5$ and $\hat{\lambda}_3 = -0.1$;
(b) the full-order state observer, so that the error in the state estimation becomes zero in at most three sampling periods;
(c) the closed-loop system block diagram incorporating the observer and the control law found in parts (a) and (b) respectively.

16. Let a discrete-time system be described in state space by

$$\mathbf{x}(k+1) = \begin{bmatrix} -2 & -1 \\ -3 & -4 \end{bmatrix} \mathbf{x}(k) + \begin{bmatrix} 1 \\ 1 \end{bmatrix} u(k)$$

$$\mathbf{y}(k) = [1 \ 3]\mathbf{x}(k)$$

Determine:

(a) the state feedback law in the form $u(k) = -\mathbf{f}^T\mathbf{x}(k)$, which yields the closed-loop system eigenvalues $\hat{\lambda}_1 = 0.1$ and $\hat{\lambda}_2 = 0.3$;

(b) the state feedback law in the form $u(k) = -\mathbf{f}^T\mathbf{u}(k)$ for deadbeat control;

(c) the full-order state observer with characteristic polynomial $p_c(z) = z^2 + 0.8z + 0.15$.

17. Consider a discrete-time system with the state-space description

$$\mathbf{x}(k+1) = \begin{bmatrix} 0 & 1 & 0 & 0 \\ 0 & 0 & 1 & 0 \\ 0 & 0 & 0 & 1 \\ 2 & -1 & 1 & -1 \end{bmatrix} \mathbf{x}(k) + \begin{bmatrix} 0 \\ 0 \\ 0 \\ 1 \end{bmatrix} u(k)$$

$$y(k) = [1\ 0\ 0\ 0]\mathbf{x}(k)$$

(a) Determine the state feedback law in the form $u(k) = -\mathbf{f}^T\mathbf{x}(k)$, so that the closed-loop system eigenvalues are $\hat{\lambda}_1 = \hat{\lambda}_2 = 0.5$ and $\hat{\lambda}_3 = \hat{\lambda}_4 = 0.6$.

(b) For the realization of the closed-loop system above, determine the deadbeat observer and draw the closed-loop system using the control law of part (a).

18. Let a discrete-time system be described in the state space by

$$\mathbf{x}(k+1) = \begin{bmatrix} 0 & 0 & 1 \\ 1 & 1 & 0 \\ -1 & 0 & 0.5 \end{bmatrix} \mathbf{x}(k) + \begin{bmatrix} 1 \\ 1 \\ -1 \end{bmatrix} u(k)$$

$$y(k) = [0\ 1\ 1]\mathbf{x}(k)$$

Find:

(a) the state feedback law in the form $u(k) = -\mathbf{f}^T\mathbf{x}(k)$, so that the closed-loop system eigenvalues are $\hat{\lambda}_1 = 0.5$, $\hat{\lambda}_2 = 0.7$ and $\hat{\lambda}_3 = -0.6$;

(b) the number of sampling periods required for the closed-loop system to reach complete standstill and the state feedback law required to do so;

(c) the full-order state observer for the above realization and draw the closed-loop system.

19. Consider the discrete-time system of problem 17. Obtain a reduced-order state observer.

6.6 **References**

Books

[1] K.J. Åström and B. Wittenmark, *Computer Controlled Systems*, Prentice Hall, Englewood Cliffs, New Jersey, 1984.

[2] G.F. Franklin, J.D. Powell and A. Emani-Naeini, *Feedback Control of Dynamic Systems*, Addison-Wesley, London, 1991 (Second Edition).

[3] G.F. Franklin, J.D. Powell and M.L. Workman. *Digital Control of Dynamic Systems*, Addison-Wesley, London, 1990 (Second Edition).

[4] T. Kailath, *Linear Systems*, Prentice Hall, Englewood Cliffs, New Jersey, 1980.

[5] J. O'Reilly, *Observers for Linear Systems*, Academic Press, London, 1983.

Papers

[6] J. Ackermann, 'Parameter Space Design of Robust Control Systems', *IEEE Transactions on Automatic Control*, Vol. AC–25, pp. 1058–1072, 1980.

[7] A.T. Alexandridis and P.N. Paraskevopoulos, 'A New Approach for Eigenstructure Assignment by Output Feedback', *IEEE Transactions on Automatic Control*, forthcoming, 1996.

[8] K.J. Åström and C.G. Kallström, 'Identification of Ship Steering Dynamics', *Automatica*, Vol. 12, pp. 9–22, 1976.

[9] J.C. Doyle and G. Stein, 'Robustness with Observers', *IEEE Transactions on Automatic Control*, Vol. AC–24, pp. 607–611, 1979.

[10] D.G. Luenberger, 'Observing the State of a Linear System', *IEEE Transactions on Military Electronics*, Vol. MIL–8, pp. 74–80, 1964.

[11] D.G. Luenberger, 'Observers for Multivariable Systems', *IEEE Transactions on Automatic Control*, Vol. AC–11, pp. 190–197, 1966.

[12] D.G. Luenberger, 'An Introduction to Observers', *IEEE Transactions on Automatic Control*, Vol. AC–16, pp. 596–602, 1971.

[13] P.N. Paraskevopoulos and S.G. Tzafestas, 'New Results on Feedback Modal-Controller Design', *International Journal of Control*, Vol. 24, pp. 209–216, 1976.

[14] W.M. Wonham, 'On Pole Assignment in Multi-Input Controllable Systems', *IEEE Transactions on Automatic Control*, Vol. AC–12, pp. 660–665, 1967.

7

Optimal control

7.1 Introduction

This chapter presents an introduction to the optimal control of discrete-time systems. This problem is of great interest from both the theoretical and the application points of view.

Optimal control aims at determining a feedback control law which minimizes a specified cost function (or performance index). The cost function involves the desired performance specifications of the closed-loop system. These specifications may include characteristics such as the energy dissipated by the system, the required energy control effort and others. Moreover, performance characteristics related to the classical techniques of Chapter 5 (such as rise time, settling time, bandwidth, phase and gain margins, etc.) may also be included in the specifications, i.e. in the cost function. The solution to the optimal control problem requires higher mathematics, which are briefly presented in the next section.

7.2 Mathematical background for the study of optimal control problems of discrete-time systems

7.2.1 *Maxima and minima using the method of calculus of variations*

The performance index or cost function of continuous-time systems is formulated as an **integral** expression [1–3], [5], [6], [8], [9], [11–13]. In the case of discrete-time systems, the cost function is formulated as a **summation** expression. Such a cost function may have the form [4], [13]:

$$J = \sum_{k=k_0}^{k_f - 1} \phi[\mathbf{x}(k), \mathbf{x}(k + 1), k] \tag{7.1}$$

Consider the problem of minimizing this cost function (7.1). A convenient way to solve this problem is to apply the method of calculus of variations, as briefly presented here. Assume that $\mathbf{x}(k) = \hat{\mathbf{x}}(k) + \varepsilon\boldsymbol{\eta}[\mathbf{x}(k)]$ and $\mathbf{x}(k+1) = \hat{\mathbf{x}}(k+1) + \varepsilon\boldsymbol{\eta}[\mathbf{x}(k+1)]$, where $\hat{\mathbf{x}}(k)$ is the optimal solution, $\boldsymbol{\eta}[\mathbf{x}(k)]$ is a deviation from $\mathbf{x}(k)$ and ε is a small real number. Let ΔJ be a small deviation of J from its optimal value. Expanding ΔJ in a Taylor series in terms of $\delta\mathbf{x}$, where $\delta\mathbf{x} = \varepsilon\boldsymbol{\eta}[\mathbf{x}(k)]$, and taking only the first term δJ of the expansion, we have

$$\delta J = \sum_{k=k_0}^{k_f - 1}\left([\delta\mathbf{x}(k)]^{\mathrm{T}}\,\frac{\partial\phi_k}{\partial\mathbf{x}(k)} + [\delta\mathbf{x}(k+1)]^{\mathrm{T}}\,\frac{\partial\phi_k}{\partial\mathbf{x}(k+1)}\right) \tag{7.2}$$

where, for simplicity, we denote $\phi[\mathbf{x}(k), \mathbf{x}(k+1), k]$ by ϕ_k and omit the symbol '^' from the optimal trajectory $\hat{\mathbf{x}}(k)$. A necessary condition for the vector $\mathbf{x}(k)$ to minimize the cost function J is that $\delta J = 0$, i.e.

$$\sum_{k=k_0}^{k_f - 1}\left([\delta\mathbf{x}(k)]^{\mathrm{T}}\,\frac{\partial\phi_k}{\partial\mathbf{x}(k)} + [\delta\mathbf{x}(k+1)]^{\mathrm{T}}\,\frac{\partial\phi_k}{\partial\mathbf{x}(k+1)}\right) = 0 \tag{7.3}$$

Setting $k = i - 1$ in the second term of the left-hand side of equation (7.3) we have

$$\sum_{k=k_0}^{k_f - 1}[\delta\mathbf{x}(k+1)]^{\mathrm{T}}\,\frac{\partial\phi[\mathbf{x}(k), \mathbf{x}(k+1), k]}{\partial\mathbf{x}(k)} = \sum_{i=k_0+1}^{k_f}[\delta\mathbf{x}(i)]^{\mathrm{T}}\,\frac{\partial\phi[\mathbf{x}(i-1), \mathbf{x}(i), i-1]}{\partial\mathbf{x}(i)} \tag{7.4a}$$

$$= \sum_{k=k_0+1}^{k_f}[\delta\mathbf{x}(k)]^{\mathrm{T}}\,\frac{\partial\phi[\mathbf{x}(k-1), \mathbf{x}(k), k-1]}{\partial\mathbf{x}(k)} \tag{7.4b}$$

$$= \sum_{k=k_0}^{k_f - 1}[\delta\mathbf{x}(k)]^{\mathrm{T}}\,\frac{\partial\phi[\mathbf{x}(k-1), \mathbf{x}(k), k-1]}{\partial\mathbf{x}(k)}$$

$$+ [\delta\mathbf{x}(k)]^{\mathrm{T}}\,\frac{\partial\phi[\mathbf{x}(k-1), \mathbf{x}(k), k-1]}{\partial\mathbf{x}(k)}\Bigg|_{k=k_0}^{k=k_f} \tag{7.4c}$$

where, going from equation (7.4a) to (7.4b) we have set $i = k$. If equation (7.4c) is substituted in (7.3) we have

$$\sum_{k=k_0}^{k_f - 1}[\delta\mathbf{x}(k)]^{\mathrm{T}}\left(\frac{\partial\phi[\mathbf{x}(k), \mathbf{x}(k+1), k]}{\partial\mathbf{x}(k)} + \frac{\partial\phi[\mathbf{x}(k-1), \mathbf{x}(k), k-1]}{\partial\mathbf{x}(k)}\right)$$

$$+ [\delta\mathbf{x}(k)]^{\mathrm{T}}\,\frac{\partial\phi[\mathbf{x}(k-1), \mathbf{x}(k), k-1]}{\partial\mathbf{x}(k)}\Bigg|_{k=k_0}^{k=k_f} = 0 \tag{7.5}$$

In order that equation (7.5) be identically zero, independently of $\delta\mathbf{x}(k)$, the following conditions must hold:

$$\frac{\partial\phi[\mathbf{x}(k),\mathbf{x}(k+1),k]}{\partial\mathbf{x}(k)} + \frac{\partial\phi[\mathbf{x}(k-1),\mathbf{x}(k),k-1]}{\partial\mathbf{x}(k)} = \mathbf{0} \tag{7.6}$$

and

$$[\delta\mathbf{x}(k)]^{\mathrm{T}}\left(\frac{\partial\phi[\mathbf{x}(k-1),\mathbf{x}(k),k-1]}{\partial\mathbf{x}(k)}\right) = 0 \quad \text{for } k = k_0 \text{ and } k_{\mathrm{f}} \tag{7.7}$$

Equation (7.6) is the **discrete-time Euler–Lagrange equation** and equation (7.7) represents the **discrete-time boundary conditions**. The solution $\mathbf{x}(k)$ of the partial differential equation (7.6) with boundary conditions given by (7.7), minimizes the cost function (7.1). Hence, the problem of determining the maxima and minima of a function using the calculus of variations reduces to one of solving a two-point boundary value problem (TPBVP).

The foregoing results can be extended to cover the more general case where constraint equations appear in the problem. In what follows, we will study the case where the vector $\mathbf{x}(k)$ of the cost function (7.1) is subject to the constraint equation

$$\mathbf{f}[\mathbf{x}(k), \mathbf{x}(k+1), k] = \mathbf{0} \tag{7.8}$$

To determine the maxima or minima of J under the constraint (7.8), we apply the method of Lagrange multipliers. To this end, a new cost function J' is formed as follows:

$$J' = \sum_{k=k_0}^{k_{\mathrm{f}}-1}\{\phi[\mathbf{x}(k),\mathbf{x}(k+1),k] + \boldsymbol{\lambda}^{\mathrm{T}}(k)\mathbf{f}[\mathbf{x}(k),\mathbf{x}(k+1),k]\} \tag{7.9}$$

or

$$J' = \sum_{k=k_0}^{k_{\mathrm{f}}-1}\psi[\mathbf{x}(k),\mathbf{x}(k+1),k]$$

where

$$\psi[\mathbf{x}(k), \mathbf{x}(k+1), k] = \phi[\mathbf{x}(k), \mathbf{x}(k+1), k] + \boldsymbol{\lambda}^{\mathrm{T}}(k)\mathbf{f}[\mathbf{x}(k), \mathbf{x}(k+1), k]$$

Clearly $J' = J$ since $\mathbf{f}[\mathbf{x}(k), \mathbf{x}(k+1), k] = \mathbf{0}$. If we extend the results of equation (7.6) to the present case, then the Euler–Lagrange equation would be

$$\frac{\partial\psi[\mathbf{x}(k),\mathbf{x}(k+1),k]}{\partial\mathbf{x}(k)} + \frac{\partial\psi[\mathbf{x}(k-1),\mathbf{x}(k),k-1]}{\partial\mathbf{x}(k)} = \mathbf{0} \tag{7.10}$$

When the cost function (7.1) is not subject to the constraints (7.8), then equation (7.10) takes the form of (7.6).

7.2.2 *The maximum principle for discrete-time systems*

The calculus of variations method, presented in Section 7.2.1, constitutes a general methodology for the study of the maxima and minima of a function. We will restrict our interest to specialized methods which facilitate the solution of optimal control design problems. Such a method is the maximum principle which was initially proposed by Pontryagin [11]. This method is based on the calculus of variations and yields a general solution to optimal control problems.

More specifically, the following problem will be studied. Consider the discrete-time system

$$\mathbf{x}(k+1) = \mathbf{f}[\mathbf{x}(k), \mathbf{u}(k), k] \tag{7.11}$$

and the cost function

$$J = \theta[\mathbf{x}(k), k]\Big|_{k=k_0}^{k=k_f} + \sum_{k=k_0}^{k_f-1} \phi[\mathbf{x}(k), \mathbf{u}(k), k] \tag{7.12}$$

where $\theta[\mathbf{x}(k), k]$ is the cost function at the boundaries k_0 and k_f and $\phi[\mathbf{x}(k), \mathbf{u}(k), k]$ is the cost function at each instant k in the interval. Find a vector $\mathbf{u}(k)$ which minimizes the cost function J under the constraint (7.11).

This problem is solved using the maximum principle method, which may be briefly presented as follows [13]. Define the **discrete-time Hamiltonian function** $H(k)$, as follows:

$$H(k) = \phi[\mathbf{x}(k), \mathbf{u}(k), k] + \boldsymbol{\lambda}^{\mathrm{T}}(k+1)\mathbf{f}[\mathbf{x}(k), \mathbf{u}(k), k] \tag{7.13}$$

where the vector $\boldsymbol{\lambda}$ is a vector of Lagrange multipliers. Note that the vector $\boldsymbol{\lambda}(k+1)$ is used in place of the vector $\boldsymbol{\lambda}(k)$ since this simplifies the final results when solving practical problems. Next, we formulate a new cost function as follows:

$$J' = \theta[\mathbf{x}(k), k]\Big|_{k=k_0}^{k=k_f} + \sum_{k=k_0}^{k_f-1} [H(k) - \boldsymbol{\lambda}^{\mathrm{T}}(k+1)\mathbf{x}(k+1)] \tag{7.14}$$

Clearly $J = J'$. Extending the well-known continuous-time technique of integrating by parts when an integral operation is involved to the case of discrete-time systems where a summation operation is involved, the last term in equation (7.14) becomes

$$\sum_{k=k_0}^{k_f-1} \boldsymbol{\lambda}^{\mathrm{T}}(k+1)\mathbf{x}(k+1) = \sum_{k=k_0+1}^{k_f} \boldsymbol{\lambda}^{\mathrm{T}}(k)\mathbf{x}(k) = \sum_{k=k_0}^{k_f-1} \boldsymbol{\lambda}^{\mathrm{T}}(k)\mathbf{x}(k) + \boldsymbol{\lambda}^{\mathrm{T}}(k_f)\mathbf{x}(k_f) - \boldsymbol{\lambda}^{\mathrm{T}}(k_0)\mathbf{x}(k_0) \tag{7.15}$$

Substituting equation (7.15) in equation (7.14) we have

$$J' = \{\theta[\mathbf{x}(k), k] - \boldsymbol{\lambda}^{\mathrm{T}}(k)\mathbf{x}(k)\}\Big|_{k=k_0}^{k=k_f} + \sum_{k=k_0}^{k_f-1} [H(k) - \boldsymbol{\lambda}^{\mathrm{T}}(k)\mathbf{x}(k)] \tag{7.16}$$

The first-order differential deviation $\delta J'$ in terms of the vectors $\mathbf{x}(k)$ and $\mathbf{u}(k)$ will be

$$\delta J' = [\delta\mathbf{x}(k)]^{\mathrm{T}}\left(\frac{\partial\theta}{\partial\mathbf{x}(k)} - \boldsymbol{\lambda}(k)\right)\Bigg|_{k=k_0}^{k=k_{\mathrm{f}}}$$

$$+ \sum_{k=k_0}^{k_{\mathrm{f}}-1}\left[[\delta\mathbf{x}(k)]^{\mathrm{T}}\left(\frac{\partial H(k)}{\partial\mathbf{x}(k)} - \boldsymbol{\lambda}(k)\right) + [\delta\mathbf{u}(k)]^{\mathrm{T}}\left(\frac{\partial H(k)}{\partial\mathbf{u}(k)}\right)\right] \tag{7.17}$$

According to the calculus of variations, presented in Section 7.2.1, a necessary condition for J' to be minimum is that $\delta J' = 0$. From (7.17) it follows that in order to have $\delta J' = 0$ for every $\delta\mathbf{x}(k)$ and $\delta\mathbf{u}(k)$ the following relations must hold:

$$\frac{\partial H(k)}{\partial\mathbf{x}(k)} = \boldsymbol{\lambda}(k) \tag{7.18a}$$

$$\frac{\partial H(k)}{\partial\mathbf{u}(k)} = \mathbf{0} \tag{7.18b}$$

$$\frac{\partial H(k)}{\partial\boldsymbol{\lambda}(k+1)} = \mathbf{x}(k+1) = \mathbf{f}[\mathbf{x}(k),\mathbf{u}(k),k] \tag{7.18c}$$

The foregoing relations must satisfy the **discrete-time boundary conditions**

$$[\delta\mathbf{x}(k)]^{\mathrm{T}}\left(\frac{\partial\theta}{\partial\mathbf{x}(k)} - \boldsymbol{\lambda}(k)\right) = 0 \quad \text{for} \quad k = k_0 \text{ and } k_{\mathrm{f}} \tag{7.18d}$$

To guarantee that the solution of equations (7.18a–d) minimizes the cost function J, it is necessary that $\delta^2 J > 0$. The set of equations (7.18) is of paramount importance in the field of automatic control systems. They are called **canonical Hamiltonian equations**.

Equation (7.18b) shows that the control signal $\mathbf{u}(k)$, which minimizes the cost function J, necessarily minimizes the Hamiltonian function, i.e.

$$H(\mathbf{x},\mathbf{u},\boldsymbol{\lambda},k) \leqslant H(\mathbf{x},\tilde{\mathbf{u}},\boldsymbol{\lambda},k)$$

where $\tilde{\mathbf{u}}(k)$ is any control signal, different from the optimal $\mathbf{u}(k)$. For this reason, the present method is known as the minimum principle method [8], [12]. However, because of a sign difference in the Hamiltonian function $H(k)$ [3], [11], the method has become known as the **maximum principle**.

■ **EXAMPLE 7.2.1** Consider the scalar system $x(k+1) = x(k) + \beta u(k)$ with $x(0) = 1$ and $x(20) = 0$. Let the cost function have the form

$$J = \tfrac{1}{2}\sum_{k=0}^{19} u^2(k)$$

Find the optimal control $u(k)$ which minimizes J.

■ **SOLUTION**   We form the cost function J' as follows:

$$J' = \sum_{k=0}^{19} \{\tfrac{1}{2} u^2(k) + \lambda(k+1)[-x(k+1) + x(k) + \beta u(k)]\} = \sum_{k=0}^{19} \psi_k$$

where

$$\psi_k = \tfrac{1}{2} u^2(k) + \lambda(k+1)[-x(k+1) + x(k) + \beta u(k)]$$

In order to determine the Euler–Lagrange equations with respect to $x(k)$ and $u(k)$, we initially determine the following:

$$\frac{\partial \psi[x(k), x(k+1), k]}{\partial x(k)} = \lambda(k+1) \qquad \frac{\partial \psi[x(k-1), x(k), k-1]}{\partial x(k)} = -\lambda(k)$$

$$\frac{\partial \psi[x(k), x(k+1), k]}{\partial u(k)} = u(k) + \beta\lambda(k+1) \qquad \frac{\partial \psi[x(k-1), x(k), k-1]}{\partial u(k)} = 0$$

whereupon

$$\lambda(k+1) - \lambda(k) = 0 \quad \text{and} \quad u(k) + \beta\lambda(k+1) = 0$$

The solution of the first equation is $\lambda(k) = \gamma = $ constant, and that of the second is $u(k) = -\beta\gamma$. From the system equation, i.e. from the equation $x(k+1) = x(k) + \beta u(k)$, we have that $x(k+1) = x(k) - \beta^2\gamma$. To solve this equation the Z-transform technique is applied. We have

$$z[X(z) - x(0)] = X(z) - \beta^2\gamma \frac{z}{z-1} \quad \text{or} \quad X(z) = \frac{z}{z-1} - \beta^2\gamma \frac{z}{(z-1)^2}$$

Applying the inverse Z-transform, we obtain that $x(k) = 1 - \beta^2\gamma k$. Based on the final value $x(20) = 0$, we obtain $\gamma = (20\beta^2)^{-1}$. Hence $x(k) = 1 - 20^{-1}k$ and the optimal control signal $u(k)$ has the form $u(k) = -\beta\lambda(k+1) = -\beta\gamma = -20^{-1}\beta^{-1}$.

■ **EXAMPLE 7.2.2**   Consider the system

$$\mathbf{x}(k+1) = \begin{bmatrix} 0 & 1 \\ -1 & 1 \end{bmatrix} \mathbf{x}(k) + \begin{bmatrix} 0 \\ 1 \end{bmatrix} u(k) \quad \text{and} \quad \mathbf{x}(0) = \begin{bmatrix} 1 \\ 1 \end{bmatrix}$$

Let the cost function have the form

$$J = \sum_{k=0}^{2} [x_1^2(k+1) + u^2(k)]$$

with $x_1(3)$ unspecified and $x_2(3) = 0$. Find the optimal sequence $u(0)$, $u(1)$, $u(2)$ which minimizes J.

■ **SOLUTION** We form the cost function J' as follows:

$$J' = \sum_{k=0}^{2} \psi_k$$

where

$$\psi_k = x_1^2(k+1) + u^2(k) + \lambda_1(k+1)[-x_1(k+1) + x_2(k)]$$
$$+ \lambda_2(k+1)[-x_2(k+1) - x_1(k) + x_2(k) + u(k)]$$

In order to determine the Euler–Lagrange equations for $\mathbf{x}(k)$ and $u(k)$ we calculate the following:

$$\frac{\partial \psi[x_1(k), x_1(k+1), k]}{\partial x_1(k)} = -\lambda_2(k+1)$$
$$\frac{\partial \psi[x_2(k), x_2(k+1), k]}{\partial x_2(k)} = \lambda_1(k+1) + \lambda_2(k+1)$$

$$\frac{\partial \psi[x_1(k-1), x_1(k), k-1]}{\partial x_1(k)} = 2x_1(k) - \lambda_1(k)$$
$$\frac{\partial \psi[x_2(k-1), x_2(k), k-1]}{\partial x_2(k)} = -\lambda_2(k)$$

$$\frac{\partial \psi[u(k), u(k-1), k]}{\partial u(k)} = 2u(k) + \lambda_2(k+1)$$
$$\frac{\partial \psi[u(k-1), u(k), k-1]}{\partial u(k)} = 0$$

The discrete-time Euler–Lagrange equations for $\mathbf{x}(k)$ and $u(k)$ are as follows:

$$-\lambda_2(k+1) + 2x_1(k) - \lambda_1(k) = 0$$
$$\lambda_1(k+1) + \lambda_2(k+1) - \lambda_2(k) = 0$$
$$2u(k) + \lambda_2(k+1) = 0$$

From all the above equations we arrive at the equation

$$2u(k) + 2x_1(k) - 2u(k-1) + 2u(k-2) = 0$$

It is easily obtained that

$$x_1(1) = 1 \qquad x_1(2) = u(0) \qquad\qquad x_1(3) = -1 + u(0) + u(1)$$
$$x_2(1) = u(0) \qquad x_2(2) = -1 + u(0) + u(1) \qquad x_2(3) = -1 + u(1) + u(2)$$

Thus, from the equation $2u(k) + 2x_1(k) - 2u(k-1) + 2u(k-2) = 0$ and for $k = 2$, we obtain

$$u(2) - u(1) + 2u(0) = 0$$

Since we know that $x_2(3) = 0$, the foregoing equation becomes

$$u(1) + u(2) = 1$$

From the two last equations we obtain

$$u(1) = u(0) + \tfrac{1}{2} \quad \text{and} \quad u(2) = -u(0) + \tfrac{1}{2}$$

Substituting the results above in the cost criterion J we obtain

$$J = 8u^2(0) - 2u(0) + \tfrac{7}{4}$$

From the relation $\partial J/\partial u(0) = 0$ it is easily shown that $u(0) = 1/8$ and further that $\partial^2 J/\partial u^2(0) = 16 > 0$. Therefore, the optimal sequence $[u(0), u(1), u(2)]$ sought is the sequence $[1/8, 5/8, 3/8]$.

7.3 The optimal linear regulator

Here, the results of Section 7.2 will be applied to the solution of the optimal linear regulator problem of discrete-time systems. The optimal linear regulator problem is a special, but very significant, optimal control problem. Simply speaking, the regulator problem may be stated as follows. Consider a linear homogeneous system with non-zero initial conditions $\mathbf{x}(k_0)$. Here the vector $\mathbf{x}(k_0)$ is the only excitation to the system. Find an optimal control signal $\mathbf{u}(k)$ capable of restoring the state vector to its equilibrium state, $\mathbf{x}(k_f) \simeq \mathbf{0}$ while minimizing a certain cost function.

As a typical practical regulator problem consider a ground antenna having fixed orientation with respect to a communications satellite. Assume that the antenna undergoes a disturbance, e.g. due to a sudden strong wind. In this case, the antenna will be forced to a new position. Let this new position be represented by the state vector $\mathbf{x}(k_0)$. Clearly, in the present situation, it is desirable to apply a control strategy which brings the antenna back from the undesired state $\mathbf{x}(k_0)$ to its initial equilibrium state. Furthermore, this must take place in a time interval $[k_0, k_f]$, while minimizing a cost function which normally includes the following three characteristics. The final value $\mathbf{x}(k_f)$ should be as close as possible to the system's equilibrium, i.e. $\mathbf{x}(k_f) \simeq \mathbf{0}$. The amplitude of the optimal control signal $\mathbf{u}(k)$ should be as small as possible, making the required control effort (control energy) for restoring the antenna to its equilibrium as small as possible. Simultaneously, the amplitude of $\mathbf{x}(k)$ should have adequately small value in order to avoid saturation or even damage (i.e. from overheating) to the system under control.

From a mathematical point of view, the problem of the linear regulator is formulated as follows. Consider a discrete-time linear system with time-varying parameters described by the vector difference equation

$$\mathbf{x}(k+1) = \mathbf{A}(k)\mathbf{x}(k) + \mathbf{B}(k)\mathbf{u}(k) \quad \text{with} \quad \mathbf{x}(k_0) = \mathbf{x}_0 \tag{7.19}$$

Find a control signal $\mathbf{u}(k)$ which minimizes the cost function

$$J = \tfrac{1}{2}\mathbf{x}^T(k_f)\mathbf{S}\mathbf{x}(k_f) + \tfrac{1}{2}\sum_{k=k_0}^{k_f-1} [\mathbf{x}^T(k)\mathbf{Q}(k)\mathbf{x}(k) + \mathbf{u}^T(k)\mathbf{R}(k)\mathbf{u}(k)] \tag{7.20}$$

The foregoing cost function is a sum of inner products of the vectors $\mathbf{x}(k)$ and $\mathbf{u}(k)$ and for this reason we say that the expression (7.20) is a **quadratic cost function**.

Here we stress that the main reason that the energy-like quadratic terms $\mathbf{x}^T(k)\mathbf{Q}(k)\mathbf{x}(k)$ and $\mathbf{u}^T(k)\mathbf{R}(k)\mathbf{u}(k)$ have been included in the cost function J is to minimize the dissipated energy in the system and the required input energy (control effort), respectively. The quadratic term $\mathbf{x}^T(k_f)\mathbf{S}\mathbf{x}(k_f)$ is included in J in order to force the final value $\mathbf{x}(k_f)$ of $\mathbf{x}(k)$ to be as near as possible to the system's equilibrium state. Note that the vector $\mathbf{x}(k_f)$ is unspecified. The matrices $\mathbf{S}$, $\mathbf{Q}(k)$ and $\mathbf{R}(k)$ are weighting matrices and are chosen to be symmetrical. The matrix $\mathbf{R}(k)$ is assumed to be positive definite, whereas the matrices $\mathbf{Q}(k)$ and $\mathbf{S}$ are assumed to be positive semi-definite.

The Hamiltonian of the system is

$$H(k) = \tfrac{1}{2}\mathbf{x}^T(k)\mathbf{Q}(k)\mathbf{x}(k) + \tfrac{1}{2}\mathbf{u}^T(k)\mathbf{R}(k)\mathbf{u}(k) + \boldsymbol{\lambda}^T(k+1)[\mathbf{A}(k)\mathbf{x}(k) + \mathbf{B}(k)\mathbf{u}(k)] \tag{7.21}$$

The canonical equations (7.18a–c) for the present problem are

$$\frac{\partial H(k)}{\partial \mathbf{x}(k)} = \boldsymbol{\lambda}(k) = \mathbf{Q}(k)\mathbf{x}(k) + \mathbf{A}^T(k)\boldsymbol{\lambda}(k+1) \tag{7.22a}$$

$$\frac{\partial H(k)}{\partial \mathbf{u}(k)} = \mathbf{0} = \mathbf{R}(k)\mathbf{u}(k) + \mathbf{B}^T(k)\boldsymbol{\lambda}(k+1) \tag{7.22b}$$

$$\frac{\partial H(k)}{\partial \boldsymbol{\lambda}(k+1)} = \mathbf{x}(k+1) = \mathbf{A}(k)\mathbf{x}(k) + \mathbf{B}(k)\mathbf{u}(k) \tag{7.22c}$$

The boundary conditions (7.18d) refer only to the final condition, i.e. for $k = k_f$, and are given by

$$\frac{\partial \theta}{\partial \mathbf{x}(k)}\bigg|_{k=k_f} = \boldsymbol{\lambda}(k_f) = \mathbf{S}\mathbf{x}(k_f) \tag{7.23}$$

From relation (7.22b) we obtain

$$\mathbf{u}(k) = -\mathbf{R}^{-1}(k)\mathbf{B}^T(k)\boldsymbol{\lambda}(k+1) \tag{7.24}$$

where $|\mathbf{R}(k)| \neq 0$ for $k = 0, 1, 2, \ldots, k_f$ since it has been assumed that the matrix $\mathbf{R}(k)$ is positive definite. Substituting equation (7.24) in equation (7.19) we obtain the closed-loop system

$$\mathbf{x}(k+1) = \mathbf{A}(k)\mathbf{x}(k) - \mathbf{B}(k)\mathbf{R}^{-1}(k)\mathbf{B}^T(k)\boldsymbol{\lambda}(k+1) \quad \text{with} \quad \mathbf{x}(0) = \mathbf{x}_0 \tag{7.25}$$

If in equation (7.25) we substitute

$$\boldsymbol{\lambda}(k) = \mathbf{P}(k)\mathbf{x}(k) \tag{7.26}$$

it follows that

$$\mathbf{x}(k+1) = \mathbf{A}(k)\mathbf{x}(k) - \mathbf{B}(k)\mathbf{R}^{-1}(k)\mathbf{B}^T(k)\mathbf{P}(k+1)\mathbf{x}(k+1) \tag{7.27}$$

From equations (7.26) and (7.22a) we obtain the relation

$$\mathbf{P}(k)\mathbf{x}(k) = \mathbf{Q}(k)\mathbf{x}(k) + \mathbf{A}^T(k)\boldsymbol{\lambda}(k+1) = \mathbf{Q}(k)\mathbf{x}(k) + \mathbf{A}^T(k)\mathbf{P}(k+1)\mathbf{x}(k+1) \tag{7.28}$$

Eliminating the vector $\mathbf{x}(k+1)$ from equations (7.27) and (7.28), we arrive at the relation

$$\{\mathbf{P}(k) - \mathbf{A}^\mathrm{T}(k)\mathbf{P}(k+1)[\mathbf{I} + \mathbf{B}(k)\mathbf{R}^{-1}(k)\mathbf{B}^\mathrm{T}(k)\mathbf{P}(k+1)]^{-1}\mathbf{A}(k) - \mathbf{Q}(k)\}\mathbf{x}(k) = \mathbf{0}$$

$$(7.29)$$

Equation (7.29) holds true for all $\mathbf{x}(k)$ if

$$\mathbf{P}(k) - \mathbf{A}^\mathrm{T}(k)\mathbf{P}(k+1)[\mathbf{I} + \mathbf{B}(k)\mathbf{R}^{-1}(k)\mathbf{B}^\mathrm{T}(k)\mathbf{P}(k+1)]^{-1}\mathbf{A}(k) = \mathbf{Q}(k) \qquad (7.30)$$

Equation (7.30) is known as the **discrete-time Riccati equation**. The final condition of equation (7.30) is

$$\mathbf{P}(k_\mathrm{f}) = \mathbf{S} \qquad (7.31)$$

Now, we return to relation (7.22a). From this relation and under the condition that $|\mathbf{A}(k)| \neq 0$, for $k = 0, 1, 2, \ldots, k_\mathrm{f}$, it follows that

$$\boldsymbol{\lambda}(k+1) = -[\mathbf{A}^\mathrm{T}(k)]^{-1}\mathbf{Q}(k)\mathbf{x}(k) + [\mathbf{A}^\mathrm{T}(k)]^{-1}\mathbf{P}(k)\mathbf{x}(k) \qquad (7.32)$$

where use was made of (7.26). Substituting equation (7.32) in equation (7.24) we arrive at the optimal control law

$$\mathbf{u}(k) = \mathbf{K}(k)\mathbf{x}(k) \qquad (7.33)$$

where

$$\mathbf{K}(k) = -\mathbf{R}^{-1}(k)\mathbf{B}^\mathrm{T}(k)[\mathbf{A}^\mathrm{T}(k)]^{-1}[\mathbf{P}(k) - \mathbf{Q}(k)] \qquad (7.34)$$

where $\mathbf{K}(k)$ is the **discrete-time Kalman matrix**. Figure 7.1 presents a simplified diagram and Figure 7.2 the block diagram of the discrete-time optimal linear system regulator.

■ **REMARK 7.3.1** An alternative solution to the optimal linear discrete-time regulator problem consists of solving two recursive equations, rather than the discrete-time Riccati equation (7.30). These two recursive equations can be obtained as follows. From relation (7.22b) we have

$$\mathbf{R}(k)\mathbf{u}(k) = -\mathbf{B}^\mathrm{T}(k)\boldsymbol{\lambda}(k+1) \qquad (7.35a)$$

$$= -\mathbf{B}^\mathrm{T}(k)\mathbf{P}(k+1)\mathbf{x}(k+1) \qquad (7.35b)$$

$$= -\mathbf{B}^\mathrm{T}(k)\mathbf{P}(k+1)[\mathbf{A}(k)\mathbf{x}(k) + \mathbf{B}(k)\mathbf{u}(k)] \qquad (7.35c)$$

where use was made of equation (7.26) in equation (7.35b) and of equation (7.19) in equation (7.35c). Solving equation (7.35c) for $\mathbf{u}(k)$, we have

$$\mathbf{u}(k) = -[\mathbf{R}(k) + \mathbf{B}^\mathrm{T}(k)\mathbf{P}(k+1)\mathbf{B}(k)]^{-1}\mathbf{B}^\mathrm{T}(k)\mathbf{P}(k+1)\mathbf{A}(k)\mathbf{x}(k)$$

or

$$\mathbf{u}(k) = \mathbf{K}(k)\mathbf{x}(k) \qquad (7.36)$$

where

$$\mathbf{K}(k) = -[\mathbf{R}(k) + \mathbf{B}^\mathrm{T}(k)\mathbf{P}(k+1)\mathbf{B}(k)]^{-1}\mathbf{B}^\mathrm{T}(k)\mathbf{P}(k+1)\mathbf{A}(k) \qquad (7.37)$$

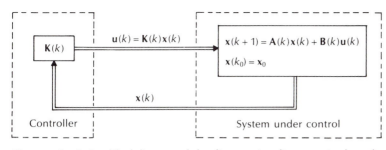

Figure 7.1 A simplified diagram of the discrete-time linear optimal regulator.

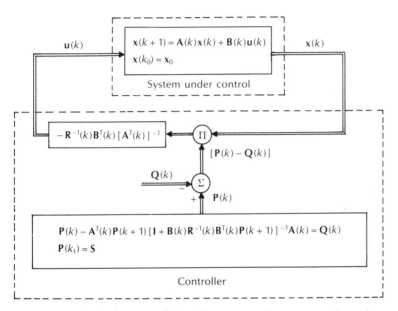

Figure 7.2 Block diagram of the discrete-time linear optimal regulator using the Riccati equation, where Π denotes multiplication.

Now, if equation (7.19) is substituted in relation (7.28), we obtain

$$\mathbf{P}(k)\mathbf{x}(k) = \mathbf{Q}(k)\mathbf{x}(k) + \mathbf{A}^{\mathrm{T}}(k)\mathbf{P}(k+1)[\mathbf{A}(k)\mathbf{x}(k) + \mathbf{B}(k)\mathbf{u}(k)]$$

Substituting equation (7.36) in the foregoing relation we have

$$[\mathbf{P}(k) - \mathbf{Q}(k) - \mathbf{A}^{\mathrm{T}}(k)\mathbf{P}(k+1)\mathbf{A}(k) + \mathbf{A}^{\mathrm{T}}(k)\mathbf{P}(k+1)\mathbf{B}(k)\mathbf{K}(k)]\mathbf{x}(k) = \mathbf{0}$$

The relation above is satisfied for every $\mathbf{x}(k)$ if

$$\mathbf{P}(k) = \mathbf{A}^{\mathrm{T}}(k)\mathbf{P}(k+1)\mathbf{A}(k) - \mathbf{A}^{\mathrm{T}}(k)\mathbf{P}(k+1)\mathbf{B}(k)\mathbf{K}(k) + \mathbf{Q}(k) \qquad (7.38)$$

Equation (7.38) can be written as two recursive equations as follows:

$$\mathbf{P}(k) = \mathbf{A}^{\mathrm{T}}(k)\mathbf{M}(k+1)\mathbf{A}(k) + \mathbf{Q}(k) \quad \text{with} \quad \mathbf{P}(k_f) = \mathbf{S} \qquad (7.39a)$$

where

$$\mathbf{M}(k+1) = \mathbf{P}(k+1) - \mathbf{P}(k+1)\mathbf{B}(k)[\mathbf{R}(k) + \mathbf{B}^{\mathrm{T}}(k)\mathbf{P}(k+1)\mathbf{B}(k)]^{-1}\mathbf{B}^{\mathrm{T}}(k)\mathbf{P}(k+1)$$
(7.39b)

The block diagram of the foregoing results is given in Figure 7.3.

It is mentioned once again that according to the above results the matrix $\mathbf{K}(k)$ is derived recursively. The recursive procedure starts from the end, i.e. for $k = k_f$, and goes back in time. The starting point is the known relation $\mathbf{P}(k_f) = \mathbf{S}$. Next, from relation (7.39b) and for $k + 1 = k_f$ we obtain the value of the matrix $\mathbf{M}(k_f)$ as follows:

$$\mathbf{M}(k_f) = \mathbf{S} - \mathbf{S}\mathbf{B}(k_f - 1)[\mathbf{R}(k_f - 1) + \mathbf{B}^{\mathrm{T}}(k_f - 1)\mathbf{S}\mathbf{B}(k_f - 1)]^{-1}\mathbf{B}^{\mathrm{T}}(k_f - 1)\mathbf{S} \quad (7.40)$$

Subsequently, the matrix $\mathbf{P}(k_f - 1)$ is obtained from equation (7.39a) as follows:

$$\mathbf{P}(k_f - 1) = \mathbf{A}^{\mathrm{T}}(k_f - 1)\mathbf{M}(k_f)\mathbf{A}(k_f - 1) + \mathbf{Q}(k_f - 1)$$
(7.41)

Up to now, a single step of the recursive procedure has been completed, going from the known value $\mathbf{P}(k_f) = \mathbf{S}$ to the next value $\mathbf{P}(k_f - 1)$ given by relation (7.41). This step is repeated until the matrix $\mathbf{P}(k_0 + 1)$ is determined. Finally, the matrix $\mathbf{K}(k)$ is formed on the basis of the matrices $\mathbf{P}(k_f)$, $\mathbf{P}(k_f - 1), \dots, \mathbf{P}(k_0 + 1)$, according to relation (7.37).

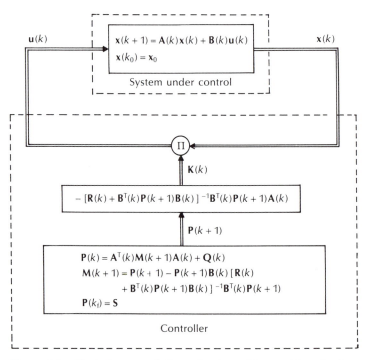

Figure 7.3 Block diagram of the optimal regulator problem for linear discrete-time systems using two recursive equations, where Π denotes multiplication.

■ **REMARK 7.3.2**   The minimum of the cost function (7.20) is proved to be given by the relation [10]

$$J_{min} = \tfrac{1}{2}\mathbf{x}^T(0)\mathbf{P}(0)\mathbf{x}(0) \tag{7.42}$$

■ **EXAMPLE 7.3.1**   Consider the system of the form (7.19), where

$$\mathbf{A}(k) = \begin{bmatrix} 1 & 1 \\ 1 & 0 \end{bmatrix}, \quad \mathbf{b}(k) = \begin{bmatrix} 1 \\ 0 \end{bmatrix} \quad \text{and} \quad \mathbf{x}(k_0) = \mathbf{x}(0) = \begin{bmatrix} 1 \\ 0 \end{bmatrix}$$

The cost function (7.20) has the particular form

$$J = \tfrac{1}{2}\mathbf{x}^T(4)\mathbf{S}\mathbf{x}(4) + \tfrac{1}{2}\sum_{k=0}^{3}[\mathbf{x}^T(k)\mathbf{Q}(k)\mathbf{x}(k) + r(k)u^2(k)]$$

with $\mathbf{S} = \mathbf{Q}(k) = \mathbf{I}_2$ and $\mathbf{R}r(k) = 1$. Find the optimal sequence $u(0)$, $u(1)$, $u(2)$, $u(3)$ and $u(4)$ which minimizes the above cost function.

■ **SOLUTION**   The matrix $\mathbf{P}(k)$, for $k = 0, 1, 2, 3$ and 4, will first be determined on the basis of the relation (7.30), i.e. on the basis of the discrete-time Riccati equation

$$\mathbf{P}(k) = \mathbf{A}^T\mathbf{P}(k+1)[\mathbf{I}_2 + \mathbf{b}r^{-1}\mathbf{b}^T\mathbf{P}(k+1)]^{-1}\mathbf{A} + \mathbf{Q}$$

To this end, the matrix $\mathbf{P}(k)$ will be determined going backwards in time, i.e. the value of $\mathbf{P}(4)$ will be calculated initially, then the matrix $\mathbf{P}(3)$ and so on, until the matrix $\mathbf{P}(0)$ is finally determined. The matrix $\mathbf{P}(4)$ is directly calculated from relation (7.31) to yield

$$\mathbf{P}(k_f) = \mathbf{P}(4) = \mathbf{S} = \mathbf{I}_2$$

The matrix $\mathbf{P}(3)$ is calculated from the recursive relation for $\mathbf{P}(k)$ given above, to yield

$$\mathbf{P}(3) = \mathbf{A}^T\mathbf{P}(4)[\mathbf{I} + \mathbf{b}r^{-1}\mathbf{b}^T\mathbf{P}(4)]^{-1}\mathbf{A} + \mathbf{Q} = \frac{1}{2}\begin{bmatrix} 5 & 1 \\ 1 & 3 \end{bmatrix}$$

Similarly, we can obtain the matrices $\mathbf{P}(2)$, $\mathbf{P}(1)$ and $\mathbf{P}(0)$. We have

$$\mathbf{P}(2) = \mathbf{A}^T\mathbf{P}(3)[\mathbf{I} + \mathbf{b}r^{-1}\mathbf{b}^T\mathbf{P}(3)]^{-1}\mathbf{A} + \mathbf{Q} = \frac{1}{7}\begin{bmatrix} 24 & 6 \\ 6 & 12 \end{bmatrix}$$

$$\mathbf{P}(1) = \mathbf{A}^T\mathbf{P}(2)[\mathbf{I} + \mathbf{b}r^{-1}\mathbf{b}^T\mathbf{P}(2)]^{-1}\mathbf{A} + \mathbf{Q} = \frac{1}{217}\begin{bmatrix} 805 & 210 \\ 210 & 385 \end{bmatrix}$$

$$\mathbf{P}(0) = \mathbf{A}^T\mathbf{P}(1)[\mathbf{I} + \mathbf{b}r^{-1}\mathbf{b}^T\mathbf{P}(1)]^{-1}\mathbf{A} + \mathbf{Q} = \begin{bmatrix} 3.77 & 0.99 \\ 0.99 & 1.79 \end{bmatrix}$$

Subsequently, the column vector $\mathbf{k}(k)$, $k = 4, 3, 2, 1$ and 0, is calculated from relation (7.34), i.e. from the relation

$$\mathbf{k}^\mathrm{T}(k) = -r^{-1}\mathbf{b}^\mathrm{T}[\mathbf{A}^\mathrm{T}]^{-1}[\mathbf{P}(k) - \mathbf{Q}]$$

Here

$$\mathbf{k}^\mathrm{T}(4) = [0 \;\; 0], \quad \mathbf{k}^\mathrm{T}(3) = [-0.5 \;\; -0.5], \quad \mathbf{k}^\mathrm{T}(2) = [-0.86 \;\; -0.71]$$

$$\mathbf{k}^\mathrm{T}(1) = [-0.97 \;\; -0.77] \quad \text{and} \quad \mathbf{k}^\mathrm{T}(0) = [-0.99 \;\; -0.79]$$

We will now determine the state vector $\mathbf{x}(k)$, $k = 0, 1, 2, 3$ and 4. The closed-loop system will be

$$\mathbf{x}(k+1) = [\mathbf{A} + \mathbf{bk}^\mathrm{T}(k)]\mathbf{x}(k) \quad \text{with} \quad \mathbf{x}(0) = \begin{bmatrix} 1 \\ 0 \end{bmatrix}$$

Therefore

$$\mathbf{x}(1) = [\mathbf{A} + \mathbf{bk}^\mathrm{T}(0)]\mathbf{x}(0) = \begin{bmatrix} 0.07 \\ 1 \end{bmatrix}, \quad \mathbf{x}(2) = [\mathbf{A} + \mathbf{bk}^\mathrm{T}(1)]\mathbf{x}(1) = \begin{bmatrix} 0.258 \\ 0.007 \end{bmatrix}$$

$$\mathbf{x}(3) = [\mathbf{A} + \mathbf{bk}^\mathrm{T}(2)]\mathbf{x}(2) = \begin{bmatrix} 0.039 \\ 0.258 \end{bmatrix} \quad \text{and} \quad \mathbf{x}(4) = [\mathbf{A} + \mathbf{bk}^\mathrm{T}(3)]\mathbf{x}(3) = \begin{bmatrix} 0.148 \\ 0.039 \end{bmatrix}$$

We are now in a position to determine the required optimal control $u(k)$ for $k = 0, 1, 2, 3$ and 4 given by relation (7.33), i.e. given by the relation $u(k) = \mathbf{k}^\mathrm{T}(k)\mathbf{x}(k)$. We have

$$u(0) = \mathbf{k}^\mathrm{T}(0)\mathbf{x}(0) = -0.993, \quad u(1) = \mathbf{k}^\mathrm{T}(1)\mathbf{x}(1) = -0.781$$

$$u(2) = \mathbf{k}^\mathrm{T}(2)\mathbf{x}(2) = -0.226, \quad u(3) = \mathbf{k}^\mathrm{T}(3)\mathbf{x}(3) = -0.148 \quad \text{and}$$

$$u(4) = \mathbf{k}^\mathrm{T}(4)\mathbf{x}(4) = 0$$

Finally, the minimum cost $J_{\min}$ is determined according to relation (7.42) to yield

$$J_{\min} = \tfrac{1}{2}\mathbf{x}^\mathrm{T}(0)\mathbf{P}(0)\mathbf{x}(0) = 1.89$$

7.4 The special case of time-invariant systems

For the case of linear time-invariant systems, the system under control is described by

$$\mathbf{x}(k+1) = \mathbf{A}\mathbf{x}(k) + \mathbf{B}\mathbf{u}(k) \quad \text{with} \quad \mathbf{x}(k_0) = \mathbf{x}_0 \tag{7.43}$$

The performance index J is

$$J = \tfrac{1}{2}\mathbf{x}^\mathrm{T}(k_\mathrm{f})\mathbf{S}\mathbf{x}(k_\mathrm{f}) + \tfrac{1}{2}\sum_{k=k_0}^{k_\mathrm{f}-1}[\mathbf{x}^\mathrm{T}(k)\mathbf{Q}\mathbf{x}(k) + \mathbf{u}^\mathrm{T}(k)\mathbf{R}\mathbf{u}(k)] \tag{7.44}$$

where the matrices $\mathbf{A}$, $\mathbf{B}$, $\mathbf{S}$, $\mathbf{Q}$ and $\mathbf{R}$ are constant. Of course, all the results of Section 7.3 apply to the present case.

When k_f is finite in relation (7.44), the feedback matrix $\mathbf{K}(k)$ is time varying. But when $k_f \to \infty$, the matrix $\mathbf{K}(k)$ becomes constant and is denoted by $\mathbf{K}$. In this case, we say that the matrix $\mathbf{K}$ is the steady state feedback matrix. When $k_f \to \infty$, the performance index J, given by equation (7.44), reduces to the following form:

$$J = \frac{1}{2} \sum_{k=k_0}^{\infty} [\mathbf{x}^T(k)\mathbf{Q}\mathbf{x}(k) + \mathbf{u}^T(k)\mathbf{R}\mathbf{u}(k)] \tag{7.45}$$

Note that the term $\frac{1}{2}\mathbf{x}^T(k_f)\mathbf{S}\mathbf{x}(k_f)$ becomes zero in the performance index (7.44) as $k_f \to \infty$, since the closed-loop system is assumed to be asymptotically stable, in which case $\mathbf{x}(\infty) = \mathbf{0}$ and therefore $\frac{1}{2}\mathbf{x}^T(\infty)\mathbf{S}\mathbf{x}(\infty) = 0$.

In the case of time-invariant systems given by (7.43) and when $k_f \to \infty$, the following interesting and very useful results have been proved:

1. The discrete-time Riccati equation (7.30) becomes

    $$\mathbf{P} - \mathbf{A}^T\mathbf{P}[\mathbf{I} + \mathbf{B}\mathbf{R}^{-1}\mathbf{B}^T\mathbf{P}]^{-1}\mathbf{A} = \mathbf{Q} \tag{7.46}$$

2. The feedback matrix $\mathbf{K}(k)$, given by equation (7.34), is constant and has the form

    $$\mathbf{K} = -\mathbf{R}^{-1}\mathbf{B}^T[\mathbf{A}^T]^{-1}[\mathbf{P} - \mathbf{Q}] \tag{7.47}$$

3. If λ_i is an eigenvalue of the matrix $\mathbf{M}$, then the eigenvalue λ_i^{-1} is also an eigenvalue of the matrix $\mathbf{M}$, where

    $$\mathbf{M} = \begin{bmatrix} \mathbf{I}_n & \mathbf{B}\mathbf{R}^{-1}\mathbf{B}^T \\ \mathbf{0} & \mathbf{A}^T \end{bmatrix}^{-1} \begin{bmatrix} \mathbf{A} & \mathbf{0} \\ -\mathbf{Q} & \mathbf{I}_n \end{bmatrix} \tag{7.48}$$

    under the condition that the matrix $\mathbf{A}$ is not singular. Therefore, n of the $2n$ eigenvalues of $\mathbf{M}$ are stable, i.e. they are located inside the unit circle, while the remaining n eigenvalues are unstable.
4. The solution of equation (7.46) for the unknown matrix $\mathbf{P}$ is

    $$\mathbf{P} = \mathbf{W}\mathbf{V}^{-1} \tag{7.49}$$

    where

    $$\mathbf{W} = [\mathbf{w}_1 \vdots \mathbf{w}_2 \vdots \cdots \vdots \mathbf{w}_n] \quad \text{and} \quad \mathbf{V} = [\mathbf{v}_1 \vdots \mathbf{v}_2 \vdots \cdots \vdots \mathbf{v}_n] \tag{7.50}$$

    where the vectors $\mathbf{w}_i$ and $\mathbf{v}_i$ are defined for the n stable eigenvalues as follows:

    $$\lambda_i \boldsymbol{\omega}_i = \mathbf{M}\boldsymbol{\omega}_i \quad \text{where} \quad \boldsymbol{\omega}_i = \begin{bmatrix} \mathbf{v}_i \\ \mathbf{w}_i \end{bmatrix} \tag{7.51}$$

    That is, the vector $\boldsymbol{\omega}_i$ is an eigenvector of the matrix $\mathbf{M}$.

■ **EXAMPLE 7.4.1**   Consider a system of the form (7.43) with a performance index of the form (7.45) (i.e. for $k_f \to \infty$) with

$$A = \begin{bmatrix} 1 & 1 \\ 1 & 0 \end{bmatrix}, \quad b = \begin{bmatrix} 1 \\ 0 \end{bmatrix}, \quad x(0) = \begin{bmatrix} 1 \\ 0 \end{bmatrix}, \quad Q = \begin{bmatrix} 1 & 0 \\ 0 & 1 \end{bmatrix} \quad \text{and} \quad R = 1$$

Find the optimal $u(k)$ and the value of J_{min}.

■ **SOLUTION**   We have

$$\begin{bmatrix} I_2 & bR^{-1}b^T \\ \hline 0 & A^T \end{bmatrix}^{-1} = \begin{bmatrix} 1 & 0 & 1 & 0 \\ 0 & 1 & 0 & 0 \\ 0 & 0 & 1 & 1 \\ 0 & 0 & 1 & 0 \end{bmatrix}^{-1} = \begin{bmatrix} 1 & 0 & 0 & -1 \\ 0 & 1 & 0 & 0 \\ 0 & 0 & 0 & 1 \\ 0 & 0 & 1 & -1 \end{bmatrix}$$

Hence, the matrix **M** would be

$$M = \begin{bmatrix} I_2 & bR^{-1}b^T \\ \hline 0 & A^T \end{bmatrix}^{-1} \begin{bmatrix} A & 0 \\ \hline -Q & I_2 \end{bmatrix} = \begin{bmatrix} 1 & 2 & 0 & -1 \\ 1 & 0 & 0 & 0 \\ 0 & -1 & 0 & 1 \\ -1 & 1 & 1 & -1 \end{bmatrix}$$

The matrix **M** has the following two stable eigenvalues: $\lambda_1 = 0.457$ and $\lambda_2 = -0.457$. For these eigenvalues the eigenvectors ω_1 and ω_2 are

$$\omega_1 = \begin{bmatrix} 0.457 \\ 1 \\ \hline 2.732 \\ 2.248 \end{bmatrix} = \begin{bmatrix} v_1 \\ \hline w_1 \end{bmatrix} \quad \text{and} \quad \omega_2 = \begin{bmatrix} -0.457 \\ 1 \\ \hline -0.732 \\ 1.334 \end{bmatrix} = \begin{bmatrix} v_2 \\ \hline w_2 \end{bmatrix}$$

From equation (7.50) we have

$$W = [w_1 \mid w_2] = \begin{bmatrix} 2.732 & -0.732 \\ 2.248 & 1.334 \end{bmatrix} \quad \text{and} \quad V = [v_1 \mid v_2] = \begin{bmatrix} 0.457 & -0.457 \\ 1 & 1 \end{bmatrix}$$

Finally, we have

$$P = WV^{-1} = \begin{bmatrix} 3.791 & 1 \\ 1 & 1.791 \end{bmatrix}$$

The column vector **k** may be determined from equation (7.47). We have

$$k^T = -R^{-1}b^T[A^T]^{-1}[P - Q] = [-1 \quad -0.791]$$

Therefore, the optimal control law is

$$u(k) = k^Tx(k) = [-1 \quad -0.791]x(k)$$

The closed-loop system is

$$\mathbf{x}(k + 1) = [\mathbf{A} + \mathbf{bk}^T]\mathbf{x}(k) = \begin{bmatrix} 0 & 0.209 \\ 1 & 0 \end{bmatrix}\mathbf{x}(k)$$

The value of J_{min}, according to (7.42), is

$$J_{min} = \tfrac{1}{2}\mathbf{x}^T(0)\mathbf{P}\mathbf{x}(0) = 1.896$$

7.5 Problems

1. Let a discrete-time system be described in the state space by

$$\mathbf{x}(k + 1) = \begin{bmatrix} 0 & 1 \\ -3 & -4 \end{bmatrix}\mathbf{x}(k) + \begin{bmatrix} 0 \\ 1 \end{bmatrix}u(k) \quad \text{with} \quad \mathbf{x}(0) = \begin{bmatrix} 1 \\ 0 \end{bmatrix}$$

Find a control signal sequence $u(0)$, $u(1)$, $u(2)$ and $u(3)$ which minimizes the cost function

$$J = 0.4 \sum_{k=0}^{4} u^2(k)$$

with $x(5) = [0 \ 0]^T$. Also, find the optimal $x(k)$.

2. Consider a discrete-time system with the state-space representation:

$$\mathbf{x}(k + 1) = \begin{bmatrix} 0 & 1 & 0 \\ 0 & 0 & 1 \\ -1 & 0 & 0.5 \end{bmatrix}\mathbf{x}(k) + \begin{bmatrix} 0 \\ 0 \\ 1 \end{bmatrix}u(k) \quad \text{with} \quad \mathbf{x}(0) = \begin{bmatrix} 0 \\ 0 \\ 1 \end{bmatrix}$$

and the cost function

$$J = \sum_{k=0}^{2} [x_1^2(k + 1) + u^2(k)]$$

with $x_1(3)$ indeterminate and $x_2(3) = x_3(3) = 0$. Find the optimal sequence $u(0)$, $u(1)$, $u(2)$ that minimizes J.

3. Consider the discrete-time system

$$\mathbf{x}(k + 1) = \begin{bmatrix} 0 & 1 & 0 & 0 \\ 0 & 0 & 1 & 0 \\ 0 & 0 & 0 & 1 \\ 2 & -1 & 1 & -1 \end{bmatrix}\mathbf{x}(k) + \begin{bmatrix} 0 \\ 0 \\ 1 \\ 1 \end{bmatrix}u(k)$$

$$\mathbf{x}(0) = \begin{bmatrix} 0 \\ 1 \\ 0 \\ 1 \end{bmatrix}$$

and the cost function

$$J = \sum_{k=0}^{6} u^2(k)$$

Also, let $x_1(7) = x_4(7) = 1$, $x_2(7) = -1$, $x_3(7) = 0$. Find the optimal sequence $u(0)$, $u(1), \ldots, u(6)$ that minimizes J.

4. Substitute $T = 0.5$ s in $G(z)$ given in problem 13 of Chapter 6.

(a) Find a state-space description for the system.
(b) In the description obtained in part (a), let $\mathbf{x}(0) = \mathbf{x}_0 = [1 \ 0]^T$. Find the optimal state feedback law of the form $u(k) = \mathbf{k}^T\mathbf{x}(k)$ such that the cost function

$$J = \frac{1}{2} \sum_{k=0}^{\infty} [\mathbf{x}^T(k)\mathbf{x}(k) + u^2(k)]$$

is minimized. Determine J_{min}.

5. The equation of motion of a satellite-tracking antenna system is given by [7]

$$J\ddot{\theta} + b\dot{\theta} = T_c + T_d$$

where T_c is the net torque from the drive motor and T_d is the disturbance torque due to wind. If we define

$$a = \frac{b}{J}, \quad u = \frac{T_c}{b} \quad \text{and} \quad w_d = \frac{T_d}{b}$$

then the equation of motion becomes

$$\frac{1}{a}\ddot{\theta} + \dot{\theta} = u + w_d$$

The transfer function of the system is

$$\frac{\Theta(s)}{U(s)} = G(s) = \frac{a}{s(s+a)}$$

In discrete time, with $u(k)$ applied through a zero-order hold (see Example 5.3.1 for detailed steps),

$$G(z) = \frac{\Theta(z)}{U(z)} = K\left(\frac{z + b}{(z - 1)(z - e^{-aT})}\right)$$

where

$$K = \frac{aT - 1 + e^{-aT}}{a} \quad \text{and} \quad b = \frac{1 - e^{-aT} - aTe^{-aT}}{aT - 1 + e^{-aT}}$$

Let $T = 0.5$ s and $a = 0.01$.

(a) Obtain a state-space description for the system.
(b) In the description derived in part (a), let $\mathbf{x}(0) = [1 \ 0]^T$. Determine the optimal state feedback law in the form $u(k) = \mathbf{k}^T\mathbf{x}(k)$ so that the cost criterion

$$J = \frac{1}{2} \sum_{k=0}^{\infty} [\mathbf{x}^T(k)\mathbf{x}(k) + u^2(k)]$$

is minimized. Determine $J_{\min}$.

6. Consider the double-mass system shown in Figure 7.4, where m_1, m_2 are two masses, β is the damping coefficient, K is the spring constant and u is the applied control force. It can be shown that [7]

$$\frac{Y(s)}{U(s)} = \frac{1}{m_1} \frac{s^2 + (\beta/m_2)s + K/m_2}{s^2\{s^2 + (1 + m_2/m_1)[(\beta/m_2)s + K/m_2]\}}$$

Let $m_1 = 1$ kg, $m_2 = 0.1$ kg, $\beta = 0.0036$ N s/m and $K = 0.091$ N/m.
(a) Obtain a discrete-time state-space system description when the sampling period is $T = 0.4$ s.
(b) In the description derived in part (a), let $\mathbf{x}_0 = [1 \ 0 \ 0 \ 0]^T$. Determine the optimal state feedback law in the form $u(k) = \mathbf{k}^T\mathbf{x}(k)$ such that the cost function

$$J = \frac{1}{2} \sum_{k=0}^{\infty} [\mathbf{x}^T(k)\mathbf{x}(k) + u^2(k)]$$

is minimized and determine $J_{\min}$.

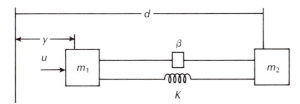

Figure 7.4 Double-mass–spring–damper system.

7. A system for advancing magnetic tape may be represented by the following state-space equations [7]:

$$\begin{bmatrix} \dot{x}_1(t) \\ \dot{x}_2(t) \\ \dot{x}_3(t) \\ \dot{x}_4(t) \end{bmatrix} = \begin{bmatrix} 0 & 0 & -10 & 0 \\ 0 & 0 & 0 & 10 \\ 3.315 & -3.315 & -0.5882 & -0.5882 \\ 3.315 & -3.315 & -0.5882 & -0.5882 \end{bmatrix} \begin{bmatrix} x_1(t) \\ x_2(t) \\ x_3(t) \\ x_4(t) \end{bmatrix} + \begin{bmatrix} 0 & 0 \\ 0 & 0 \\ 8.553 & 1 \\ 0 & 8.533 \end{bmatrix} \begin{bmatrix} u_1(t) \\ u_2(t) \end{bmatrix}$$

$$\begin{bmatrix} y_1(t) \\ y_2(t) \end{bmatrix} = \begin{bmatrix} 0.5 & 0.5 & 0 & 0 \\ -2.113 & 2.133 & 0.375 & 0.375 \end{bmatrix} \begin{bmatrix} x_1(t) \\ x_2(t) \\ x_3(t) \\ x_4(t) \end{bmatrix}$$

It can be shown that the system eigenvalues are $s_1 = s_2 \cong 0$ and $s_{3,4} \cong \pm j8$. Hence, there are two oscillatory roots for this system. Use a sampling period $T = 0.5$ s and

(a) find a discrete-time state-space description for the system;

(b) In the description derived in part (a), let $\mathbf{x}_0 = [1 \ 0 \ 0 \ 0]^T$. Find the optimal state feedback law in the form

$$\begin{bmatrix} u_1(t) \\ u_2(t) \end{bmatrix} = \mathbf{K} \begin{bmatrix} x_1(t) \\ x_2(t) \\ x_3(t) \\ x_4(t) \end{bmatrix}$$

so that the cost function

$$J = \frac{1}{2} \sum_{k=0}^{\infty} [x_1^2(k) + x_2^2(k) + x_3^2(k) + x_4^2(k) + u_1^2(k) + u_2^2(k)]$$

is minimized and determine J_{min}.

8. Consider the continuous-time model for lateral helicopter motion given in Problem 3 of Chapter 6. Let the sampling period be $T = 0.3$ s. Find the optimal feedback control law in the form $u(k) = \mathbf{k}^T \mathbf{x}(k)$ so that the cost function

$$J = \frac{1}{2} \sum_{k=0}^{\infty} [\mathbf{x}^T(k)\mathbf{x}(k) + u^2(k)]$$

is minimized when the initial condition vector for the discrete-time system is $\mathbf{x}_0 = [1 \ -1 \ 0]^T$. Determine J_{min}.

9. Consider the continuous-time model for the yaw motion of a tanker given in Problem 4 of Chapter 6. Let the sampling period be $T = 0.05$ s. Find the optimal state feedback law in the form $u(k) = \mathbf{k}^T\mathbf{x}(k)$ such that the cost function

$$J = \tfrac{1}{2}\sum_{k=0}^{\infty}[\mathbf{x}^T(k)\mathbf{x}(k) + 10^{-3}u^2(k)]$$

is minimized when the initial condition vector for the discrete-time system is $\mathbf{x}_0 = [1 \;\; -1 \;\; 0]^T$. Determine $J_{\min}$.

10. The control of a pressurized flow box is analyzed in [7] where the dynamical equations of a multivariable system take the form

$$\begin{bmatrix} \dot{x}_1 \\ \dot{x}_2 \\ \dot{x}_3 \end{bmatrix} = \begin{bmatrix} -0.2 & 0.1 & 1 \\ -0.05 & 0 & 0 \\ 0 & 0 & -1 \end{bmatrix}\begin{bmatrix} x_1 \\ x_2 \\ x_3 \end{bmatrix} + \begin{bmatrix} 0 & 1 \\ 0 & 0.7 \\ 1 & 0 \end{bmatrix}\begin{bmatrix} u_1(t) \\ u_2(t) \end{bmatrix}$$

For this system:

(a) Find a discrete-time state-space description when the sampling period is $T = 0.5$ s.

(b) In the description derived in part (a), let $\mathbf{x}_0 = [1 \;\; -1 \;\; 1]^T$. Find the optimal state feedback law in the form $u(k) = \mathbf{f}^T\mathbf{x}(k)$ such that the cost function

$$J = \tfrac{1}{2}\sum_{k=0}^{\infty}[\mathbf{x}^T(k)\mathbf{x}(k) + \mathbf{u}^T(k)\mathbf{u}(k)]$$

is minimized and determine $J_{\min}$.

7.6 References

Books

[1] B.D.O. Anderson and J.B. Moore, *Linear Optimal Control*, Prentice Hall, Englewood Cliffs, New Jersey, 1971.

[2] B.D.O. Anderson and J.B. Moore, *Optimal Control: Linear Quadratic Methods*, Prentice Hall, Englewood Cliffs, New Jersey, 1990.

[3] K.J. Åström, *Introduction to Stochastic Control Theory*, Academic Press, New York, 1970.

[4] K.J. Åström and B. Wittenmark, *Computer Controlled Systems*, Prentice Hall, Englewood Cliffs, New Jersey, 1984.

[5] M. Athans and P.L. Falb, *Optimal Control*, McGraw-Hill, New York, 1966.

[6] A.E. Bryson Jr and Y.C. Ho, *Applied Optimal Control*, John Wiley, New York, 1975.

[7] G.F. Franklin, J.D. Powell and M.L. Workman, *Digital Control of Dynamic Systems*, Addison-Wesley, London, 1990 (Second Edition).

[8] D.E. Kirk, *Optimal Control Theory, An Introduction*, Prentice Hall, Englewood Cliffs, New Jersey, 1970.

[9] H. Kwakernaak and R. Sivan, *Linear Optimal Control Systems*, Wiley-Interscience, New York, 1972.

[10] K. Ogata, *Discrete-Time Control Systems*, Prentice Hall, Englewood Cliffs, New Jersey, 1987.

[11] L.S. Pontryagin *et al.*, *The Mathematical Theory of Optimal Processes*, Interscience Publishers, New York, 1962.

[12] H.A. Prime, *Modern Concepts in Control Theory*, McGraw-Hill, New York, 1969.

[13] A.D. Sage and C.C. White, *Optimum Systems Control*, Prentice Hall, Englewood Cliffs, New Jersey, 1977 (Second Edition).

8

System identification

8.1 Introduction

A fundamental concept in science and technology is that of mathematical modelling. A mathematical model is a very useful, as well as a very compact, way of describing what we know about a process or system. The determination of a mathematical model of a system is known as **system identification**. In control systems, a mathematical model of the process is necessary in most cases for the design of the controller. This becomes apparent if we recall that all the controller design methods covered in previous chapters are based on the model of the system under control which was assumed known. The model is also necessary for the design of adaptive control systems presented in Chapter 9.

A process or system may be described by several models, ranging from necessarily very detailed and complex microscopic models to simplistic macroscopic models which facilitate an understanding of the gross characteristics of a system's performance. Complex microscopic models require a long time to determine and they are mostly used for the detailed control of a system's performance. Between these two extremes, there exist several different types of models. Clearly, one must be able to choose a suitable type of model for each specific application.

There are basically two ways of determining a mathematical model of a system: by implementing known laws of nature or by experimenting on the process. A popular approach to obtain a model is to combine both ways.

Mathematical models may be distinguished as parametric or non-parametric models. Parametric models obviously involve parameters, as for example the coefficients of difference equations, state equations and transfer functions. Non-parametric models do not involve parameters and are usually graphical representations, such as the Nyquist or Bode diagrams of a transfer function or impulse response function. This chapter covers parametric models. Non-parametric models are covered in, for example, [6]. Overall, the parametric identification problem reduces to the development of methods which give a good estimate of the parameters of the system model.

Here we deal with the problem of determining mathematical models of linear, time-invariant, single-input single-output, discrete-time systems, described by difference equations. The proposed method for the identification (estimation) of the coefficients (parameters) of a difference equation is experimental and may be briefly described as follows. First, a set of N linear algebraic equations is formulated, where N is the number of measurements. From these equations, one may easily derive the **canonical equation** whose solution yields the parameter estimate $\boldsymbol{\theta}(N)$, where $\boldsymbol{\theta}$ is the vector parameter under identification. If an estimate of the initial conditions of the dynamic equation is also required, then $N + n$ measurements are taken and hence $N + n$ equations are produced, where n is the order of the difference equation. Minor alterations in the procedure for determining $\boldsymbol{\theta}(N)$ are then made, resulting in an estimate for both $\boldsymbol{\theta}(N)$ and the initial conditions (Section 8.2).

An interesting feature, which is also included in this chapter, is the determination of a recursive algorithm which allows estimation of the vector parameter $\boldsymbol{\theta}$ for $N + 1$ measurements, based on the following formula:

$$\boldsymbol{\theta}(N+1) = \boldsymbol{\theta}(N) + \Delta\boldsymbol{\theta} = \boldsymbol{\theta}(N) + \boldsymbol{\gamma}(N)[y_{N+1} - \boldsymbol{\phi}^{\mathrm{T}}(N+1)\boldsymbol{\theta}(N)] \tag{8.1}$$

where $\boldsymbol{\gamma}(N)$ and $\boldsymbol{\phi}(N+1)$ are known vector quantities and y_{N+1} is the $(N+1)$th measurement of the output y of the system. This formula shows that for the determination of $\boldsymbol{\theta}(N+1)$ one can use the previous estimate $\boldsymbol{\theta}(N)$ plus a corrective term $\Delta\boldsymbol{\theta}$, which is due to the new $(N+1)$th measurement, instead of starting the estimation procedure right from the beginning. This algorithm facilitates the numerical part of the problem and constitutes the cornerstone of and tool for the solution of ON-LINE identification, i.e. of identification which takes place in real time while the system is operating under normal conditions (Section 8.3). In contrast, when the identification procedure is required to take place only for the first N measurements, it is carried out off-line once only, and for this reason it is called OFF-LINE identification or parameter estimation.

For relevant references on system identification see [1–23].

8.2 OFF-LINE parameter estimation

8.2.1 First-order systems

The simple case of a first-order discrete-time system is studied first. Assume that the system under consideration is described by the difference equation

$$y(k) + a_1 y(k-1) = b_1 u(k-1) \tag{8.2}$$

with initial condition $y(-1)$. Assume also that the system (8.2) is excited with an input sequence $u(-1)$, $u(0)$, $u(1)$, ... As a result, the output of the system has a sequence $y(0)$, $y(1)$, ... The identification problem may now be defined as follows. Given the known input sequence $u(-1)$, $u(0)$, $u(1)$, ..., as well as the measured output sequence $y(0)$, $y(1)$, ..., find an estimate of the system's parameters a_1 and b_1 and of the initial condition $y(-1)$.

To solve the problem, we begin by writing down equation (8.2) for $N + 1$ measurements, i.e. for $k = 0, 1, 2, \ldots, N$. In this way, we arrive at the following set of linear algebraic equations:

$$y(0) + a_1 y(-1) = b_1 u(-1)\} \tag{8.3a}$$

$$
\left.\begin{aligned}
y(1) + a_1 y(0) &= b_1 u(0) \\
y(2) + a_1 y(1) &= b_1 u(1) \\
&\vdots \\
y(N) + a_1 y(N-1) &= b_1 u(N-1)
\end{aligned}\right\} \tag{8.3b}
$$

The last N equations, i.e. equations (8.3b), are used for the estimation of the parameters a_1 and b_1. Then equation (8.3a) can be used for the estimation of the initial condition $y(-1)$. To this end, we define

$$
\boldsymbol{\theta} = \begin{bmatrix} a_1 \\ b_1 \end{bmatrix}, \quad \mathbf{y} = \begin{bmatrix} y(1) \\ y(2) \\ \vdots \\ y(N) \end{bmatrix} \quad \text{and} \quad \boldsymbol{\Phi} = \begin{bmatrix} \boldsymbol{\phi}^{\mathrm{T}}(0) \\ \boldsymbol{\phi}^{\mathrm{T}}(1) \\ \vdots \\ \boldsymbol{\phi}^{\mathrm{T}}(N-1) \end{bmatrix} = \begin{bmatrix} -y(0) & u(0) \\ -y(1) & u(1) \\ \vdots & \vdots \\ -y(N-1) & u(N-1) \end{bmatrix} \tag{8.4}
$$

Using these definitions, equations (8.3b) can be written compactly as

$$\mathbf{y} = \boldsymbol{\Phi}\boldsymbol{\theta} \tag{8.5}$$

Equation (8.5) is an algebraic system of N equations with two unknowns. It is clear that if the known input and output sequences involve errors due to measurement or noise, then, for every input–output pair $\{u(k), y(k)\}$, there exists an error $e(k)$; thus, equations (8.3) will take on the form

$$y(k) + a_1 y(k-1) = b_1 u(k-1) + e(k) \quad \text{for} \quad k = 0, 1, 2, \ldots, N \tag{8.6}$$

Consequently (8.5) becomes

$$\mathbf{y} = \boldsymbol{\Phi}\boldsymbol{\theta} + \mathbf{e} \tag{8.7}$$

where $\mathbf{e}$ is the N-dimensional error vector $\mathbf{e}^{\mathrm{T}} = [e(1)\ e(2)\ \ldots\ e(N)]$. For the minimization of the error vector $\mathbf{e}$, the least-squares method can be applied. To this end, we define the following cost function:

$$J = \mathbf{e}^{\mathrm{T}}\mathbf{e} = \sum_{k=1}^{N} e^2(k) \tag{8.8}$$

If (8.7) is substituted in (8.8), we obtain

$$J = (\mathbf{y} - \boldsymbol{\Phi}\boldsymbol{\theta})^{\mathrm{T}}(\mathbf{y} - \boldsymbol{\Phi}\boldsymbol{\theta})$$

Hence

$$\frac{\partial J}{\partial \boldsymbol{\theta}} = -2\boldsymbol{\Phi}^{\mathrm{T}}(\mathbf{y} - \boldsymbol{\Phi}\boldsymbol{\theta})$$

where the following formula was used (see Appendix A):

$$\frac{\partial}{\partial\theta}[\mathbf{A}\theta] = \frac{\partial}{\partial\theta}[\theta^{\mathsf{T}}\mathbf{A}^{\mathsf{T}}] = \mathbf{A}^{\mathsf{T}}$$

If we set $\partial J/\partial\theta$ equal to zero, we obtain

$$\mathbf{\Phi}^{\mathsf{T}}\mathbf{\Phi}\theta = \mathbf{\Phi}^{\mathsf{T}}\mathbf{y} \tag{8.9}$$

Relation (8.9) is known as the **canonical equation** and has a solution when the matrix $\mathbf{\Phi}^{\mathsf{T}}\mathbf{\Phi}$ is invertible, in which case we have

$$\theta = (\mathbf{\Phi}^{\mathsf{T}}\mathbf{\Phi})^{-1}\mathbf{\Phi}^{\mathsf{T}}\mathbf{y} = \mathbf{\Phi}^{*}\mathbf{y} \tag{8.10}$$

where $\mathbf{\Phi}^{*}$ is the pseudo-inverse of $\mathbf{\Phi}$. The solution (8.10) minimizes the cost function (8.8).

The matrix $\mathbf{\Phi}^{\mathsf{T}}\mathbf{\Phi}$ is symmetrical and has the following form:

$$\mathbf{\Phi}^{\mathsf{T}}\mathbf{\Phi} = \begin{bmatrix} \sum_{k=0}^{N-1} y^{2}(k) & -\sum_{k=0}^{N-1} y(k)u(k) \\ -\sum_{k=0}^{N-1} y(k)u(k) & \sum_{k=0}^{N-1} u^{2}(k) \end{bmatrix} \tag{8.11a}$$

Moreover, the vector $\mathbf{\Phi}^{\mathsf{T}}\mathbf{y}$ has the form

$$\mathbf{\Phi}^{\mathsf{T}}\mathbf{y} = \begin{bmatrix} -\sum_{k=1}^{N} y(k)y(k-1) \\ \sum_{k=1}^{N} y(k)u(k-1) \end{bmatrix} \tag{8.11b}$$

The estimate of the parameters a_1 and a_2 is based on equation (8.10). The initial condition $y(-1)$ is estimated on the basis of equation (8.3a), which gives

$$y(-1) = \frac{1}{a_1}[b_1 u(-1) - y(0)]$$

where it is assumed that $a_1 \neq 0$ and $u(-1)$ is known.

■ **EXAMPLE 8.2.1**   A discrete-time system is described by the first-order difference equation

$$y(k) + a_1 y(k-1) = b_1 u(k-1)$$

Table 8.1   Input–output measurements for Example 8.2.1.

k	-1	0	1	2	3	4	5	6
$u(k)$	1	1	1	1	1	1	1	1
$y(k)$		1	1/2	3/4	5/8	11/16	21/32	43/64

The input and output sequences $u(k)$ and $y(k)$, for $N = 6$, are given in Table 8.1. Estimate the parameters a_1 and b_1, as well as the initial condition $y(-1)$.

■ **SOLUTION**   From equations (8.11) and for $N = 6$ we have that

$$\boldsymbol{\Phi}^T\boldsymbol{\Phi} = \begin{bmatrix} \sum_{k=0}^{5} y^2(k) & -\sum_{k=0}^{5} y(k)u(k) \\ -\sum_{k=0}^{5} y(k)u(k) & \sum_{k=0}^{5} u^2(k) \end{bmatrix} = \begin{bmatrix} 3.106\,445 & -4.218\,75 \\ -4.218\,75 & 6 \end{bmatrix}$$

$$\boldsymbol{\Phi}^T\mathbf{y} = \begin{bmatrix} -\sum_{k=1}^{6} y(k)y(k-1) \\ \sum_{k=1}^{6} y(k)u(k-1) \end{bmatrix} = \begin{bmatrix} -2.665\,527 \\ 3.890\,625 \end{bmatrix}$$

Hence

$$\boldsymbol{\theta} = \begin{bmatrix} a_1 \\ b_1 \end{bmatrix} = (\boldsymbol{\Phi}^T\boldsymbol{\Phi})^{-1}\boldsymbol{\Phi}^T\mathbf{y} = \frac{1}{0.840\,82} \begin{bmatrix} 6 & 4.218\,75 \\ 4.218\,75 & 3.106\,445 \end{bmatrix} \begin{bmatrix} -2.665\,527 \\ 3.890\,625 \end{bmatrix} = \begin{bmatrix} 0.5 \\ 1 \end{bmatrix}$$

Finally, the estimate of $y(-1)$, derived using equation (8.3a), is

$$y(-1) = \frac{1}{a_1}[b_1 u(-1) - y(0)] = 2[1 - 1] = 0$$

■ **EXAMPLE 8.2.2**   For the estimation of the acceleration constant due to gravity, g, the following experiment is carried out. A steel ball is allowed to free-fall with zero initial speed. During the free fall, the height h between the ball and the ground is measured at different time instants. The measurements are given in Table 8.2. The measurements in the table are assumed accurate, but there exists a height error. Estimate g using the least-squares method.

■ **SOLUTION**   Using Newton's law for a free-falling body with zero initial speed, the height h is given by the formula:

$$h = \frac{g}{2} t^2$$

Table 8.2   Position measurements for Example 8.2.2.

t (s)	1	2	3	4	5	6
h (m)	8.49	20.05	50.56	72.19	129.85	171.56

Consequently, the model of the system has the form $h(k) = ak^2 + e(k)$, where $a = g/2$ and $e(k)$ is the height error. To determine a, the least-squares method is used. The cost function has the form

$$J = \sum_{k=1}^{6} e^2(k) = \sum_{k=1}^{6} [h(k) - ak^2]^2$$

Minimizing J, we obtain

$$\frac{\partial J}{\partial a} = 2 \sum_{k=1}^{6} [h(k) - ak^2]k^2 = 0$$

Hence

$$a = \frac{\sum_{k=1}^{6} h(k)k^2}{\sum_{k=1}^{6} k^4} = 4.888\ 786\ 813 \text{ m/s}^2$$

Therefore, the estimate of the acceleration constant g has the value

$$g = 2a = 9.777\ 573\ 626 \text{ m/s}^2$$

8.2.2 *Higher-order systems*

In this section, the results of Section 8.2.1 will be extended to cover the general case where the difference equation is of order n and has the form

$$y(k) + a_1 y(k-1) + \cdots + a_n y(k-n) = b_1 u(k-1) + \cdots + b_n u(k-n) \qquad (8.12)$$

with initial conditions $y(-1)$, $y(-2)$, ..., $y(-n)$. In this case, the unknowns are the parameters $a_1, a_2, ..., a_n$, $b_1, b_2, ..., b_n$ and the initial conditions $y(-1)$, $y(-2), ..., y(-n)$. As before we take $(N+n)$ measurements. For $k = 0, 1, ..., N + n - 1$, the difference equation (8.12) yields the following equations:

$$\left.\begin{array}{l} y(0) + a_1 y(-1) + \cdots + a_n y(-n) = b_1 u(-1) + \cdots + b_n u(-n) \\ y(1) + a_1 y(0) + \cdots + a_n y(-n+1) = b_1 u(0) + \cdots + b_n u(-n+1) \\ \qquad\qquad\vdots \\ y(n-1) + a_1 y(n-2) + \cdots + a_n y(-1) = b_1 u(n-2) + \cdots + b_n u(-1) \end{array}\right\} \qquad (8.13a)$$

$$\left.\begin{array}{l} y(n) + a_1 y(n-1) + \cdots + a_n y(0) = b_1 u(n-1) + \cdots + b_n u(0) \\ y(n+1) + a_1 y(n) + \cdots + a_n y(1) = b_1 u(n) + \cdots + b_n u(1) \\ \qquad\qquad\vdots \\ y(n+N-1) + a_1 y(n+N-2) + \cdots + a_n y(N-1) \\ \qquad = b_1 u(n+N-2) + \cdots + b_n u(N-1) \end{array}\right\} \qquad (8.13b)$$

Relations (8.13) have a total of $N + n$ algebraic equations: the first n equations are in (8.13a), whereas the remaining N equations are in (8.13b). We define

$$\boldsymbol{\theta}^T = [a_1 \quad a_2 \quad \ldots \quad a_n \quad b_1 \quad b_2 \quad \ldots \quad b_n]$$

$$\mathbf{y}^T = [y(n) \quad y(n+1) \quad \ldots \quad y(n+N-1)]$$

(8.14a)

$$\boldsymbol{\Phi} = \begin{bmatrix} \boldsymbol{\phi}^T(0) \\ \boldsymbol{\phi}^T(1) \\ \vdots \\ \boldsymbol{\phi}^T(n+N-1) \end{bmatrix} = \begin{bmatrix} -y(n-1) & \ldots & -y(0) & u(n-1) & \ldots & u(0) \\ -y(n) & \ldots & -y(1) & u(n) & \ldots & u(1) \\ & \vdots & & & \vdots & \\ -y(n+N-2) & \ldots & -y(N-1) & u(n+N-2) & \ldots & u(N-1) \end{bmatrix}$$

(8.14b)

Using the foregoing definitions, equations (8.13b) can be written compactly as follows:

$$\mathbf{y} = \boldsymbol{\Phi}\boldsymbol{\theta}$$

(8.15)

Based on the formula (8.15), the results derived for the first-order systems in Section 8.2.1 can easily be extended to the higher-order case. Hence, the canonical equation for (8.15) takes the form

$$\boldsymbol{\Phi}^T\boldsymbol{\Phi}\boldsymbol{\theta} = \boldsymbol{\Phi}^T\mathbf{y}$$

(8.16)

and therefore

$$\boldsymbol{\theta} = (\boldsymbol{\Phi}^T\boldsymbol{\Phi})^{-1}\boldsymbol{\Phi}^T\mathbf{y} = \boldsymbol{\Phi}^*\mathbf{y}$$

(8.17)

under the assumption that the matrix $\boldsymbol{\Phi}^T\boldsymbol{\Phi}$ is invertible.

The estimate of the initial conditions $y(-1)$, $y(-2), \ldots, y(-n)$ is based on (8.13a), making use of the known estimates of the parameters $a_1, a_2, \ldots, a_n$, b_1, $b_2, \ldots, b_n$ given in (8.17). We define

$$\tilde{\mathbf{A}} = \begin{bmatrix} a_1 & a_2 & a_3 & \ldots & a_{n-1} & a_n \\ a_2 & a_3 & a_4 & \ldots & a_n & 0 \\ \vdots & \vdots & \vdots & & \vdots & \vdots \\ a_n & 0 & 0 & \ldots & 0 & 0 \end{bmatrix}, \quad \mathbf{A}^* = \begin{bmatrix} 1 & 0 & 0 & \ldots & 0 & 0 \\ a_1 & 1 & 0 & \ldots & 0 & 0 \\ \vdots & \vdots & \vdots & & \vdots & \vdots \\ a_{n-1} & a_{n-2} & a_{n-3} & \ldots & a_1 & 1 \end{bmatrix}$$

$$\tilde{\mathbf{B}} = \begin{bmatrix} b_n & b_{n-1} & \ldots & b_1 & 0 & \ldots & 0 \\ 0 & b_n & \ldots & b_2 & b_1 & \ldots & 0 \\ \vdots & \vdots & & \vdots & \vdots & & \vdots \\ 0 & 0 & \ldots & b_n & b_{n-1} & \ldots & b_1 \end{bmatrix}, \quad \boldsymbol{\eta} = \begin{bmatrix} y(-1) \\ y(-2) \\ \vdots \\ y(-n) \end{bmatrix}, \quad \mathbf{d} = \begin{bmatrix} u(-n) \\ u(-n+1) \\ \vdots \\ u(n-2) \end{bmatrix}, \quad \mathbf{c} = \begin{bmatrix} y(0) \\ y(1) \\ \vdots \\ y(n-1) \end{bmatrix}$$

Equations (8.13a) may now be written compactly as

$$\tilde{\mathbf{A}}\boldsymbol{\eta} + \mathbf{A}^*\mathbf{c} = \tilde{\mathbf{B}}\mathbf{d}$$

(8.18)

The only unknown in (8.18) is the initial condition vector $\boldsymbol{\eta}$. Hence

$$\boldsymbol{\eta} = \tilde{\mathbf{A}}^{-1}[\tilde{\mathbf{B}}\mathbf{d} - \mathbf{A}^*\mathbf{c}] \tag{8.19}$$

It is assumed that $a_1 \neq 0$ (so that $\det \tilde{\mathbf{A}} \neq 0$) and that the values of $u(-1)$, $u(-2), \ldots, u(-n)$ are known.

Note that in identifying a diskrete-time system, there is usually no particular interest in estimating the initial conditions $y(-1)$, $y(-2), \ldots, y(-n)$. For this reason, they are not usually estimated (see Section 8.3 below). Equation (8.19) was derived above mainly for the sake of completeness of the presentation of the identification procedure.

■ **EXAMPLE 8.2.3** Consider a discrete-time system described by the second-order difference equation:

$$y(k+2) + \omega^2 y(k) = bu(k)$$

The input $u(k)$ and the output $y(k)$, for $N = 5$, are presented in Table 8.3. Estimate the parameters ω and b.

■ **SOLUTION** For $k = 0, 1, 2, 3$ we have

$$y(2) + \omega^2 y(0) = bu(0) + e(0)$$
$$y(3) + \omega^2 y(1) = bu(1) + e(1)$$
$$y(4) + \omega^2 y(2) = bu(2) + e(2)$$
$$y(5) + \omega^2 y(3) = bu(3) + e(3)$$

where $e(k)$ is the measurement error at time k. These equations can be grouped as follows:

$$\mathbf{y} = \boldsymbol{\Phi}\boldsymbol{\theta} + \mathbf{e}$$

where

$$\mathbf{y}^T = [y(2) \quad y(3) \quad y(4) \quad y(5)] = [1/2 \quad 1/3 \quad 1/4 \quad 1/5]$$

$$\boldsymbol{\Phi} = \begin{bmatrix} -y(0) & 1 \\ -y(1) & 1 \\ -y(2) & 1 \\ -y(3) & 1 \end{bmatrix} = \begin{bmatrix} 0 & 1 \\ 0 & 1 \\ -1/2 & 1 \\ -1/3 & 1 \end{bmatrix}, \quad \boldsymbol{\theta} = \begin{bmatrix} \omega^2 \\ b \end{bmatrix} \quad \text{and} \quad \mathbf{e} = \begin{bmatrix} e(0) \\ e(1) \\ e(2) \\ e(3) \end{bmatrix}$$

Table 8.3 Input–output data sequence for Example 8.2.3.

k	0	1	2	3	4	5
$u(k)$	1	1	1	1	1	1
$y(k)$	0	0	1/2	1/3	1/4	1/5

Using the above, we have

$$\Phi^T\Phi = \begin{bmatrix} 13/36 & -5/6 \\ -5/6 & 4 \end{bmatrix}, \quad \Phi^T\mathbf{y} = \begin{bmatrix} -23/120 \\ 77/60 \end{bmatrix} \quad \text{and} \quad (\Phi^T\Phi)^{-1} = \frac{36}{27}\begin{bmatrix} 4 & 5/6 \\ 5/6 & 13/36 \end{bmatrix}$$

The optimum estimates of ω^2 and b are obtained from

$$\theta = \begin{bmatrix} \omega^2 \\ b \end{bmatrix} = (\Phi^T\Phi)^{-1}\Phi^T\mathbf{y} = \begin{bmatrix} 0.404 \\ 0.405 \end{bmatrix}$$

whence $\omega = (0.404)^{1/2} = 0.635$ and $b = 0.405$.

■ **EXAMPLE 8.2.4** The mathematician L.F. Richardson suggested the following simple model for the presentation of an arms race between two countries [1]:

$$x(k+1) = ax(k) + by(k) + f$$
$$y(k+1) = cx(k) + dy(k) + g$$

where $x(k)$ and $y(k)$ are the expenditures on arms of the two countries and a, b, c, d, f and g are constants. In Table 8.4 the expenditures on arms by Iran, Iraq, NATO and the Warsaw Treaty Organization (WTO) are given (from World Armaments and Disarmaments, SIPRI Yearbook, 1982). Estimate the parameters of the model and check its stability.

■ **SOLUTION** Using the least-squares method, the following matrix Φ is formulated first:

$$\Phi = \begin{bmatrix} x_1 & y_1 & 1 \\ x_2 & y_2 & 1 \\ \vdots & \vdots & \vdots \\ x_9 & y_9 & 1 \end{bmatrix}$$

Table 8.4 Armaments expenditures in millions of US dollars for the countries and organizations in Example 8.2.4.

Year	Iran	Iraq	NATO	WTO
1972	2891	909	216478	112893
1973	3982	1123	211146	115020
1974	8801	2210	212267	117169
1975	11230	2247	210525	119612
1976	12178	2204	205717	121461
1977	9867	2203	212009	123561
1978	9165	2179	215988	125498
1979	5080	2675	218561	127185
1980	4040		225411	129000
1981			233957	131595

where

$$[x_1, x_2, \ldots, x_9, x_{10}] = [x(1972), x(1973), \ldots, x(1980), x(1981)]$$

$$[y_1, y_2, \ldots, y_9, y_{10}] = [y(1972), y(1973), \ldots, y(1980), y(1981)]$$

Hence

$$\mathbf{\Phi}^T\mathbf{\Phi} = \begin{bmatrix} \sum_{k=1}^{9} x_k^2 & \sum_{k=1}^{9} x_k y_k & \sum_{k=1}^{9} x_k \\ \sum_{k=1}^{9} x_k y_k & \sum_{k=1}^{9} y_k^2 & \sum_{k=1}^{9} y_k \\ \sum_{k=1}^{9} x_k & \sum_{k=1}^{9} y_k & 9 \end{bmatrix}$$

and

$$(\mathbf{\Phi}^T\mathbf{\Phi})^{-1} = \frac{1}{|\mathbf{\Phi}^T\mathbf{\Phi}|} \begin{bmatrix} \phi_{11} & \phi_{12} & \phi_{13} \\ \phi_{21} & \phi_{22} & \phi_{23} \\ \phi_{31} & \phi_{32} & \phi_{33} \end{bmatrix}$$

where

$$|\mathbf{\Phi}^T\mathbf{\Phi}| = \left[\sum_{k=1}^{9} x_k^2\right]\left[9\left[\sum_{k=1}^{9} y_k^2\right] - \left[\sum_{k=1}^{9} y_k\right]^2\right] - \left[\sum_{k=1}^{9} x_k y_k\right]\left[9\left[\sum_{k=1}^{9} x_k y_k\right] - \left[\sum_{k=1}^{9} x_k\right]\left[\sum_{k=1}^{9} y_k\right]\right]$$

$$+ \left[\sum_{k=1}^{9} x_k\right]\left[\left[\sum_{k=1}^{9} y_k\right]\left[\sum_{k=1}^{9} x_k y_k\right] - \left[\sum_{k=1}^{9} y_k^2\right]\left[\sum_{k=1}^{9} x_k\right]\right]$$

$$\phi_{11} = 9\left[\sum_{k=1}^{9} y_k^2\right] - \left[\sum_{k=1}^{9} y_k\right]^2, \quad \phi_{12} = \phi_{21} = \left[\sum_{k=1}^{9} x_k\right]\left[\sum_{k=1}^{9} y_k\right] - 9\left[\sum_{k=1}^{9} x_k y_k\right]$$

$$\phi_{13} = \phi_{31} = \left[\sum_{k=1}^{9} y_k\right]\left[\sum_{k=1}^{9} x_k y_k\right] - \left[\sum_{k=1}^{9} y_k^2\right]\left[\sum_{k=1}^{9} x_k\right]$$

$$\phi_{22} = 9\left[\sum_{k=1}^{9} x_k^2\right] - \left[\sum_{k=1}^{9} x_k\right]^2, \quad \phi_{23} = \phi_{32} = \left[\sum_{k=1}^{9} x_k\right]\left[\sum_{k=1}^{9} x_k y_k\right] - \left[\sum_{k=1}^{9} x_k^2\right]\left[\sum_{k=1}^{9} y_k\right]$$

$$\phi_{33} = \left[\sum_{k=1}^{9} x_k^2\right]\left[\sum_{k=1}^{9} y_k^2\right] - \left[\sum_{k=1}^{9} x_k y_k\right]^2$$

Moreover, we have

$$\mathbf{\Phi}^T\tilde{\mathbf{x}} = \begin{bmatrix} \sum_{k=1}^{9} x_{k+1} x_k \\ \sum_{k=1}^{9} x_{k+1} y_k \\ \sum_{k=1}^{9} x_{k+1} \end{bmatrix} \quad \text{and} \quad \mathbf{\Phi}^T\tilde{\mathbf{y}} = \begin{bmatrix} \sum_{k=1}^{9} y_{k+1} x_k \\ \sum_{k=1}^{9} y_{k+1} y_k \\ \sum_{k=1}^{9} y_{k+1} \end{bmatrix}$$

where

$$
\tilde{\mathbf{x}} = \begin{bmatrix} x_2 \\ x_3 \\ \vdots \\ x_{10} \end{bmatrix} \quad \text{and} \quad \tilde{\mathbf{y}} = \begin{bmatrix} y_2 \\ y_3 \\ \vdots \\ y_{10} \end{bmatrix}
$$

The optimum estimates of the parameters a, b, f and c, d, g are given by

$$
\begin{bmatrix} a \\ b \\ f \end{bmatrix} = (\mathbf{\Phi}^T \mathbf{\Phi})^{-1} \mathbf{\Phi}^T \tilde{\mathbf{x}} \quad \text{and} \quad \begin{bmatrix} c \\ d \\ g \end{bmatrix} = (\mathbf{\Phi}^T \mathbf{\Phi})^{-1} \mathbf{\Phi}^T \tilde{\mathbf{y}}
$$

The numerical solution of the problem may be calculated by substituting the data of Table 8.4 in the foregoing equations.

With regard to the stability of the model, we note that in order for it to be stable, it suffices that the roots of the equation

$$
\begin{vmatrix} z - a & -b \\ -c & z - d \end{vmatrix} = 0
$$

be inside the unit circle. That is, the roots of the polynomial $z^2 - (a + d)z + (ad - cb)$ must have an amplitude of less than 1. In accordance with the Jury stability criterion, for the system to be stable, the following relations should hold (see Example 4.3.3):

$$
ad - cb < 1, \quad ad - cb > -1 - a - d \quad \text{and} \quad ad - cb > -1 + a + d
$$

8.3   ON-LINE parameter estimation

In many practical cases, it is necessary that parameter estimation takes place concurrently with operation of the system. This parameter estimation problem is called **ON-LINE identification** and its methodology usually leads to a recursive procedure for every new measurement (or data entry). For this reason, it is also called **recursive identification**. In simple words, ON-LINE identification is based on the following idea. Assume that we have an estimate of the parameter vector $\boldsymbol{\theta}$ based on N pairs of input–output data entries. Let this estimate be denoted by $\boldsymbol{\theta}(N)$. Assume also that $\boldsymbol{\theta}(N)$ is not accurate enough and we wish to improve its accuracy by using the new (the next) $N + 1$ data entry. Clearly, using $N + 1$ data entries, we will obtain a new estimate for $\boldsymbol{\theta}$, denoted by $\boldsymbol{\theta}(N + 1)$, which is expected to be an improved estimate on the previous estimate $\boldsymbol{\theta}(N)$.

Now, it is natural to ask the following question. For the calculation of $\boldsymbol{\theta}(N + 1)$, do we have to estimate $\boldsymbol{\theta}(N + 1)$ right from the beginning, based on equation (8.17), or is there an easier way by taking advantage of the already known parameter vector

$\theta(N)$? The answer to this question is that the estimate $\theta(N+1)$ may indeed be determined in terms $\theta(N)$, in accordance with the following general expression:

$$\theta(N+1) = \theta(N) + \Delta\theta \tag{8.20}$$

where $\Delta\theta$ is the change in $\theta(N)$ as a consequence of the new $(N+1)$th measurement. Expression (8.20) is computationally attractive, since for each new measurement we do not have to compute $\theta(N+1)$ from the beginning, a fact which requires a great deal of computation, but only the correction term $\Delta\theta$, which requires much less computation. Even though the calculation of the correction term $\Delta\theta$ is not always simple, for the case of linear time-invariant systems a computationally simple expression for $\Delta\theta$ may be found, as is shown below. To this end, we return to the results of Section 8.2.2. Since the initial conditions $y(-1)$, $y(-2)$, ..., $y(-n)$ are not of interest in ON-LINE identification, they are dropped from the identification procedure. We are therefore left with the equation (8.15) which, for simplicity, will be stated in the rest of the chapter as follows:

$$\mathbf{\Phi}(N)\theta(N) = \mathbf{y}(N) \tag{8.21}$$

Working as in Section 8.2, we obtain the estimate for $\theta(N)$:

$$\theta(N) = [\mathbf{\Phi}^T(N)\mathbf{\Phi}(N)]^{-1}\mathbf{\Phi}^T(N)\mathbf{y}(N) \tag{8.22}$$

We may partition $\mathbf{\Phi}(N+1)$ and $\mathbf{y}(N+1)$ as follows:

$$\mathbf{\Phi}(N+1) = \begin{bmatrix} \mathbf{\Phi}(N) \\ \hline \boldsymbol{\phi}^T(N+1) \end{bmatrix} \quad \text{and} \quad \mathbf{y}(N+1) = \begin{bmatrix} \mathbf{y}(N) \\ \hline y(N+1) \end{bmatrix} = \begin{bmatrix} \mathbf{y}(N) \\ \hline y_{N+1} \end{bmatrix} \tag{8.23}$$

where y_{N+1} indicates the last measurement $y(N+1)$ in order to avoid any confusion between the vector $\mathbf{y}(N+1)$ and the data entry y_{N+1}. Then, the equation (8.21) for the $N+1$ measurements takes the form

$$\mathbf{\Phi}(N+1)\theta(N+1) = \mathbf{y}(N+1) \tag{8.24}$$

Hence

$$\theta(N+1) = [\mathbf{\Phi}^T(N+1)\mathbf{\Phi}(N+1)]^{-1}\mathbf{\Phi}^T(N+1)\mathbf{y}(N+1)$$
$$= [\mathbf{\Phi}^T(N)\mathbf{\Phi}(N) + \boldsymbol{\phi}(N+1)\boldsymbol{\phi}^T(N+1)]^{-1}[\mathbf{\Phi}^T(N)\mathbf{y}(N) + \boldsymbol{\phi}(N+1)y_{N+1}] \tag{8.25}$$

Equation (8.25) may take the form of equation (8.20) by using the following formula (known as the **matrix inversion lemma**):

$$[\mathbf{A} + \mathbf{BCD}]^{-1} = \mathbf{A}^{-1} - \mathbf{A}^{-1}\mathbf{B}[\mathbf{C}^{-1} + \mathbf{DA}^{-1}\mathbf{B}]^{-1}\mathbf{DA}^{-1} \tag{8.26}$$

The foregoing equation can be easily verified. To this end, the matrix $[\mathbf{A} + \mathbf{BCD}]$ is multiplied from the left on the right-hand side of equation (8.26) to yield

$$[\mathbf{A} + \mathbf{BCD}][\mathbf{A}^{-1} - \mathbf{A}^{-1}\mathbf{B}[\mathbf{C}^{-1} + \mathbf{DA}^{-1}\mathbf{B}]^{-1}\mathbf{DA}^{-1}]$$
$$= \mathbf{I} - \mathbf{B}[\mathbf{C}^{-1} + \mathbf{DA}^{-1}\mathbf{B}]^{-1}\mathbf{DA}^{-1} + \mathbf{BCDA}^{-1} - \mathbf{BCDA}^{-1}\mathbf{B}[\mathbf{C}^{-1} + \mathbf{DA}^{-1}\mathbf{B}]^{-1}\mathbf{DA}^{-1}$$
$$= \mathbf{I} + \mathbf{BCDA}^{-1} - \mathbf{BC}[\mathbf{C}^{-1} + \mathbf{DA}^{-1}\mathbf{B}][\mathbf{C}^{-1} + \mathbf{DA}^{-1}\mathbf{B}]^{-1}\mathbf{DA}^{-1}$$
$$= \mathbf{I} + \mathbf{BCDA}^{-1} - \mathbf{BCDA}^{-1} = \mathbf{I}$$

Using (8.26), we obtain

$$[\mathbf{\Phi}^T(N)\mathbf{\Phi}(N) + \boldsymbol{\phi}(N+1)\boldsymbol{\phi}^T(N+1)]^{-1} = [\mathbf{\Phi}^T(N)\mathbf{\Phi}(N)]^{-1}$$
$$- [\mathbf{\Phi}^T(N)\mathbf{\Phi}(N)]^{-1}\boldsymbol{\phi}(N+1)[1 + \boldsymbol{\phi}^T(N+1)[\mathbf{\Phi}^T(N)\mathbf{\Phi}(N)]^{-1}$$
$$\times \boldsymbol{\phi}(N+1)]^{-1}\boldsymbol{\phi}^T(N+1)[\mathbf{\Phi}^T(N)\mathbf{\Phi}(N)]^{-1} \quad (8.27)$$

Substituting (8.27) in (8.25) we obtain

$$\boldsymbol{\theta}(N+1) = [\mathbf{\Phi}^T(N)\mathbf{\Phi}(N)]^{-1}\mathbf{\Phi}^T(N)y(N) + [[\mathbf{\Phi}^T(N)\mathbf{\Phi}(N) + \boldsymbol{\phi}(N+1)\boldsymbol{\phi}^T(N+1)]^{-1}$$
$$- [\mathbf{\Phi}^T(N)\mathbf{\Phi}(N)]^{-1}]^{-1}\mathbf{\Phi}^T(N)\mathbf{y}(N)$$
$$+ [\mathbf{\Phi}^T(N)\mathbf{\Phi}(N) + \boldsymbol{\phi}(N+1)\boldsymbol{\phi}^T(N+1)]^{-1}\boldsymbol{\phi}(N+1)y_{N+1} \quad (8.28)$$

The following holds true:

$$[[\mathbf{\Phi}^T(N)\mathbf{\Phi}(N) + \boldsymbol{\phi}(N+1)\boldsymbol{\phi}^T(N+1)]^{-1} - [\mathbf{\Phi}^T(N)\mathbf{\Phi}(N)]^{-1}]\mathbf{\Phi}^T(N)\mathbf{y}(N)$$
$$= [\mathbf{\Phi}^T(N)\mathbf{\Phi}(N) + \boldsymbol{\phi}(N+1)\boldsymbol{\phi}^T(N+1)]^{-1}[[\mathbf{\Phi}^T(N)\mathbf{\Phi}(N)] - [\mathbf{\Phi}^T(N)\mathbf{\Phi}(N)$$
$$+ \boldsymbol{\phi}(N+1)\boldsymbol{\phi}^T(N+1)]][\mathbf{\Phi}^T(N)\mathbf{\Phi}(N)]^{-1}\mathbf{\Phi}^T(N)\mathbf{y}(N) = -[\mathbf{\Phi}^T(N)\mathbf{\Phi}(N)$$
$$+ \boldsymbol{\phi}(N+1)\boldsymbol{\phi}^T(N+1)]^{-1}\boldsymbol{\phi}(N+1)\boldsymbol{\phi}^T(N+1)[\mathbf{\Phi}^T(N)\mathbf{\Phi}(N)]^{-1}\mathbf{\Phi}^T(N)\mathbf{y}(N)$$
$$= -[\mathbf{\Phi}^T(N)\mathbf{\Phi}(N) + \boldsymbol{\phi}(N+1)\boldsymbol{\phi}^T(N+1)]^{-1}\boldsymbol{\phi}(N+1)\boldsymbol{\phi}^T(N+1)\boldsymbol{\theta}(N)$$

where, in deriving the final step in the foregoing equation, use was made of the expression $[\mathbf{\Phi}^T(N)\mathbf{\Phi}(N)]^{-1}\mathbf{\Phi}^T(N)\mathbf{y}(N) = \boldsymbol{\theta}(N)$. Using this result, equation (8.28) can be written as

$$\boldsymbol{\theta}(N+1) = \boldsymbol{\theta}(N) - [\mathbf{\Phi}^T(N)\mathbf{\Phi}(N) + \boldsymbol{\phi}(N+1)\boldsymbol{\phi}^T(N+1)]^{-1}\boldsymbol{\phi}(N+1)\boldsymbol{\phi}^T(N+1)\boldsymbol{\theta}(N)$$
$$+ [\mathbf{\Phi}^T(N)\mathbf{\Phi}(N) + \boldsymbol{\phi}(N+1)\boldsymbol{\phi}^T(N+1)]^{-1}\boldsymbol{\phi}(N+1)y_{N+1} \quad (8.29)$$

where use was made of equation (8.22). Finally, defining

$$\boldsymbol{\gamma}(N) = [\mathbf{\Phi}^T(N)\mathbf{\Phi}(N) + \boldsymbol{\phi}(N+1)\boldsymbol{\phi}^T(N+1)]^{-1}\boldsymbol{\phi}(N+1)$$
$$= [\mathbf{\Phi}^T(N+1)\mathbf{\Phi}(N+1)]^{-1}\boldsymbol{\phi}(N+1) \quad (8.30)$$

equation (8.29) is transformed as follows:

$$\boldsymbol{\theta}(N+1) = \boldsymbol{\theta}(N) + \boldsymbol{\gamma}(N)[y_{N+1} - \boldsymbol{\phi}^T(N+1)\boldsymbol{\theta}(N)] \quad (8.31)$$

Equation (8.31) is of the general form (8.20). Unfortunately, the determination of the vector $\boldsymbol{\gamma}(N)$ of equation (8.30) is numerically cumbersome since it requires inversion of the matrix $\mathbf{\Phi}^T(N+1)\mathbf{\Phi}(N+1)$ at every step. To overcome this difficulty, we define

$$\mathbf{P}(N) = [\mathbf{\Phi}^T(N)\mathbf{\Phi}(N)]^{-1} \quad (8.32)$$

Hence

$$\mathbf{P}(N+1) = [\mathbf{\Phi}^T(N+1)\mathbf{\Phi}(N+1)]^{-1} = [\mathbf{\Phi}^T(N)\mathbf{\Phi}(N) + \boldsymbol{\phi}(N+1)\boldsymbol{\phi}^T(N+1)]^{-1}$$

Using equation (8.27), the matrix $\mathbf{P}(N+1)$ can be written as

$$\mathbf{P}(N+1) = \mathbf{P}(N) - \mathbf{P}(N)\boldsymbol{\phi}(N+1)[1 + \boldsymbol{\phi}^T(N+1)\mathbf{P}(N)\boldsymbol{\phi}(N+1)]^{-1}\boldsymbol{\phi}^T(N+1)\mathbf{P}(N)$$
$$(8.33)$$

Equation (8.33) offers a convenient way of calculating $\mathbf{P}(N+1)$. It is noted that the term $1 + [\boldsymbol{\phi}^T(N+1)\mathbf{P}(N)\boldsymbol{\phi}(N+1)]$ is scalar, whereupon calculation of its inverse is simple. The calculation of the matrix $\mathbf{P}(N+1)$, according to the recursive formula (8.33), requires a matrix inversion only once at the beginning of the procedure to obtain

$$\mathbf{P}(N_0) = [\boldsymbol{\Phi}^T(N_0)\boldsymbol{\Phi}(N_0)]^{-1}$$

where N_0 is the starting number of data entries. Upon computing $\mathbf{P}(N+1)$, the vector $\boldsymbol{\gamma}(N)$ can easily be determined from equation (8.30), i.e. from the expression

$$\boldsymbol{\gamma}(N) = \mathbf{P}(N+1)\boldsymbol{\phi}(N+1) \tag{8.34}$$

In summary, the proposed recursive algorithm is given by the following theorem.

■ **THEOREM 8.3.1**   Suppose that $\boldsymbol{\theta}(N)$ is an estimate of the parameters of the nth-order system (8.12) for N data entries. Then, the estimate of the parameter vector $\boldsymbol{\theta}(N+1)$ for $N+1$ data entries is given by the expression

$$\boldsymbol{\theta}(N+1) = \boldsymbol{\theta}(N) + \boldsymbol{\gamma}(N)[y_{N+1} - \boldsymbol{\phi}^T(N+1)\boldsymbol{\theta}(N)] \tag{8.35}$$

where

$$\boldsymbol{\gamma}(N) = \mathbf{P}(N+1)\boldsymbol{\phi}(N+1) \tag{8.36a}$$

and the matrix $\mathbf{P}(N+1)$ is calculated from the recursive formula

$$\mathbf{P}(N+1) = \mathbf{P}(N) - \mathbf{P}(N)\boldsymbol{\phi}(N+1)[1 + \boldsymbol{\phi}^T(N+1)\mathbf{P}(N)\boldsymbol{\phi}(N+1)]^{-1}\boldsymbol{\phi}^T(N+1)\mathbf{P}(N) \tag{8.36b}$$

with initial conditions

$$\mathbf{P}(N_0) = [\boldsymbol{\Phi}^T(N_0)\boldsymbol{\Phi}(N_0)]^{-1} \tag{8.36c}$$

$$\boldsymbol{\theta}(N_0) = \mathbf{P}(N_0)\boldsymbol{\Phi}^T(N_0)\mathbf{y}(N_0) \tag{8.36d}$$

In Figure 8.1, a block diagram of the ON-LINE algorithm is given. It is clear that at every step the (known) inputs are y_{N+1}, $\boldsymbol{\phi}^T(N+1)$, $\boldsymbol{\theta}(N)$ and $\mathbf{P}(N)$ and the algorithm produces the new estimate $\boldsymbol{\theta}(N+1)$. The algorithm also produces the matrix $\mathbf{P}(N+1)$, which is used in the next step. In Figure 8.2, a more detailed block diagram of the ON-LINE algorithm is given.

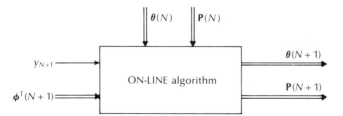

Figure 8.1   Block diagram presentation of the ON-LINE algorithm.

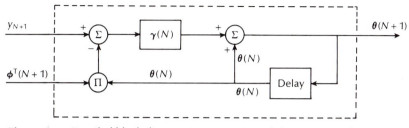

Figure 8.2 Detailed block diagram representation of the ON-LINE algorithm.

■ **REMARK 8.3.1**   In equation (8.35) we observe that the correction term $\Delta\boldsymbol{\theta}$, defined in (8.20), is proportional to the difference $y_{N+1} - \boldsymbol{\phi}^T(N+1)\boldsymbol{\theta}(N)$, where y_{N+1} is the new data entry and $\boldsymbol{\phi}^T(N+1)\boldsymbol{\theta}(N)$ is an estimate of this new entry, based on equation (8.24) and using the latest estimate of the system parameters. Had there not been any error in the measurements, the expected value $\boldsymbol{\phi}^T(N+1)\boldsymbol{\theta}(N)$ of y_{N+1} would have been equal to the respective measurement value y_{N+1} and the difference between them would have been zero, in which case $\boldsymbol{\theta}(N+1) = \boldsymbol{\theta}(N)$. In other words, when there is no error in the data entries, a new entry does not add any new information, so the new estimate of $\boldsymbol{\theta}$ has exactly the same value as that of the previous estimate. Finally, the term $\boldsymbol{\gamma}(N)$ may be considered as a weighting factor of the difference term $y_{N+1} - \boldsymbol{\phi}^T(N+1)\boldsymbol{\theta}(N)$.

■ **EXAMPLE 8.3.1**   Consider the simple case of a resistive network given in Figure 8.3. Estimate the parameter a when $u(k)$ is the unit step function, i.e. when $u(k) = 1$, for $k = 1, 2, 3, \dots$.

■ **SOLUTION**   The difference equation for the system is $y(k) = au(k)$. Since $u(k) = 1$, for all k, it follows that

$$y(k) = a + e(k) \quad k = 1, 2, \dots, N$$

To start with, we will solve the problem using the following very simple technique. We define the cost function

$$J = \sum_{k=1}^{N} e^2(k) = \sum_{k=1}^{N} [y(k) - a]^2$$

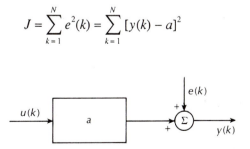

Figure 8.3 Simple case of a resistive network.

Then

$$\frac{\partial J}{\partial a} = -2 \sum_{k=1}^{N} [y(k) - a] = 0$$

The foregoing equation can be written as

$$\sum_{k=1}^{N} [y(k) - a] = \left[\sum_{k=1}^{N} y(k) \right] - Na = 0$$

Hence, the estimate $a(N)$ of the parameter a is

$$a(N) = \frac{1}{N} \sum_{k=1}^{N} y(k)$$

The expression above for $a(N)$ is the mean value of the measurements $y(k)$, as was anticipated. Assume now that we have a new measurement. Then

$$a(N + 1) = \frac{1}{N + 1} \sum_{k=1}^{N+1} y(k)$$

$$= \frac{1}{N} \sum_{k=1}^{N} y(k) - \frac{1}{N} \sum_{k=1}^{N} y(k) + \frac{1}{N + 1} \sum_{k=1}^{N} [y(k) + y(N + 1)]$$

$$= \frac{1}{N} \sum_{k=1}^{N} y(k) + \frac{N - (N + 1)}{N(N + 1)} \sum_{k=1}^{N} y(k) + \frac{1}{N + 1} y(N + 1)$$

$$= a(N) + \frac{1}{N + 1} [y_{N+1} - a(N)]$$

The foregoing equation represents the recursive algorithm (8.35), where $\boldsymbol{\phi}^{\mathrm{T}}(N + 1) = 1$, $y_{N+1} = y(N + 1)$ and $\boldsymbol{\gamma}(N) = 1/(N + 1)$.

Now, we will solve the problem using the method presented in the present section. As the first step, we formulate the equation

$$\boldsymbol{\Phi}(N)\boldsymbol{\theta}(N) = \mathbf{y}(N)$$

where $\boldsymbol{\theta}(N) = a(N)$, $\boldsymbol{\Phi}(N) = [1 \ 1 \ \dots \ 1]^{\mathrm{T}}$, and $\mathbf{y}(N) = [y(1) \ y(2) \ \dots \ y(N)]^{\mathrm{T}}$. The solution of the canonical equation is

$$a(N) = [\boldsymbol{\Phi}^{\mathrm{T}}(N)\boldsymbol{\Phi}(N)]^{-1}\boldsymbol{\Phi}^{\mathrm{T}}(N)\mathbf{y}(N) = N^{-1}\boldsymbol{\Phi}^{\mathrm{T}}(N)\mathbf{y}(N) = N^{-1} \sum_{k=1}^{N} y(k)$$

We observe that $\mathbf{P}(N) = N^{-1}$. Thus equation (8.36b) yields

$$\mathbf{P}(N + 1) = N^{-1} - N^{-1}[1 + N^{-1}]N^{-1} = \frac{1}{N + 1}$$

whereas equation (8.36a) gives

$$\gamma(N) = \mathbf{P}(N+1)\boldsymbol{\phi}(N+1) = \frac{1}{N+1}$$

Hence, equation (8.35) becomes

$$a(N+1) = a(N) + \frac{1}{N+1}\, [y_{N+1} - a(N)]$$

Figure 8.4 shows the block diagram of the ON-LINE algorithm for the present example.

■ **EXAMPLE 8.3.2** Consider the system of Example 8.3.1 where the input is not a unit step function but any other type of function. Estimate the parameter a.

■ **SOLUTION** The difference equation is $y(k) = au(k)$. We formulate the equation

$$\boldsymbol{\Phi}(N)\boldsymbol{\theta}(N) = \mathbf{y}(N)$$

where

$$\boldsymbol{\theta}(N) = a(N), \quad \boldsymbol{\Phi}(N) = [u(1)\ u(2)\ \dots\ u(N)]^{\mathrm{T}} \quad \text{and} \quad \mathbf{y}(N) = [y(1)\ y(2)\ \dots\ y(N)]^{\mathrm{T}}$$

The solution of the canonical equation is

$$a(N) = [\boldsymbol{\Phi}^{\mathrm{T}}(N)\boldsymbol{\Phi}(N)]^{-1}\boldsymbol{\Phi}^{\mathrm{T}}(N)\mathbf{y}(N) = \left[\sum_{k=1}^{N} u^2(k)\right]^{-1} \sum_{k=1}^{N} u(k)\phi(k)$$

We observe that

$$\mathbf{P}(N) = \left[\sum_{k=1}^{N} u^2(k)\right]^{-1}$$

Hence, equation (8.36b) becomes

$$\begin{aligned}
\mathbf{P}(N+1) &= \mathbf{P}(N) - \mathbf{P}(N)u(N+1)[1 + u^2(N+1)\mathbf{P}(N)]^{-1}u(N+1)\mathbf{P}(N) \\
&= [[1 + u^2(N+1)\mathbf{P}(N)] - u^2(N+1)\mathbf{P}(N)]\mathbf{P}(N)[1 + u^2(N+1)\mathbf{P}(N)]^{-1} \\
&= \mathbf{P}(N)[1 + u^2(N+1)\mathbf{P}(N)]^{-1}
\end{aligned}$$

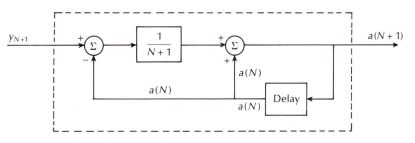

Figure 8.4 Block diagram of the ON-LINE algorithm of Example 8.3.1.

Therefore

$$a(N+1) = a(N) + \gamma(N)[y_{N+1} - \phi^T(N+1)a(N)]$$
$$= a(N) + P(N)u(N+1)[1 + u^2(N+1)P(N)]^{-1}[y_{N+1} - u(N+1)a(N)]$$

If $u(k) = 1$, $k = 1, 2, \ldots$, the results of Example 8.3.1 can readily be derived as a special case of the foregoing results. Indeed, since $P(N) = N^{-1}$ and $u(k) = 1$, $k = 1, 2, \ldots$, the above expression for $a(N+1)$ becomes:

$$a(N+1) = a(N) + N^{-1}[1 + N^{-1}][y_{N+1} - a(N)]$$

$$= a(N) + \frac{1}{N+1}[y_{N+1} - a(N)]$$

■ EXAMPLE 8.3.3   Consider a discrete-time system which is described by the difference equation:

$$y(k+1) + ay^2(k) = bu(k)$$

If the input $u(k)$ is a unit step sequence, it results in the following output measurements:

k	0	1	2	3	4	5	6
$y(k)$	0	0.01	1.05	1.69	3.02	7.4	39.3

(a) Determine an estimate of the parameters a and b.
(b) Assume that a new output measurement $y(7) = 1082$ is available. Find the new estimates for a and b using the recursive formula.

■ SOLUTION
(a) For $k = 1, 2, 3, 4, 5$ we have

$$y(2) + ay^2(1) = bu(1) + e(2)$$
$$y(3) + ay^2(2) = bu(2) + e(3)$$
$$y(4) + ay^2(3) = bu(3) + e(4)$$
$$y(5) + ay^2(4) = bu(4) + e(5)$$
$$y(6) + ay^2(5) = bu(5) + e(6)$$

where $e(k)$ is the measurement error at time k. The above equations can be written compactly as follows:

$$\mathbf{y}(N) = \mathbf{\Phi}(N)\boldsymbol{\theta}(N) + \mathbf{e}$$

where

$$\mathbf{y}^{\mathrm{T}}(N) = [y(2) \quad y(3) \quad \cdots \quad y(6)] = [1.05 \quad 1.69 \quad 3.02 \quad 7.4 \quad 39.3]$$

$$\Phi(N) = \begin{bmatrix} -y^2(1) & u(1) \\ -y^2(2) & u(2) \\ -y^2(3) & u(3) \\ -y^2(4) & u(4) \\ -y^2(5) & u(5) \end{bmatrix} = \begin{bmatrix} -0.0001 & 1 \\ -1.1025 & 1 \\ -2.8561 & 1 \\ -9.1204 & 1 \\ -54.76 & 1 \end{bmatrix}, \quad \theta(N) = \begin{bmatrix} a \\ b \end{bmatrix} \quad \text{and} \quad \mathbf{e} = \begin{bmatrix} e(2) \\ e(3) \\ e(4) \\ e(5) \\ e(6) \end{bmatrix}$$

Using the above results, we obtain

$$\Phi^{\mathrm{T}}(N)\Phi(N) = \begin{bmatrix} 3091.12 & -67.839 \\ -67.839 & 5 \end{bmatrix}, \quad \Phi^{\mathrm{T}}(N)\mathbf{y}(N) = \begin{bmatrix} -2230.047\,71 \\ 52.46 \end{bmatrix}$$

$$[\Phi^{\mathrm{T}}(N)\Phi(N)]^{-1} = \begin{bmatrix} 0.0005 & 0.0063 \\ 0.0063 & 0.2848 \end{bmatrix}$$

The optimal estimate of the parameters a and b is

$$\theta(N) = \begin{bmatrix} a \\ b \end{bmatrix} = [\Phi^{\mathrm{T}}(N)\Phi(N)]^{-1}\Phi^{\mathrm{T}}(N)\mathbf{y}(N) = \begin{bmatrix} -0.7845 \\ 0.891 \end{bmatrix}$$

(b) Using the recursive equation (8.35) we have

$$\theta(7) = \theta(6) + \gamma(6)[y_7 - \phi^{\mathrm{T}}(7)\theta(6)]$$

where

$$\theta(6) = \begin{bmatrix} -0.7845 \\ 0.891 \end{bmatrix}, \quad y_7 = y(7) = 1082, \quad \phi^{\mathrm{T}}(7) = [-1544.5 \quad 1]$$

$$\gamma(6) = \mathbf{P}(7)\phi(7) = \begin{bmatrix} -0.0006 \\ -0.008 \end{bmatrix}$$

Hence

$$\theta(7) = \begin{bmatrix} a \\ b \end{bmatrix} = \begin{bmatrix} -0.7062 \\ 1.9354 \end{bmatrix}$$

8.4 Problems

1. A system is described by the following difference equation:

$$y(k + 3) + a_1 y(k + 2) + a_2 y(k + 1) + a_3 y(k) = u(k)$$

$$y(k+3) + a_1 y(k+2) + a_2 y(k+1) + a_3 y(k) = u(k)$$

If the input to the system is the impulse function $u(k) = \delta(k)$, it results in the following output measurements:

k	0	1	2	3	4	5	6
$y(k)$	1	0.2	−0.6	−1.2	−1.6	−1.7	−1.6

Estimate the parameters a_1, a_2 and a_3.

2. Estimate the parameters a, b and ω of a system described by the difference equation

$$y(k) + \omega^2 y(k-2) = au(k-1) + bu(k-2)$$

given that the following measurements are available:

k	1	2	3	4	5
$u(k)$	1	1	1	1	1
$y(k)$	0	1/2	1/3	1/4	1/5

Are the estimates unique? Explain your results.

3. Estimate the parameters a, b and c of the system

$$y(k) + ay(k-1) = bu^2(k) + cu(k-1)$$

given the information shown in Table 8.5 relating its input and output.

Make use of all the information provided by the table. The resulting identification must be unique.

4. Harmonic retrieval in digital signal processing involves parameter estimation (frequency, amplitude, phase) of unknown sinusoids contaminated with noise. Consider, for example, the signal

$$y(k) = a \cos(\omega k + \phi) + e(k)$$

Table 8.5 Output sequences $y(k)$ for several inputs.

Input	$y(1)$	$y(2)$	$y(3)$	$y(4)$
$u(k) = k$	0	0	1	1.5
$u(k) = k^2$	0	−0.5	1	2.5
$u(k) = 1$	0	0.5	1	4

In practice, however, it very often happens that not all state variables of a system are accessible to measurement. This obstacle can be circumvented if a mathematical model for the system is available, in which case it is possible to estimate (reconstruct) the state vector [4], [5], [9–12].

We will present two methods for estimating the state vector. The first method is the direct estimation of the state vector based on input and output measurements (Section 6.4.1), while the second is the estimation of the state vector using a dynamic system called a state observer (Sections 6.4.2–6.4.4). The system model is assumed known and of the following form:

$$\mathbf{x}(k+1) = \mathbf{A}\mathbf{x}(k) + \mathbf{B}\mathbf{u}(k), \mathbf{x} \in \mathbb{R}^n \quad \text{and} \quad \mathbf{u} \in \mathbb{R}^m \tag{6.28a}$$

$$\mathbf{y}(k) = \mathbf{C}\mathbf{x}(k), \mathbf{y} \in \mathbb{R}^p \tag{6.28b}$$

6.4.1 *Direct state vector estimation*

The direct state vector estimate will be obtained on the basis of the measured output sequences $\mathbf{y}(k)$, $\mathbf{y}(k-1)$, … and input sequences $\mathbf{u}(k)$, $\mathbf{u}(k-1)$, …. To facilitate the presentation, we assume that the system (6.28) has only one output. Then, using (6.28), we obtain

$$y(k-n+1) = \mathbf{c}^T\mathbf{x}(k-n+1)$$
$$y(k-n+2) = \mathbf{c}^T\mathbf{A}\mathbf{x}(k-n+1) + \mathbf{c}^T\mathbf{B}\mathbf{u}(k-n+1)$$
$$\vdots$$
$$y(k) = \mathbf{c}^T\mathbf{A}^{n-1}\mathbf{x}(k-n+1) + \mathbf{c}^T\mathbf{A}^{n-2}\mathbf{B}\mathbf{u}(k-n+1) + \cdots + \mathbf{c}^T\mathbf{B}\mathbf{u}(k-1)$$

These equations can be written more compactly as

$$\begin{bmatrix} y(k-n+1) \\ y(k-n+2) \\ \vdots \\ y(k) \end{bmatrix} = \mathbf{R}\mathbf{x}(k-n+1) + \mathbf{\Omega} \begin{bmatrix} \mathbf{u}(k-n+1) \\ \mathbf{u}(k-n+2) \\ \vdots \\ \mathbf{u}(k-1) \end{bmatrix} \tag{6.29}$$

where

$$\mathbf{R} = \begin{bmatrix} \mathbf{c}^T \\ \mathbf{c}^T\mathbf{A} \\ \vdots \\ \mathbf{c}^T\mathbf{A}^{n-1} \end{bmatrix} \quad \text{and} \quad \mathbf{\Omega} = \begin{bmatrix} \mathbf{0} & \mathbf{0} & \cdots & \mathbf{0} \\ \mathbf{c}^T\mathbf{B} & \mathbf{0} & \cdots & \mathbf{0} \\ \vdots & \vdots & & \vdots \\ \mathbf{c}^T\mathbf{A}^{n-2}\mathbf{B} & \mathbf{c}^T\mathbf{A}^{n-3}\mathbf{B} & \cdots & \mathbf{c}^T\mathbf{B} \end{bmatrix} \tag{6.30}$$

where $\mathbf{R}$ is the well-known observability matrix of the system (6.28). If the system (6.28) is observable, then the matrix $\mathbf{R}$ is non-singular and the state vector

$\mathbf{x}(k - n + 1)$ can consequently be estimated by using equation (6.29) as follows:

$$\mathbf{x}(k - n + 1) = \mathbf{R}^{-1}\begin{bmatrix} y(k - n + 1) \\ y(k - n + 2) \\ \vdots \\ y(k) \end{bmatrix} - \mathbf{R}^{-1}\Omega\begin{bmatrix} \mathbf{u}(k - n + 1) \\ \mathbf{u}(k - n + 2) \\ \vdots \\ \mathbf{u}(k - 1) \end{bmatrix}$$

Repeated use of equation (6.28) leads to

$$\mathbf{x}(k - n + 2) = \mathbf{A}\mathbf{x}(k - n + 1) + \mathbf{B}\mathbf{u}(k - n + 1)$$

$$= \mathbf{A}\left[\mathbf{R}^{-1}\begin{bmatrix} y(k - n + 1) \\ y(k - n + 2) \\ \vdots \\ y(k) \end{bmatrix} - \mathbf{R}^{-1}\Omega\begin{bmatrix} \mathbf{u}(k - n + 1) \\ \mathbf{u}(k - n + 2) \\ \vdots \\ \mathbf{u}(k - 1) \end{bmatrix}\right] + \mathbf{B}\mathbf{u}(k - n + 1)$$

$$= \mathbf{A}\mathbf{R}^{-1}\begin{bmatrix} y(k - n + 1) \\ y(k - n + 2) \\ \vdots \\ y(k) \end{bmatrix} - \mathbf{A}\mathbf{R}^{-1}\Omega\begin{bmatrix} \mathbf{u}(k - n + 1) \\ \mathbf{u}(k - n + 2) \\ \vdots \\ \mathbf{u}(k - 1) \end{bmatrix} + \mathbf{B}\mathbf{u}(k - n + 1)$$

Furthermore

$$\mathbf{x}(k - n + 3) = \mathbf{A}\mathbf{x}(k - n + 2) + \mathbf{B}\mathbf{u}(k - n + 2)$$

$$= \mathbf{A}^2\mathbf{R}^{-1}\begin{bmatrix} y(k - n + 1) \\ y(k - n + 2) \\ \vdots \\ y(k) \end{bmatrix} - \mathbf{A}^2\mathbf{R}^{-1}\Omega\begin{bmatrix} \mathbf{u}(k - n + 1) \\ \mathbf{u}(k - n + 2) \\ \vdots \\ \mathbf{u}(k - 1) \end{bmatrix} + \mathbf{A}\mathbf{B}\mathbf{u}(k - n + 1) + \mathbf{B}\mathbf{u}(k - n + 2)$$

and so on. Finally, we arrive at the desired expression

$$\mathbf{x}(k) = \mathbf{A}^{n-1}\mathbf{R}^{-1}\begin{bmatrix} y(k - n + 1) \\ y(k - n + 2) \\ \vdots \\ y(k) \end{bmatrix} + \Psi\begin{bmatrix} \mathbf{u}(k - n + 1) \\ \mathbf{u}(k - n + 2) \\ \vdots \\ \mathbf{u}(k - 1) \end{bmatrix} \qquad (6.31)$$

where

$$\Psi = [\mathbf{A}^{n-2}\mathbf{B} \ \ \mathbf{A}^{n-3}\mathbf{B} \ \cdots \ \mathbf{B}] - \mathbf{A}^{n-1}\mathbf{R}^{-1}\Omega \qquad (6.32)$$

From equation (6.31) it is clear that the state vector is given as a linear combination of $y(k)$, $y(k-1), \ldots, y(k-n+1)$ and $\mathbf{u}(k)$, $\mathbf{u}(k-1), \ldots, \mathbf{u}(k-n+1)$. Equation (6.31) shows that the estimation of $\mathbf{x}(k)$ will be achieved after n steps at most. For this reason, an observer of the form of equation (6.31) is called a **deadbeat observer**. The following theorem holds.

■ **THEOREM 6.4.1**   The state vector of system (6.28) can be estimated on the basis of input and output measurements, if and only if the system is observable.

■ **EXAMPLE 6.4.1**   Consider the double-integrator system of Example 6.3.1. A discrete-time state-space representation of this system, derived in Example 6.3.1, has the form

$$\mathbf{x}[(k+1)T] = \mathbf{A}(T)\mathbf{x}(kT) + \mathbf{b}(T)u(kT), \quad y(kT) = \mathbf{c}^\mathsf{T}\mathbf{x}(kT)$$

where

$$\mathbf{A}(T) = \begin{bmatrix} 1 & T \\ 0 & 1 \end{bmatrix}, \quad \mathbf{b}(T) = \begin{bmatrix} T^2/2 \\ T \end{bmatrix} \quad \text{and} \quad \mathbf{c} = \begin{bmatrix} 1 \\ 0 \end{bmatrix}$$

Estimate the state variables in terms of $u(kT)$ and $y(kT)$.

■ **SOLUTION**   The observability matrix $\mathbf{R}$ is

$$\mathbf{R} = \begin{bmatrix} \mathbf{c}^\mathsf{T} \\ \mathbf{c}^\mathsf{T}\mathbf{A} \end{bmatrix} = \begin{bmatrix} 1 & 0 \\ 1 & T \end{bmatrix} \quad \text{and} \quad \mathbf{R}^{-1} = \frac{1}{T}\begin{bmatrix} T & 0 \\ -1 & 1 \end{bmatrix} = \begin{bmatrix} 1 & 0 \\ -1/T & 1/T \end{bmatrix}$$

We also have

$$\mathbf{R}^{-1}\Omega = \mathbf{R}^{-1}\begin{bmatrix} 0 \\ \mathbf{c}^\mathsf{T}\mathbf{b} \end{bmatrix} = \begin{bmatrix} 1 & 0 \\ -1/T & 1/T \end{bmatrix}\begin{bmatrix} 0 \\ T^2/2 \end{bmatrix} = \begin{bmatrix} 0 \\ T/2 \end{bmatrix}$$

and therefore

$$\Psi = \mathbf{B} - \mathbf{A}\mathbf{R}^{-1}\Omega = \begin{bmatrix} T^2/2 \\ T \end{bmatrix} - \begin{bmatrix} 1 & T \\ 0 & 1 \end{bmatrix}\begin{bmatrix} 0 \\ T/2 \end{bmatrix} = \begin{bmatrix} 0 \\ T/2 \end{bmatrix}$$

We further have

$$\mathbf{A}^{n-1}\mathbf{R}^{-1} = \mathbf{A}\mathbf{R}^{-1} = \begin{bmatrix} 1 & T \\ 0 & 1 \end{bmatrix}\begin{bmatrix} 1 & 0 \\ -1/T & 1/T \end{bmatrix} = \begin{bmatrix} 0 & 1 \\ -1/T & 1/T \end{bmatrix}$$

Hence, equation (6.31) finally becomes

$$\mathbf{x}(kT) = \begin{bmatrix} 0 & 1 \\ -1/T & 1/T \end{bmatrix}\begin{bmatrix} y(kT-T) \\ y(kT) \end{bmatrix} + \begin{bmatrix} 0 \\ T/2 \end{bmatrix}u(kT-T)$$

or

$$\mathbf{x}(kT) = \begin{bmatrix} 1 \\ 1/T \end{bmatrix}y(kT) + \begin{bmatrix} 0 \\ -1/T \end{bmatrix}y(kT-T) + \begin{bmatrix} 0 \\ T/2 \end{bmatrix}u(kT-T)$$

From the output equation $y(kT) = \mathbf{c}^{\mathsf{T}}\mathbf{x}(kT) = [1,0]\mathbf{x}(kT) = x_1(kT)$ it follows that $x_1(kT)$ is directly measured since $x_1(kT) = y(kT)$. The state variable $x_2(kT)$ is obtained by using the foregoing expression for $\mathbf{x}(kT)$ which gives

$$x_2(kT) = \frac{y(kT) - y(kT - T)}{T} + \frac{T}{2}u(kT - T)$$

The direct state vector estimate has the advantage of determining the state variables in n steps, at most. The drawback of this method is its susceptibility to disturbances. For this reason it is important to have other alternative methods. A widely known alternative method is the state vector estimation (or reconstruction) using an observer, first proposed by Luenberger [10–12], which we present in the following section.

6.4.2 *State vector reconstruction using a Luenberger observer*

Consider the system (6.28). Assume that state vector $\mathbf{x}(k)$ is given approximately by the state vector $\hat{\mathbf{x}}(k)$ of the model

$$\hat{\mathbf{x}}(k + 1) = \hat{\mathbf{A}}\hat{\mathbf{x}}(k) + \hat{\mathbf{B}}\mathbf{u}(k) + \mathbf{K}\mathbf{y}(k) \tag{6.33}$$

where $\hat{\mathbf{x}} \in \mathbb{R}^n$ and the matrices $\hat{\mathbf{A}}$, $\hat{\mathbf{B}}$ and $\mathbf{K}$ are unknown. System (6.33) is called the **state observer** of system (6.28). In particular, the observer is a dynamic system having two inputs, the input vector $\mathbf{u}(k)$ and the output vector $\mathbf{y}(k)$ of the initial system (6.28) (see Figure 6.3) and whose matrices $\hat{\mathbf{A}}$, $\hat{\mathbf{B}}$ and $\hat{\mathbf{K}}$ are such that $\hat{\mathbf{x}}(k)$ is as close as possible to $\mathbf{x}(k)$. In cases where $\hat{\mathbf{x}}(k)$ and $\mathbf{x}(k)$ are of equal dimension, then the observer is referred to as a **full-order** observer. This case is studied in the present section. When the dimension of $\hat{\mathbf{x}}(k)$ is smaller than that of $\mathbf{x}(k)$, then the observer is referred to as a **reduced-order** observer which is studied in Section 6.4.3.

Define the state error

$$\mathbf{e}(k) = \mathbf{x}(k) - \hat{\mathbf{x}}(k) \tag{6.34}$$

The definition of the problem of designing the observer (6.33) is to determine appropriate matrices $\hat{\mathbf{A}}$, $\hat{\mathbf{B}}$ and $\mathbf{K}$, such that the error $\mathbf{e}(k)$ tends to zero as fast as possible.

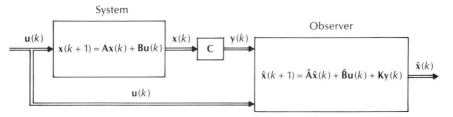

Figure 6.3 Simplified presentation of the system (6.28) and the observer (6.33).

ϕ is to assume that this signal is the output of a linear system which is excited by an impulse and $e(k)$ is the output measurement noise (see Figure 8.5).

(a) Determine the particular form of the difference equation

$$w(k) + a_1 w(k-1) + \cdots + a_n w(k-n) = b_0 u(k) + b_1 u(k-1) + \cdots + b_m u(k-m)$$

which describes $H(z)$.

(b) Write the equations for the determination of the unknown parameters a_i and b_i using the least-squares method, provided that $u(0), \ldots, u(N)$, $y(0), \ldots, y(N)$ are known.

(c) Indicate how these parameters can be used to obtain the original parameters a, ω and ϕ given that $0 < \omega < \pi/2$.

5. Let a system be described by

$$y(k+2) = ay(k+1) + y(k) + bu(k)$$

(a) Estimate the parameters a and b given the following measurements:

k	0	1	2	3	4
$u(k)$	1	1	1	1	1
$y(k)$	1	0.9	0.9	0.8	0.6

(b) What are the new parameter estimates in view of the additional measurements $u(5) = 1$ and $y(5) = 0.4$?

6. The output of a given system $H(z)$ is compared with the output of a known system $H_2(z)$, as shown in Figure 8.6.

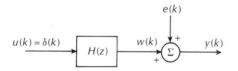

Figure 8.5 Linear system and measurement noise.

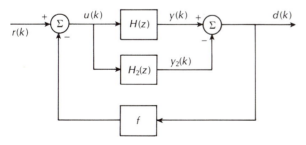

Figure 8.6 Closed-loop configuration.

It is known that the difference equation that describes $H(z)$ is of the form

$$y(k) = au(k) + bu(k-1) + cu(k-2)$$

The equation of $H_2(z)$ is

$$y(k) = u(k) + 2u(k-1) - 3u(k-2)$$

and the feedback coefficient $f = 1$. Find the unknown parameters a, b and c given the following measurements:

k	0	1	2	3	4	5
$r(k)$	1	1	1	1	1	1
$d(k)$	0	0	−0.5	−0.8	−1.1	−1.4

7. Let θ be the parameter estimate obtained from the identification of a linear system. Then, an estimate of the output is $\hat{y} = \phi^T(k)\theta$. Show that the error $e(k) = y(k) - \hat{y}(k)$ in the output estimation is orthogonal to $\hat{y}(k)$ in the following sense:

$$\sum_k e(k)\hat{y}(k) = 0$$

Also, show that

$$\sum_k y(k)\hat{y}(k) \geq 0$$

8. If a matrix Φ has more rows than columns, then the matrix $\Phi^* = (\Phi^T\Phi)^{-1}\Phi^T$ is called the pseudo-inverse of Φ because it has the property $\Phi^*\Phi = I$. (In particular, if Φ is square and invertible, then $\Phi^* = \Phi^{-1}$.)

(a) Show that a vector $\mathbf{x}$ is an eigenvector of $\Phi\Phi^T$ if and only if it is an eigenvector of $(\Phi^*)^T\Phi^*$.

(b) Show that a vector $\mathbf{x}$ is an eigenvector of $\Phi^T\Phi$ if and only if it is an eigenvector of $\Phi^*(\Phi^*)^T$.

9. Consider a discrete-time system which is described in state space by the equation

$$\mathbf{x}(k+1) = \mathbf{A}\mathbf{x}(k) + \mathbf{b}u(k)$$

where $\mathbf{x}(k) \in \mathbb{R}^2$ and $u(k) \in \mathbb{R}^1$. If

$$u(k) = \begin{cases} 0 & \forall\, k < 0 \\ 1 & \forall\, k \geq 0 \end{cases}$$

we have the following measurements:

$$\mathbf{x}(0) = \begin{bmatrix} 0 \\ 1 \end{bmatrix}, \quad \mathbf{x}(1) = \begin{bmatrix} 1 \\ 1.5 \end{bmatrix}, \quad \mathbf{x}(2) = \begin{bmatrix} 1.2 \\ 2 \end{bmatrix} \quad \text{and} \quad \mathbf{x}(3) = \begin{bmatrix} 1.5 \\ 7 \end{bmatrix}$$

(a) Find an estimate of the matrix **A** and vector **b**.

(b) Assume that a new measurement

$$\mathbf{x}(4) = \begin{bmatrix} 4 \\ 12 \end{bmatrix}$$

is available. Find the new estimates for the matrix **A** and the vector **b**.

10. Consider the system

$$\mathbf{y}(k) = \begin{bmatrix} y_1(k) \\ y_2(k) \end{bmatrix} = \begin{bmatrix} a_{11} & a_{12} \\ a_{12} & a_{22} \end{bmatrix} \begin{bmatrix} 1 \\ u_2(k) \end{bmatrix} + \begin{bmatrix} y_1(k-1) \\ y_2(k-1) \end{bmatrix}$$

The following measurements are given:

k	1	2	3	4	5	6
$y_1(k)$	1	0.5	3	0.5	4.5	4
$y_2(k)$	1	−1.5	−2	−5.5	−5	−7.5
$u_2(k)$	1	−1	2	−2	1	−1

Estimate the parameters a_{11}, a_{12} and a_{22}.

8.5 References

Books

[1] K.J. Åström and B. Wittenmark, *Computer Controlled Systems*, Prentice Hall, Englewood Cliffs, New Jersey, 1984.

[2] C.I. Byrnes and A. Lindquist (eds), *Modelling, Identification and Robust Control*, North-Holland, Amsterdam, 1986.

[3] R. Calaba and K. Spingarn, *Control Identification and Input Optimization*, Plenum Press, New York, 1982.

[4] R.H. Cannon Jr, *Dynamics of Physical Systems*, McGraw-Hill, New York, 1967.

[5] P. Eykhoff, *System Identification. Parameter and State Estimation*, John Wiley, New York, 1977.

[6] P. Eykhoff, *Trends and Progress in System Identification*, Pergamon Press, Oxford, 1981.

[7] G.C. Goodwin and R.L. Payne, *Dynamic System Identification: Experimental Design and Data Analysis*, Academic Press, New York, 1977.

[8] T.C. Hsia, *System Identification*, Lexington, Boston, Massachusetts, 1977.

[9] L. Ljung and T. Söderström, *Theory and Practice of Recursive Identification*, MIT Press, Cambridge, Massachusetts, 1983.

[10] L. Ljung, *System Identification – Theory for the User*, Prentice Hall, Englewood Cliffs, New Jersey, 1987.

[11] R.K. Mehra and D.G. Lainiotis (eds), *System Identification. Advances and Case Studies*, Academic Press, New York, 1976.

[12] J.M. Mendel, *Discrete Techniques of Parameter Estimation*, Marcel Dekker, New York, 1973.

[13] M. Milanese, R. Tempo and A. Vicino (eds), *Robustness in Identification and Control*, Plenum Press, New York, 1989.

[14] J.P. Norton, *An Introduction to Identification*, Academic Press, London, 1986.

[15] M.S. Rajbman and V.M. Chadeer, *Identification of Industrial Processes*, North Holland, Amsterdam, 1980.

[16] A.D. Sage and J.L. Melsa, *System Identification*, Academic Press, New York, 1971.

[17] N.K. Sinha and B. Kuszta, *Modelling and Identification of Dynamic Systems*, Van Nostrand Reinhold, New York, 1983.

[18] T. Söderström and P. Stoica, *System Identification*, Prentice Hall, London, 1989.

[19] H.W. Sorenson, *Parameter Estimation. Principles and Problems*, Marcel Dekker, New York, 1980.

[20] C.B. Speedy, R.F. Brown and G.C. Goodwin, *Control Theory: Identification and Optimal Control*, Oliver & Boyd, Edinburgh, 1970.

[21] E. Walter, *Identification of State Space Models*, Springer Verlag, Berlin, 1982.

Papers

[22] K.J. Åström and P.E. Eykhoff, 'System Identification – A Survey', *Automatica*, Vol. 7, pp. 123–162, 1971.

[23] 'Special Issue on Identification and System Parameter Estimation', *Automatica*, Vol. 17, 1981.

9

Adaptive control

9.1 Introduction

An **adaptive control system** is a system which automatically adjusts on-line the parameters of its controller, so as to maintain a satisfactory level of performance when the parameters of the system under control are unknown and/or time varying.

Generally speaking, the performance of a system is affected either by external perturbations or by parameter variations. Closed-loop systems involving feedback (top portion of Figure 9.1) are used in order to cope with external perturbations. In this case, the measured value of the output $y(kT)$ is compared with the desired value of the reference signal $\omega(kT)$. The difference $e(kT)$ between the two signals is applied to the controller, which in turn provides the appropriate control action $u(kT)$ to the plant under control. A somewhat similar approach can be used when parametric uncertainties (unknown parameters) appear in the system model of Figure 9.1. In this case the controller involves adjustable parameters. A performance index is defined reflecting the actual performance of the system. This index is then measured and compared with a desired performance index (see Figure 9.1) and the error between the two performance indices activates the controller adaptation mechanism. This mechanism is suitably designed so as to adjust the parameters of the controller (or modify the input signals in a more general case), so that the error between the two performance indices lies within acceptable bounds.

Closer examination of Figure 9.1 reveals that two closed loops are involved: the 'inner' feedback closed loop, whose controller involves adjustable parameters (upper portion of the figure), and the supplementary 'outer' feedback closed loop (or adaptation loop) which involves the performance indices and the adaptation mechanism (lower portion of the figure). The role of the adaptation loop is to find appropriate estimates for the adjustable controller parameters at each sampling instant.

It is mentioned that a general definition, on the basis of which one could characterize a system as being adaptive or not, is still missing. However, it is clear that constant feedback systems are not adaptive systems. The existence of a

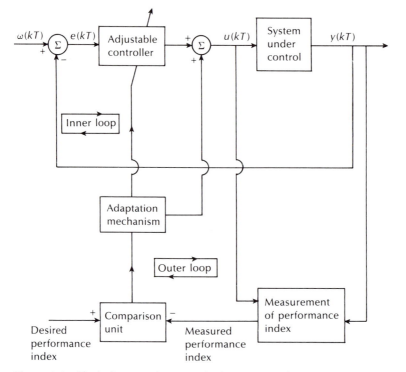

Figure 9.1 Block diagram of a general adaptive control system.

feedback loop involving the performance index of the closed-loop system is a safe rule for characterizing a system as adaptive or not.

An adaptive control system is inherently non-linear since the controller parameters are non-linear functions of the measured signals through the adaptation mechanism. This is true even for the control of linear systems with unknown parameters, a fact which makes the analysis of adaptive systems very difficult. This analysis involves the stability characteristics of the closed-loop system, satisfaction of the performance requirements and convergence of the parameter estimates.

Adaptive control has been under investigation for many years. Major breakthroughs in the area have been reported in the last two decades [1–26]. Adaptive control schemes have been applied in the paper industries, rolling mills, power plants, motor drives, chemical reactors, cement mills, autopilots for aircraft, missiles and ships, etc. Advances in microprocessors have made it quite easy to implement adaptive controllers at low cost. The use of adaptive controllers may lead to improvements in product quality, increases in production rates, fault detection and energy saving.

The two basic techniques for controlling discrete-time systems with unknown parameters are the **model reference adaptive control** (MRAC) scheme [8],

[19–23] and **self-tuning regulators** (STRs) [2], [7], [14–17], [25], [26]. These two techniques are presented in Sections 9.4 and 9.5, respectively.

In MRAC, a reference model is used explicitly in the control scheme and sets the desired performance. Then, an appropriate on-line adaptation mechanism is designed to adjust the controller parameters at each step so that the output of the system converges asymptotically to the output of the reference model, while simultaneously the stability of the closed-loop system is secured. In STRs, the control design and the adaptation procedure are separate. Different parameter estimators can be combined with appropriate control schemes to yield a variety of STRs. Restrictions on the structure of the models under which both methods can be applied are discussed on several appropriate occasions in this chapter.

MRAC schemes can be either **direct** or **indirect**. The essential difference lies in the fact that in direct MRAC the controller parameters are directly adjusted by the adaptation mechanism, while in indirect MRAC the adjustment of the controller parameters is made in two steps. In the first step, the control law is reparametrized so that the plant parameters appear explicitly in the control law. A relation between the controller parameters and the plant parameters is thus established. The plant parameters are adjusted by the adaptation mechanism. In the second step, the controller parameters are calculated from the estimates of the plant parameters. Direct MRAC, using the hyperstability approach for proving stability, is discussed in Section 9.4.

STRs can be either **explicit** or **implicit**. In explicit STRs an estimate of the explicit plant model parameters is obtained. The explicit plant model is the actual plant model. In implicit STRs the parameters of an implicit model are estimated. The implicit model is a reparametrization of the explicit plant model. The parameters of the implicit model and those of the controller are the same. In view of this fact, we call the plant parameters explicit or indirect and the controller parameters implicit or direct. Though they are of different nature and origin, a close relation between MRAC systems and STRs has been established [17], [19]. It is clear that explicit self-tuners correspond to indirect model MRAC schemes, while implicit self-tuners correspond to direct MRAC schemes. Self-tuners based on minimum variance control and pole-placement control are presented in Sections 9.5.2 and 9.5.3, respectively.

Another approach to discrete-time MRAC is that of using Lyapunov functions to prove asymptotic stability and satisfy performance requirements. An expression for the error between the output of the reference model and that of the plant is formed and then the adaptation mechanism is chosen in order to make the increments of a Lyapunov candidate function negative. The method is demonstrated in Section 9.3, using a simple example. The difficulty of finding an appropriate Lyapunov candidate function in the general discrete-time case restricts the use of this method. The hyperstability approach of Section 9.4 is preferable for discrete-time MRAC systems, while for continuous time systems the Lyapunov design has mainly been used.

A first approach to MRAC was based on the gradient method. The parameter adaptation scheme obtained for synthesizing the adaptive loop was developed heuristically, initially for continuous-time systems, and is known as the **MIT rule** [3]. A version of MRAC for discrete-time systems, based on the gradient method, is presented below.

9.2   Adaptive control with the gradient method (MIT rule)

9.2.1   *Introduction*

Consider a system with a single output $y(kT, \boldsymbol{\theta})$, where T is the sampling period and $\boldsymbol{\theta}$ is the vector of unknown parameters which parametrizes the adjustable controller (hence the system's input signal is a function of $\boldsymbol{\theta}$, i.e. $u(kT, \boldsymbol{\theta})$) and the output of the system. The control objective is to follow the output $y_m(kT)$ of a reference model, in the sense that a particular performance index, involving the error $e(kT, \boldsymbol{\theta}) = y(kT, \boldsymbol{\theta}) - y_m(kT)$, is a minimum.

Consider the quadratic performance index

$$J(kT, \boldsymbol{\theta}) = \tfrac{1}{2} e^2(kT, \boldsymbol{\theta}) \tag{9.1}$$

It is obvious that in order to minimize $J(kT, \boldsymbol{\theta})$, the parameter vector $\boldsymbol{\theta}$ of the adjustable controller should change in the opposite direction to that of the gradient $\partial J/\partial \boldsymbol{\theta}$. Consequently, the adaptation rule for $\boldsymbol{\theta}$, i.e. the difference equation giving the time evolution of $\boldsymbol{\theta}$ at the sampling instants, is

$$\boldsymbol{\theta}(kT + T) = \boldsymbol{\theta}(kT) - \gamma \left(\frac{\partial J(kT, \boldsymbol{\theta})}{\partial \theta} \right) = \boldsymbol{\theta}(kT) - \gamma e(kT, \boldsymbol{\theta}) \left(\frac{\partial e(kT, \boldsymbol{\theta})}{\partial \theta} \right) \tag{9.2}$$

where γ is a constant positive adaptation gain. More precisely, when $\partial J(kT, \boldsymbol{\theta})/\partial \boldsymbol{\theta}$ is negative, i.e. when J decreases while $\boldsymbol{\theta}$ increases, then $\boldsymbol{\theta}$ should increase in order for J to decrease further. In the case where $\partial J(kT, \boldsymbol{\theta})/\partial \boldsymbol{\theta}$ is positive, i.e. when J and $\boldsymbol{\theta}$ increase simultaneously, then $\boldsymbol{\theta}$ should decrease in order for J to decrease further. This is achieved by the heuristic adaptation mechanism of equation (9.2). The partial derivative $\partial e(kT, \boldsymbol{\theta})/\partial \boldsymbol{\theta}$ appearing in (9.2) is called the system's **sensitivity derivative**. In order for the MIT rule to perform well, the adaptation gain γ should be small since its value influences the convergence rate significantly. Moreover, it is possible for the MIT rule to lead to an unstable closed-loop system, since it is only a heuristic algorithm not rigidly based on stability requirements.

Other performance indices are possible. For example, by choosing

$$J(kT, \boldsymbol{\theta}) = | e(kT, \boldsymbol{\theta}) | \tag{9.3}$$

the adaptation rule becomes

$$\boldsymbol{\theta}(kT + T) = \boldsymbol{\theta}(kT) - \gamma \left(\frac{\partial e(kT, \boldsymbol{\theta})}{\partial \theta} \right) (\text{sign}\{ e(kT, \boldsymbol{\theta}) \}) \tag{9.4}$$

where

$$\text{sign}\{x\} = \begin{cases} +1 & \text{when } x > 0 \\ 0 & \text{when } x = 0 \\ -1 & \text{when } x < 0 \end{cases}$$

An even simpler implementation, often used in telecommunications where fast computations are required, is the sign–sign algorithm

$$\boldsymbol{\theta}(kT + T) = \boldsymbol{\theta}(kT) - \gamma\left(\text{sign}\left\{\frac{\partial e(kT, \boldsymbol{\theta})}{\partial \boldsymbol{\theta}}\right\}\right)(\text{sign}\{e(kT, \boldsymbol{\theta})\}) \tag{9.5}$$

Before the results are presented for the general case of linear systems in Section 9.2.2, the following example will illustrate the application of the MIT rule. This example will reveal the main problem in applying this method, namely the necessity to use approximations in order to calculate the sensitivity derivatives of a certain system.

■ **EXAMPLE 9.2.1**   Consider a first-order system described by the difference equation

$$y(kT + T) = -ay(kT) + bu(kT) \tag{9.6}$$

where $u(kT)$ is the input and $y(kT)$ is the output. It is desired to obtain a closed-loop system of the form

$$y_{\text{m}}(kT + T) = -a_{\text{m}}y_{\text{m}}(kT) + b_{\text{m}}\omega(kT) \tag{9.7}$$

where $\omega(kT)$ is a bounded reference sequence and $y_{\text{m}}(kT)$ is the output of the reference model. To this end, an output feedback control law of the following form is used:

$$u(kT) = -fy(kT) + g\omega(kT) \tag{9.8}$$

Assume that the system parameters a and b are unknown. Determine the appropriate adaptation mechanism for the controller parameters f and g, using the MIT rule.

■ **SOLUTION**   Combining (9.6) and (9.8) we obtain the closed-loop system

$$y(kT + T) = -ay(kT) + b[-fy(kT) + g\omega(kT)] = -(a + bf)y(kT) + bg\omega(kT)$$

or

$$y(kT) = -(a + bf)y(kT - T) + bg\omega(kT - T)$$

or

$$y(kT) = \left(\frac{bgq^{-1}}{1 + (a + bf)q^{-1}}\right)\omega(kT) \tag{9.9}$$

where q^{-1} is the backward shift operator such that $q^{-1}y(kT) \equiv y(kT - T)$. Comparing (9.7) and (9.9), we have that in the case of known parameters a and b, the particular choice $f = (a_m - a)/b$ and $g = b_m/b$ leads to satisfaction of the control objective. This case is called **perfect model following**.

In the case of uncertain system parameters, we will use the MIT rule. Here, the controller is parametrized by the adjustable parameters $f(k)$ and $g(k)$. The error $e(kT)$ between the outputs of the system and the reference model is

$$e(kT) = y(kT) - y_m(kT) = \left(\frac{bgq^{-1}}{1 + (a + bf)q^{-1}}\right)\omega(kT) - y_m(kT) \tag{9.10}$$

Using the foregoing expression for $e(kT)$, the system's sensitivity derivatives $\partial e/\partial g$ and $\partial e/\partial f$ can be easily determined. We have

$$\frac{\partial e(kT)}{\partial g} = \left(\frac{bq^{-1}}{1 + (a + bf)q^{-1}}\right)\omega(kT) \tag{9.11}$$

$$\frac{\partial e(kT)}{\partial f} = -\left(\frac{b^2 g q^{-2}}{[1 + (a + bf)q^{-1}]^2}\right)\omega(kT) = -\left(\frac{bq^{-1}}{1 + (a + bf)q^{-1}}\right)y(kT) \tag{9.12}$$

These expressions for the sensitivity derivatives cannot be used in the adaptation mechanism, since the unknown parameters a and b appear explicitly. For the present system, when perfect model following is achieved, we have that $a + bf = a_m$. Taking advantage of this fact, the following approximate forms can be used for the sensitivity derivatives (still containing the unknown b):

$$\frac{\partial e(kT)}{\partial g} \simeq \left(\frac{bq^{-1}}{1 + a_m q^{-1}}\right)\omega(kT) \tag{9.13}$$

$$\frac{\partial e(kT)}{\partial f} \simeq -\left(\frac{bq^{-1}}{1 + a_m q^{-1}}\right)y(kT) \tag{9.14}$$

These sensitivity derivatives lead to the following parameter adaptation laws (MIT rule):

$$g(kT + T) = g(kT) - \gamma\left(\frac{q^{-1}}{1 + a_m q^{-1}}\omega(kT)\right)e(kT) \tag{9.15}$$

$$f(kT + T) = f(kT) + \gamma\left(\frac{q^{-1}}{1 + a_m q^{-1}}y(kT)\right)e(kT) \tag{9.16}$$

Notice here that the adaptation laws were obtained by absorbing the parameter b in the adaptation gain γ. This is done because b is unknown and should not appear in the adaptation laws, but this requires that the sign of b is known. Therefore the sign of γ depends on the sign of b. The foregoing laws are initialized with arbitrary $g(0)$ and $f(0)$, which should reflect our *a priori* knowledge on the appropriate controller parameters which achieve model following.

Finally, the adjustable controller is given by

$$u(kT) = -f(kT)y(kT) + g(kT)\omega(kT) \tag{9.17}$$

Equations (9.15), (9.16) and (9.17) specify the dynamic adaptive controller being sought.

9.2.2 *Results for the general case of linear systems*

Consider a single-input single-output linear system described by the difference equation

$$A(q^{-1})y(kT) = q^{-d}B(q^{-1})u(kT) \tag{9.18}$$

where $d \geq 1$ is the system's delay and $A(q^{-1})$ and $B(q^{-1})$ are polynomials in the backward shift operator having the form

$$A(q^{-1}) = 1 + a_1 q^{-1} + \cdots + a_{n_A}q^{-n_A} \tag{9.19}$$

$$B(q^{-1}) = b_0 + b_1 q^{-1} + \cdots + b_{n_B}q^{-n_B} \tag{9.20}$$

The polynomial $A(q^{-1})$ is monic (i.e. the first coefficient is unity), the system is causal (i.e. $n_A = \deg A(q^{-1}) \geq \deg B(q^{-1}) = n_B$) and $A(q^{-1})$ and $B(q^{-1})$ are relatively prime (i.e. have no common factors). It is desired to find a controller which minimizes $J = \frac{1}{2}e^2(kT)$, where $e(kT) = y(kT) - y_m(kT)$ and where the relation between the command signal $\omega(kT)$ and the desired output signal $y_m(kT)$ is given by

$$A_m(q^{-1})y_m(kT) = q^{-d}B_m(q^{-1})\omega(kT) \tag{9.21}$$

with

$$A_m(q^{-1}) = 1 + a_1^m q^{-1} + \cdots + a_{n_{A_m}}^m q^{-n_{A_m}} \tag{9.22}$$

$$B_m(q^{-1}) = b_0^m + b_1^m q^{-1} + \cdots + b_{n_{B_m}}^m q^{-n_{B_m}} \tag{9.23}$$

Here $A_m(q^{-1})$ is a monic, asymptotically stable polynomial (i.e. with all its roots lying inside the unit circle). The following canonical structure (called R–S–T structure) for the controller is used:

$$R(q^{-1})u(kT) = T(q^{-1})\omega(kT) - S(q^{-1})y(kT) \tag{9.24}$$

where

$$R(q^{-1}) = r_0 + r_1 q^{-1} + \cdots + r_{n_R}q^{-n_R} \tag{9.25}$$

$$S(q^{-1}) = s_0 + s_1 q^{-1} + \cdots + s_{n_S}q^{-n_S} \tag{9.26}$$

$$T(q^{-1}) = t_0 + t_1 q^{-1} + \cdots + t_{n_T}q^{-n_T} \tag{9.27}$$

The output $y(kT)$ of the closed-loop system is given by

$$y(kT) = \left(\frac{q^{-d}B(q^{-1})T(q^{-1})}{A(q^{-1})R(q^{-1}) + q^{-d}B(q^{-1})S(q^{-1})} \right) \omega(kT) \qquad (9.28)$$

Moreover, the signals $u(kT)$ and $\omega(kT)$ are related as follows:

$$u(kT) = \left(\frac{A(q^{-1})T(q^{-1})}{A(q^{-1})R(q^{-1}) + q^{-d}B(q^{-1})S(q^{-1})} \right) \omega(kT) \qquad (9.29)$$

Using the error expression $e(kT) = y(kT) - y_m(kT)$ and relations (9.28) and (9.29) we obtain the following sensitivity derivatives:

$$\frac{\partial e(kT)}{\partial r_i} = -\left(\frac{q^{-d}B(q^{-1})T(q^{-1})A(q^{-1})q^{-i}}{[A(q^{-1})R(q^{-1}) + q^{-d}B(q^{-1})S(q^{-1})]^2} \right) \omega(kT)$$

$$= -\left(\frac{B(q^{-1})q^{-(d+i)}}{A(q^{-1})R(q^{-1}) + q^{-d}B(q^{-1})S(q^{-1})} \right) u(kT) \quad \text{for} \quad i = 0, 1, \dots, n_R \qquad (9.30)$$

$$\frac{\partial e(kT)}{\partial s_i} = -\left(\frac{q^{-d}B(q^{-1})T(q^{-1})q^{-d}B(q^{-1})q^{-i}}{[A(q^{-1})R(q^{-1}) + q^{-d}B(q^{-1})S(q^{-1})]^2} \right) \omega(kT)$$

$$= -\left(\frac{B(q^{-1})q^{-(d+i)}}{A(q^{-1})R(q^{-1}) + q^{-d}B(q^{-1})S(q^{-1})} \right) y(kT) \quad \text{for} \quad i = 0, 1, \dots, n_S \qquad (9.31)$$

$$\frac{\partial e(kT)}{\partial t_i} = \left(\frac{B(q^{-1})q^{-(d+i)}}{A(q^{-1})R(q^{-1}) + q^{-d}B(q^{-1})S(q^{-1})} \right) \omega(kT) \quad \text{for} \quad i = 0, 1, \dots, n_T \qquad (9.32)$$

The foregoing expressions for the sensitivity derivatives are not implementable, since they involve the unknown polynomials $A(q^{-1})$ and $B(q^{-1})$ which contain the unknown plant parameters. Several approximations can be applied in order to yield approximate implementable forms for the sensitivity derivatives. Two such approaches are presented below:

1. The polynomial $B(q^{-1})$ can be factored out uniquely as $B(q^{-1}) = B^+(q^{-1})B^-(q^{-1})$, where $B^+(q^{-1})$ is monic and includes well-damped stable zeros (which may be cancelled out) and $B^-(q^{-1})$ includes unstable and poorly damped zeros (which should not be cancelled out). One can then use the approximation

$$A(q^{-1})R(q^{-1}) + q^{-d}B(q^{-1})S(q^{-1}) \cong A_o(q^{-1})A_m1(q^{-1})B^+(q^{-1}) \qquad (9.33)$$

where $A_o(q^{-1})$ is an asymptotically stable polynomial of our choice. This approximation will be clarified in Section 9.5.3. Now, the sensitivity derivatives are approximated by

$$\frac{\partial e(kT)}{\partial r_i} \simeq -\left(\frac{B^-(q^{-1})q^{-(d+i)}}{A_o(q^{-1})A_m(q^{-1})}\right)u(kT) \quad \text{for} \quad i = 0, 1, \ldots, n_R \tag{9.34}$$

$$\frac{\partial e(kT)}{\partial s_i} \simeq -\left(\frac{B^-(q^{-1})q^{-(d+i)}}{A_o(q^{-1})A_m(q^{-1})}\right)y(kT) \quad \text{for} \quad i = 0, 1, \ldots, n_S \tag{9.35}$$

$$\frac{\partial e(kT)}{\partial t_i} \simeq \left(\frac{B^-(q^{-1})q^{-(d+i)}}{A_o(q^{-1})A_m(q^{-1})}\right)w(kT) \quad \text{for} \quad i = 0, 1, \ldots, n_T \tag{9.36}$$

Note that these expressions still involve the unknown $B^-(q^{-1})$. This does not allow use of the approximations for the sensitivity derivatives. However, consider the special case of minimum-phase systems, i.e. systems where all zeros are stable. In this case $B^-(q^{-1}) = b_0$ and b_0 can be absorbed by the adaptation gain γ in the adaptation laws, provided that the sign of b_0 is known. In this case of minimum-phase systems with known sign for b_0, the following adaptation laws for the controller parameters can be applied:

$$r_i(kT + T) = r_i(kT) + \gamma\left(\frac{q^{-(d+i)}}{A_o(q^{-1})A_m(q^{-1})}\right)u(kT)e(kT) \quad \text{for} \quad i = 0, 1, \ldots, n_R \tag{9.37}$$

$$s_i(kT + T) = s_i(kT) + \gamma\left(\frac{q^{-(d+i)}}{A_o(q^{-1})A_m(q^{-1})}\right)y(kT)e(kT) \quad \text{for} \quad i = 0, 1, \ldots, n_S \tag{9.38}$$

$$t_i(kT + T) = t_i(kT) - \gamma\left(\frac{q^{-(d+i)}}{A_o(q^{-1})A_m(q^{-1})}\right)w(kT)e(kT) \quad \text{for} \quad i = 0, 1, \ldots, n_T \tag{9.39}$$

Any *a priori* knowledge of the system parameters a_i and b_i can be used in order to define the initial values $r_i(0)$, $s_i(0)$ and $t_i(0)$ of the controller parameters.

2. Another approximation can be used if a parameter estimator of the system parameters in $A(q^{-1})$ and $B(q^{-1})$ is added to the control scheme. This estimator yields the estimated polynomials $\hat{A}(q^{-1}, kT)$ and $\hat{B}(q^{-1}, kT)$ at each step. In this case, $B(q^{-1})$ and $A(q^{-1})R(q^{-1}) + q^{-d}B(q^{-1})S(q^{-1})$ are replaced by $\hat{B}(q^{-1}, kT)$ and $\hat{A}(q^{-1}, kT)R(q^{-1}) + q^{-d}\hat{B}(q^{-1}, kT)S(q^{-1})$, respectively, in (9.30), (9.31) and (9.32). The polynomials $R(q^{-1})$, $S(q^{-1})$ and $T(q^{-1})$ are also calculated, at each step, by using the values obtained from the adaptation laws. The adaptation laws use the sensitivity derivatives of (9.30) to (9.32). It is then possible to obtain an algorithm applicable to non-minimum-phase systems and systems with unknown sign for b_0.

■ **REMARK 9.2.1** The algorithm presented in paragraph 1 above is a direct algorithm, since the controller parameters are updated directly. In contrast, the algorithm presented in paragraph 2 is an indirect algorithm, since the system model parameters are initially estimated, and the controller parameters are subsequently updated using these estimates.

9.3 Adaptive control using a Lyapunov design

The Lyapunov design method will be illustrated by considering the first-order system

$$y(k + 1) = ay(k) + u(k) \tag{9.40}$$

where a is an unknown parameter (open-loop pole). It is desired to find an appropriate control law and an adaptation law in order that the pole a is moved to a new desirable position a_m.

It is obvious that in the case of a known parameter a, the control law

$$u(k) = (a_m - a)y(k) \tag{9.41}$$

achieves the control objective. When the parameter a is unknown, we cannot use (9.41). However, we may use the same controller structure and replace the unknown parameter a in (9.41) with an estimate $\hat{a}'(k)$. This procedure is widely known in the literature as the **certainty equivalence principle**. In this case of unknown parameter a, $u(k)$ is given by

$$u(k) = [a_m - \hat{a}'(k)]y(k) = -\hat{a}(k)y(k) \tag{9.42}$$

A suitable adaptation mechanism for $\hat{a}(k)$ is sought in what follows. The closed-loop system will be

$$y(k + 1) = [a - \hat{a}(k)]y(k) \tag{9.43}$$

We define $\tilde{a}(k) = \hat{a}(k) - a + a_m$. Then (9.43) may be rewritten as

$$y(k + 1) = [a_m - \tilde{a}(k)]y(k) \tag{9.44}$$

By using a Lyapunov design, we will prove that the adaptation mechanism

$$\hat{a}(k + 1) = \hat{a}(k) - \left(\frac{y^2(k)}{\gamma^2 + y^2(k)} \right)\tilde{a}(k) = \hat{a}(k) + \left(\frac{y(k)}{\gamma^2 + y^2(k)} \right)[y(k + 1) - a_m y(k)] \tag{9.45}$$

where the gain is $\gamma > 0$, associated with the control law (9.42), yields

$$\lim_{k \to +\infty} y(k + 1) = a_m y(k) \tag{9.46}$$

Indeed, from (9.45)

$$\tilde{a}(k + 1) = \tilde{a}(k) - \left(\frac{y^2(k)}{\gamma^2 + y^2(k)} \right)\tilde{a}(k) = \left(\frac{\gamma^2}{\gamma^2 + y^2(k)} \right)\tilde{a}(k) \tag{9.47}$$

Equations (9.44) and (9.47) define the closed-loop system. From the structure of (9.47), it is obvious that, irrespective of the time evolution of $y(k)$, $\bar{a}(k)$ is a decreasing function of time. It is also clear that if $y(k)$ does not converge to zero, then $\lim_{k \to +\infty} \bar{a}(k) = 0$. The identification of the unknown system parameter a is thus guaranteed in this case by the controller (9.42) and (9.45). In contrast, if $y(k)$ converges to zero, then $\bar{a}(k)$ converges to a constant. The convergence rate of $\bar{a}(k)$ increases as γ decreases.

We will prove now that $\lim_{k \to +\infty} y(k) = 0$. To this end, consider the Lyapunov candidate function

$$V(k) = y^2(k) + \bar{a}^2(k) \tag{9.48}$$

This function is continuous with $V[y(k) = 0, \bar{a}(k) = 0] = 0$ and $V(k) > 0$ for $y(k)$ and $\bar{a}(k) \neq 0$ (i.e. $V(k)$ is a positive definite function). By using (9.44) and (9.47), its increments are calculated as follows:

$$V(k + 1) - V(k) = y^2(k + 1) - y^2(k) + \bar{a}^2(k + 1) - \bar{a}^2(k)$$

$$= \{[a_m - \bar{a}(k)]^2 - 1\}y^2(k) - \frac{y^2(k)[2\gamma^2 + y^2(k)]\bar{a}^2(k)}{[\gamma^2 + y^2(k)]^2} \tag{9.49}$$

Here we guarantee that $\bar{a}(k)$ is a decreasing function. It is then clear from (9.47) that as long as $y(k)$ remains different from zero, it is always possible, for a given $\varepsilon > 0$, to find a $\gamma > 0$ and a k_0 such that $|\bar{a}(k)| < \varepsilon \ \forall \ k > k_0$. Note also that since the desired pole a_m is supposed to be stable, then $a_m^2 - 1 < 0$. Hence, there exists a k_0 such that for $k > k_0$ the term $\{[a_m - \bar{a}(k)]^2 - 1\}y^2(k)$ becomes and remains strictly negative as long as $y(k) \neq 0$. That is, $V(k + 1) - V(k) < 0$ as long as $y(k) \neq 0$, for $k > k_0$. Moreover, $V(k)$ tends to infinity when $y(k)$ and $\bar{a}(k)$ tend to infinity. We conclude that $V(k)$ is a Lyapunov function for the overall system and, consequently, $\lim_{k \to +\infty} y(k) = 0$ for all initial conditions $\bar{a}(0)$ and $y(0)$. Hence, relation (9.46) may easily be derived from (9.44). The control objective is thus satisfied in the case of unknown plant parameter a, by using the control law (9.42) together with the adaptation mechanism (9.45).

■ **EXAMPLE 9.3.1** Consider the system (9.40) with $a = 1.5$, i.e. consider the following unstable system:

$$y(k + 1) = 1.5y(k) + u(k) \tag{9.50}$$

This system is initialized at $y(0) = 5$. The desired closed-loop behaviour is given by the stable system

$$y_m(k + 1) = a_m y_m(k) = 0.7y_m(k) \tag{9.51}$$

initialized at $y_m(0) = 5$. Simulate the adaptive Lyapunov design by choosing $\gamma = 4.5$ and $\hat{a}(0) = 0.6$.

■ **SOLUTION**   By applying the Lyapunov design approach given in the present section, one arrives at the following simulation results. In Figure 9.2 the time evolution of $y(k)$ is given, together with the time evolution of (9.51). In Figures 9.3 to 9.5, the time evolution of $u(k)$, $\hat{a}(k)$ and $V(k+1) - V(k)$ is presented. Note that the unknown parameter a is not identified in this example. This does not have any influence on the satisfaction of the control objective, as it was deduced from the Lyapunov analysis carried out just after (9.49).

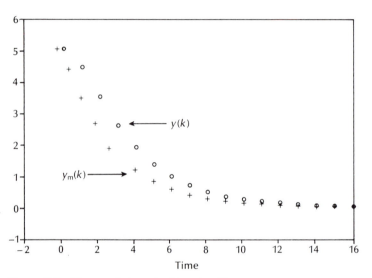

Figure 9.2   Time evolution of $y(k)$ and $y_m(k)$.

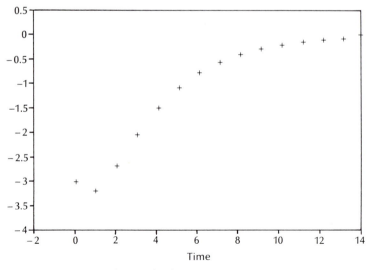

Figure 9.3   Time evolution of $u(k)$.

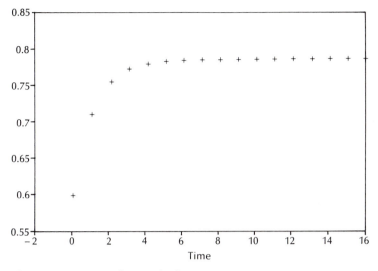

Figure 9.4   Time evolution of $\hat{a}(k)$.

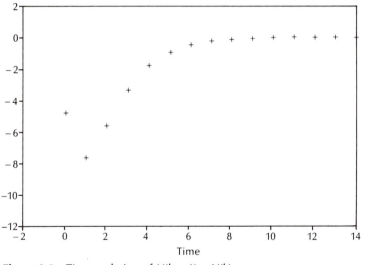

Figure 9.5   Time evolution of $V(k + 1) - V(k)$.

Adaptive control using Lyapunov design has the advantage that it guarantees the stability of the closed-loop system [18], [24]. However, it involves the very difficult task of determining a suitable Lyapunov function and, for this reason, the use of this method is limited.

9.4 Model reference adaptive control – hyperstability design

9.4.1 Introduction

Model reference adaptive control (MRAC) is a systematic method for controlling plants with unknown parameters. The basic scheme of an MRAC system is presented in Figure 9.6. In comparison with the general structure of an adaptive control system given in Figure 9.1, here the desired performance index is generated by means of a reference model.

The reference model is a dynamic system whose behaviour is considered to be the desired (ideal) one and it is a part of the control system itself, since it appears explicitly in the control scheme. The output $y_m(kT)$ of the reference model indicates how the output $y(kT)$ of the plant should behave. Both systems are excited by the same command signal $\omega(kT)$. Comparing Figures 9.1 and 9.6, we observe that the desired performance index is now replaced by $y_m(kT)$ and the measured performance index by $y(kT)$.

We distinguish two control loops: the 'inner' loop and the 'outer' loop. The 'inner' loop consists of the plant which involves unknown parameters and the adjustable controller. The 'outer' loop is designed appropriately to adjust the controller's para-

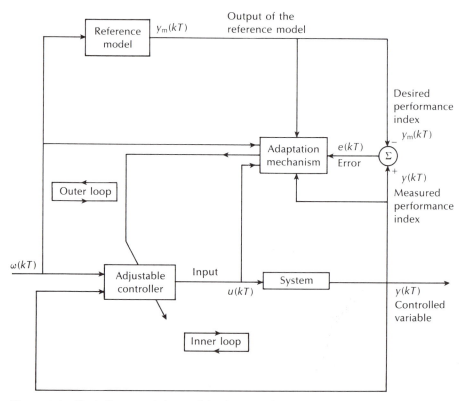

Figure 9.6 Block diagram of the model reference adaptive control scheme.

meters so that the error $e(kT) = y(kT) - y_m(kT)$ approaches zero asymptotically, while the stability of the overall system can be proved using the hyperstability approach.

Compared with techniques which involve other kinds of performance indices, the MRAC technique is characterized by high speed of adaptation. This is due to the fact that a simple subtractor is needed to form the error $e(kT) = y(kT) - y_m(kT)$. This error, together with other available on-line data, is then fed to the adaptation mechanism. The parameters of the adjustable controller are modified accordingly, in order to minimize the difference between the two performance indices, namely the desired performance index and the measured performance index.

9.4.2 *Definition of the model reference control problem*

Consider a deterministic, single-input single-output, discrete-time, linear, time-invariant system described by

$$A(q^{-1})y(k+d) = B(q^{-1})u(k) \quad \text{or} \quad A(q^{-1})y(k) = q^{-d}B(q^{-1})u(k) \tag{9.52}$$

with initial condition $y(0) \neq 0$. Here

$$A(q^{-1}) = 1 + a_1 q^{-1} + \cdots + a_{n_A} q^{-n_A} = 1 + q^{-1}A^*(q^{-1}) \tag{9.53}$$

$$B(q^{-1}) = b_0 + b_1 q^{-1} + \cdots + b_{n_B} q^{-n_B} = b_0 + q^{-1}B^*(q^{-1}) \quad \text{with} \quad b_0 \neq 0 \tag{9.54}$$

where q^{-1} is the backward shift operator, $d > 0$ represents the system's time delay, and $u(k)$ and $y(k)$ represent the system's input and output signals, respectively. The following three assumptions are made:

1. The roots of $B(z^{-1})$, which are the system zeros, are all inside the unit circle $|z| < 1$, i.e. $z_i^{n_B}B(z_i^{-1}) = 0$ with $|z_i| < 1$. Thus, the system zeros are stable and can be cancelled out without leading to an unbounded control signal.
2. The system delay d is known (this implies $b_0 \neq 0$).
3. An upper limit for the orders n_A and n_B of the polynomials $A(q^{-1})$ and $B(q^{-1})$, respectively, is given.

According to the foregoing assumptions, any change in the system characteristics should not affect the delay d, whereas the system zeros can move only inside the unit circle. The method is therefore valid only for minimum-phase systems.

The control objective is: linear model following during tracking and elimination of any initial output disturbance during regulation. It is desirable to be able to specify the tracking and regulation objectives independently. This flexibility is crucial for certain applications. The control objectives are specified as follows.

1. Tracking

During tracking, it is necessary for the plant output $y(k)$ to satisfy the equation

$$A_m(q^{-1})y(k) = q^{-d}B_m(q^{-1})\omega(k) \tag{9.55}$$

where

$$A_m(q^{-1}) = 1 + a_1^m q^{-1} + \cdots + a_{n_{A_m}}^m q^{-n_{A_m}} \tag{9.56}$$

$$B_m(q^{-1}) = b_0^m + b_1^m q^{-1} + \cdots + b_{n_{B_m}}^m q^{-n_{B_m}} \tag{9.57}$$

Here, $\omega(k)$ is a bounded reference sequence and the polynomial $A_m(q^{-1})$ is chosen to be asymptotically stable.

2. Regulation $(\omega(k) \equiv 0, y_m(k) \equiv 0)$

In regulation, the influence of any initial non-zero output $y(0) \neq 0$ (which corresponds to an impulse perturbation) should be eliminated via the dynamics defined by

$$\Gamma(q^{-1})y(k+d) = 0 \quad \text{for} \quad k \geqslant 0 \tag{9.58}$$

where $\Gamma(q^{-1})$ is an asymptotically stable polynomial of our choice having the form

$$\Gamma(q^{-1}) = 1 + \gamma_1 q^{-1} + \cdots + \gamma_{n_r} q^{-n_r} \tag{9.59}$$

Consider the following explicit reference model:

$$A_m(q^{-1})y_m(k) = q^{-d}B_m(q^{-1})\omega(k) \tag{9.60}$$

with input $\omega(k)$ and output $y_m(k)$. Note here that the sequence $y_m(k)$, apart from being calculated by means of the reference model (9.60), can also be a predefined sequence stored in memory.

It is obvious that both control objectives (i.e. tracking and regulation) can be accomplished if the control law $u(k)$ is such that

$$\bar{e}(k+d) = \Gamma(q^{-1})e(k+d) = \Gamma(q^{-1})[y(k+d) - y_m(k+d)] \equiv 0 \quad \text{for} \quad k \geqslant 0 \tag{9.61}$$

The error $e(k)$ is the difference between the plant and reference model outputs (**plant–model error**), i.e.

$$e(k) = y(k) - y_m(k) \tag{9.62}$$

and $\bar{e}(k) = \Gamma(q^{-1})e(k)$ is the **filtered error** between the plant and the reference model outputs. The error $\bar{e}(k)$ is also called the **a priori adaptation error**. The foregoing objectives will be satisfied below in the case of known or unknown parameters in the polynomials $A(q^{-1})$ and $B(q^{-1})$.

If equation (9.61) is satisfied, then any initial plant–model error or any initial output disturbance will converge to zero, i.e. $\lim_{k \to +\infty} e(k) = 0$. Note that if $\Gamma(q^{-1}) = 1$, one has $e(k+d) \equiv 0$, $k \geqslant 0$, which means that in this case the plant–model error vanishes d steps after the control input is applied. The polynomial $\Gamma(q^{-1})$ is a filtering polynomial. As is made clear in what follows, adaptive control performance depends critically on the choice of $\Gamma(q^{-1})$.

9.4.3 Design in the case of known parameters

In this section we assume that the plant parameters appearing in the polynomials $A(q^{-1})$ and $B(q^{-1})$ are known. Consider the general case where $d > 1$. We wish to obtain a controller satisfying the control objectives and being causal, i.e. not depending on future values of the input and output. This controller will therefore have the form

$$u(k) = f(y(k), y(k-1), \ldots, u(k-1), u(k-2), \ldots)$$

To this end, consider the following equation:

$$y(k+d) = -A^*(q^{-1})y(k+d-1) + B(q^{-1})u(k) \tag{9.63}$$

Equation (9.63) is equivalent to equation (9.52). Next, we express (9.63) in the form

$$\Gamma(q^{-1})y(k+d) = g(y(k), y(k-1), \ldots, u(k), u(k-1), \ldots) \tag{9.64}$$

The specific form of g can be determined in two ways: either by repeatedly substituting $y(k+d-1), \ldots, y(k+1)$ in (9.63) generated by the same equation (9.63) delayed in time, or more easily by directly considering the following polynomial identity:

$$\Gamma(q^{-1}) = A(q^{-1})S(q^{-1}) + q^{-d}R(q^{-1}) \tag{9.65}$$

with $R(q^{-1})$ and $S(q^{-1})$ as appropriate polynomials. We adopt the second method for simplicity and, to this end, the results of the following remark will be useful.

■ **REMARK 9.4.1** The above identity (9.65) is a special case of what is referred to as the **Diophantine equation** or the **Bezout identity**. It can be proven that it has a unique solution for the polynomials $S(q^{-1})$ and $R(q^{-1})$ (i.e. $\Gamma(q^{-1})$ is uniquely factorized as in (9.65)), where

$$S(q^{-1}) = 1 + s_1 q^{-1} + \cdots + s_{n_S} q^{-n_S} \tag{9.66}$$

$$R(q^{-1}) = r_0 + r_1 q^{-1} + \cdots + r_{n_R} q^{-n_R} \tag{9.67}$$

when $n_S = d-1$ and $n_R = \max(n_A - 1, n_\Gamma - d)$ (see Section 9.5.3 for the uniqueness conditions). This results from the unique solution of the following algebraic equation:

$$
\begin{bmatrix}
1 & & & & & & & & & & \\
a_1 & 1 & & & & & & & & & \\
a_2 & a_1 & 1 & & & & & & & & \\
\vdots & \vdots & \vdots & & & & \mathbf{0} & & & & \\
a_{d-1} & a_{d-2} & \cdots & a_1 & 1 & & & & & & \\
a_d & a_{d-1} & \cdots & a_2 & a_1 & 1 & & & & & \\
a_{d+1} & a_d & \cdots & a_3 & a_2 & 0 & 1 & & & & \\
a_{d+2} & a_{d+1} & \cdots & a_4 & a_3 & 0 & 0 & 1 & & & \\
\vdots & \vdots & & \vdots & \vdots & 0 & 0 & 0 & 1 & & \\
& & & & & \vdots & \vdots & \vdots & \vdots & & \\
& & & & & 0 & 0 & 0 & \cdots & 1 & \\
& & & & & 0 & 0 & 0 & \cdots & 0 & 1
\end{bmatrix}
\begin{bmatrix}
1 \\ s_1 \\ s_2 \\ \vdots \\ s_{d-1} \\ r_0 \\ r_1 \\ r_2 \\ \vdots \\ \\ \\ r_{n_R}
\end{bmatrix}
=
\begin{bmatrix}
1 \\ \gamma_1 \\ \gamma_2 \\ \vdots \\ \gamma_{d-1} \\ \vdots \\ \\ \\ \\ \\ \\
\end{bmatrix}
\tag{9.68}
$$

Returning to (9.64) and using (9.65), we express $\Gamma(q^{-1})y(k+d)$ as follows:

$$\Gamma(q^{-1})y(k+d) = A(q^{-1})S(q^{-1})y(k+d) + q^{-d}R(q^{-1})y(k+d)$$
$$= B(q^{-1})S(q^{-1})u(k) + R(q^{-1})\,y(k) \tag{9.69}$$

Let $\Psi(q^{-1}) = B(q^{-1})S(q^{-1})$. Then

$$\Psi(q^{-1}) = b_0 + q^{-1}\Psi^*(q^{-1}) = b_0 + \psi_1 q^{-1} + \cdots + \psi_{d+n_B-1}q^{-(d+n_B-1)} \tag{9.70}$$

where

$$\psi_1 = b_0 s_1 + b_1, \quad \psi_2 = b_0 s_2 + b_1 s_1 + b_2, \ldots \quad \text{and} \quad \psi_{d+n_B-1} = b_{n_B}s_{d-1} \tag{9.71}$$

Finally, we have

$$\Gamma(q^{-1})y(k+d) = b_0 u(k) + \Psi^*(q^{-1})u(k-1) + R(q^{-1})y(k) \tag{9.72}$$

Note that the right-hand side of (9.72) is the function g appearing in (9.64). Equation (9.72) can also be written as

$$\Gamma(q^{-1})y(k+d) = \boldsymbol{\theta}^{\mathrm{T}}\boldsymbol{\phi}(k) = b_0 u(k) + \boldsymbol{\theta}_0^{\mathrm{T}}\boldsymbol{\phi}_0(k) \tag{9.73}$$

where

$$\boldsymbol{\theta}^{\mathrm{T}} = [b_0 \mid \psi_1, \ldots, \psi_{d+n_B-1}, r_0, r_1, \ldots, r_{n_R}] = [b_0 \mid \boldsymbol{\theta}_0^{\mathrm{T}}] \tag{9.74}$$

and $\boldsymbol{\phi}(k)$ is the so-called **regression vector** having the form

$$\boldsymbol{\phi}^{\mathrm{T}}(k) = [u(k) \mid u(k-1), \ldots, u(k-d-n_B+1), y(k), y(k-1), \ldots, y(k-n_R)]$$
$$= [u(k) \mid \boldsymbol{\phi}_0^{\mathrm{T}}(k)] \tag{9.75}$$

In what follows, we seek to find a control law $u(k)$ which drives the filtered plant–model error $\bar{e}(k+d) = \Gamma(q^{-1})e(k+d)$ to zero. Using (9.72) and (9.73) we have

$$\Gamma(q^{-1})e(k+d) = \Gamma(q^{-1})y(k+d) - \Gamma(q^{-1})y_{\mathrm{m}}(k+d)$$
$$= b_0 u(k) + \Psi^*(q^{-1})u(k-1) + R(q^{-1})y(k) - \Gamma(q^{-1})y_{\mathrm{m}}(k+d)$$
$$= b_0 u(k) + \boldsymbol{\theta}_0^{\mathrm{T}}\boldsymbol{\phi}_0(k) - \Gamma(q^{-1})y_{\mathrm{m}}(k+d) \tag{9.76}$$

Solving for $u(k)$ which drives the filtered error to zero, i.e. $\bar{e}(k+d) = \Gamma(q^{-1})e(k+d) = 0$, we arrive at the desired control law

$$u(k) = \frac{\Gamma(q^{-1})y_{\mathrm{m}}(k+d) - R(q^{-1})y(k) - \Psi^*(q^{-1})u(k-1)}{b_0} \tag{9.77}$$

or equivalently

$$u(k) = \frac{\Gamma(q^{-1})y_{\mathrm{m}}(k+d) - \boldsymbol{\theta}_0^{\mathrm{T}}\boldsymbol{\phi}_0(k)}{b_0} \tag{9.78}$$

where use has been made of the condition that $b_0 \neq 0$. Finally, using the fact that

$\Psi(q^{-1}) = B(q^{-1})S(q^{-1})$, the expression for $u(k)$ becomes

$$u(k) = \frac{1}{B(q^{-1})S(q^{-1})} [\Gamma(q^{-1})y_m(k+d) - R(q^{-1})y(k)] \tag{9.79}$$

From this last expression for $u(k)$, it is readily seen why the process should be minimum phase as the system zeros appear in the denominator of the control law.

It can be seen that the control law (9.77), which satisfies the control objective $\Gamma(q^{-1})e(k+d) = 0$, also minimizes the quadratic performance index

$$J(k+d) = \{\Gamma(q^{-1})[y(k+d) - y_m(k+d)]\}^2 \tag{9.80}$$

thereby assuring that $J(k+d) \equiv 0$, for $k \geqslant 0$.

The control scheme analyzed above for the case of known parameters is shown in Figure 9.7.

■ **REMARK 9.4.2** A more general control strategy, for the case of known parameters, results in the control law

$$u(k) = \frac{\Gamma(q^{-1})y_m(k+d) - R(q^{-1})y(k) - \Psi^*(q^{-1})u(k-1)}{b_0 + \beta} \tag{9.81}$$

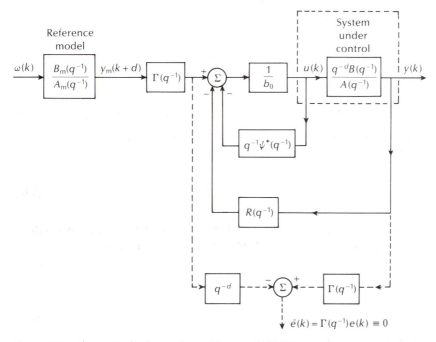

Figure 9.7 The control scheme for tracking and regulation with independent dynamics, for the case of known parameters.

This expression for $u(k)$ minimizes the performance index

$$J(k+d) = \{\Gamma(q^{-1})[y(k+d) - y_m(k+d)]\}^2 + \beta b_0 u^2(k) \quad \text{for} \quad \beta b_0 > 0 \tag{9.82}$$

Here, the input $u(k)$ is also penalized, i.e. $u(k)$ appears in the function $J(k+d)$ which is to be minimized. In this case, the control objective is given by the equation

$$\Gamma(q^{-1})[y(k+d) - y_m(k+d)] + \beta u(k) \equiv 0 \quad \text{for} \quad k \geq 0 \tag{9.83}$$

which is to be compared with (9.61).

By using the control law (9.81), the plant zeros are not cancelled out. Hence, with an appropriate choice of β, it is possible to control non-minimum-phase plants, since now the control law may be rewritten as

$$u(k) = \frac{\Gamma(q^{-1})y_m(k+d) - R(q^{-1})y(k)}{\beta + B(q^{-1})S(q^{-1})} \tag{9.84}$$

In this case, $\beta + B(q^{-1})S(q^{-1})$ should be asymptotically stable and this can be achieved by appropriately selecting β.

9.4.4 *Hyperstability design in the case of unknown parameters*

1. The adaptation algorithm

When the system parameters appearing in the polynomials $A(q^{-1})$ and $B(q^{-1})$ are unknown, we keep the same structure for the controller but replace the constant b_0 and the vector θ_0 in equation (9.78) with the time-varying adjustable parameters (this is the certainty equivalence principle first presented in Section 9.3)

$$\hat{b}_0(k) \quad \text{and} \quad \hat{\theta}_0^T(k) = [\hat{\psi}_1(k), \ldots, \hat{\psi}_{d+n_B-1}(k), \hat{r}_0(k), \hat{r}_1(k), \ldots, \hat{r}_{n_R}(k)] \tag{9.85}$$

These adjustable parameters will be updated by the adaptation mechanism. The control law becomes

$$u(k) = \frac{\Gamma(q^{-1})y_m(k+d) - \hat{\theta}_0^T(k)\phi_0(k)}{\hat{b}_0(k)} \tag{9.86}$$

For convenience, we keep the same notation $u(k)$ for the certainty equivalence control law (9.86) as well. Expression (9.86) may also be written as

$$\Gamma(q^{-1})y_m(k+d) = \hat{\theta}^T(k)\phi(k) \tag{9.87}$$

where

$$\hat{\theta}(k) = [\hat{b}_0(k) \,\vdots\, \hat{\theta}_0^T(k)] \tag{9.88}$$

In the case of unknown plant parameters, it is not possible to keep the filtered error $\bar{e}(k+d) = \Gamma(q^{-1})e(k+d)$ identically equal to zero. The design objective now becomes one of finding a suitable adaptation mechanism for the adjustable parameters in (9.88), which will secure the asymptotic convergence of $\bar{e}(k+d)$ to zero, with bounded input and output signals. Consequently, the control objective becomes

$$\lim_{k \to +\infty} \bar{e}(k+d) = \lim_{k \to +\infty} \Gamma(q^{-1})[y(k+d) - y_m(k+d)] = 0 \quad \forall \; \bar{e}(0) \neq 0, \hat{\theta}(0) \in \mathbb{R}^{d+n_B+n_R+1}$$

(9.89)

with $\| \boldsymbol{\phi}(k) \|$ bounded for all k. Then, since $\Gamma(q^{-1})$ is an asymptotically stable polynomial, one concludes that $\lim_{k \to +\infty} e(k) = 0$. That is, the plant–model error vanishes asymptotically. Using (9.73) and (9.87), the filtered plant–model error (or *a priori* adaptation error) $\bar{e}(k+d)$ is expressed as

$$\bar{e}(k+d) = \Gamma(q^{-1})e(k+d) = \Gamma(q^{-1})[y(k+d) - y_m(k+d)] = \boldsymbol{\theta}^{\mathrm{T}}\boldsymbol{\phi}(k) - \hat{\boldsymbol{\theta}}^{\mathrm{T}}(k)\boldsymbol{\phi}(k)$$

or

$$\bar{e}(k+d) = (\boldsymbol{\theta} - \hat{\boldsymbol{\theta}}(k))^{\mathrm{T}}\boldsymbol{\phi}(k)$$

(9.90)

Equivalently

$$\bar{e}(k) = \Gamma(q^{-1})e(k) = [\boldsymbol{\theta} - \hat{\boldsymbol{\theta}}(k-d)]^{\mathrm{T}}\boldsymbol{\phi}(k-d)$$

(9.91)

We define the **auxiliary error** $\bar{\varepsilon}(k)$ as

$$\bar{\varepsilon}(k) = [\hat{\boldsymbol{\theta}}(k-d) - \hat{\boldsymbol{\theta}}(k)]^{\mathrm{T}}\boldsymbol{\phi}(k-d)$$

(9.92)

and the **a posteriori filtered plant–model error** or **augmented error** $\varepsilon(k)$ as

$$\varepsilon(k) = \bar{e}(k) + \bar{\varepsilon}(k) = [\boldsymbol{\theta} - \hat{\boldsymbol{\theta}}(k)]^{\mathrm{T}}\boldsymbol{\phi}(k-d)$$

(9.93)

As explained in the hyperstability approach presented in the second part of this section, it can be proved that the adaptation algorithm

$$\hat{\boldsymbol{\theta}}(k) = \hat{\boldsymbol{\theta}}(k-1) + \mathbf{F}(k)\boldsymbol{\phi}(k-d)\varepsilon(k)$$

(9.94)

assures that, for all $\bar{e}(0) \neq 0$ and $\hat{\boldsymbol{\theta}}(0) \in \mathbb{R}^{d+n_B+n_R+1}$, we have

$$\lim_{k \to +\infty} \varepsilon(k) = 0 \quad \text{and} \quad \lim_{k \to +\infty} \bar{e}(k) = \lim_{k \to +\infty} e(k) = 0$$

(9.95)

The gain matrix $\mathbf{F}(k)$ is positive definite and is generated by

$$\mathbf{F}^{-1}(k+1) = \lambda_1(k)\mathbf{F}^{-1}(k) + \lambda_2(k)\boldsymbol{\phi}(k-d)\boldsymbol{\phi}^{\mathrm{T}}(k-d) \quad \text{when} \quad \mathbf{F}(0) > 0$$

(9.96)

with

$$0 < \lambda_1(k) \leqslant 1 \quad \text{and} \quad 0 \leqslant \lambda_2(k) < 2, \quad \forall k$$

(9.97)

Clearly, relation (9.95) states that the control objective is satisfied asymptotically.

Note that the algorithm presented above is a special case of the algorithm given by Ionescu and Monopoli [18], who introduced the notion of the augmented error for the first time for discrete-time systems.

It is further noted that a more general design procedure, which uses filtered variables, can be applied [21], [22]. The motivation is threefold: to obtain more flexibility in the conception of the MRAC system, to assure improved performance, and to give a general framework in which many known design procedures can be viewed as particular cases.

■ **REMARK 9.4.3** An implementable form for the *a posteriori* filtered plant–model error $\varepsilon(k)$ may be derived using (9.87) and (9.94), as follows:

$$\varepsilon(k) = \bar{e}(k) + \tilde{\varepsilon}(k) = \Gamma(q^{-1})[y(k) - y_m(k)] + [\hat{\boldsymbol{\theta}}(k-d) - \hat{\boldsymbol{\theta}}(k)]^T \boldsymbol{\phi}(k-d)$$
$$= \Gamma(q^{-1})y(k) - \hat{\boldsymbol{\theta}}^T(k-d)\boldsymbol{\phi}(k-d) + [\hat{\boldsymbol{\theta}}(k-d) - \hat{\boldsymbol{\theta}}(k)]^T \boldsymbol{\phi}(k-d)$$
$$= \Gamma(q^{-1})y(k) - \hat{\boldsymbol{\theta}}^T(k)\boldsymbol{\phi}(k-d)$$
$$= \Gamma(q^{-1})y(k) - \hat{\boldsymbol{\theta}}^T(k-1)\boldsymbol{\phi}(k-d) - \boldsymbol{\phi}^T(k-d)\mathbf{F}(k)\boldsymbol{\phi}(k-d)\varepsilon(k) \qquad (9.98)$$

Hence

$$\varepsilon(k) = \frac{\tilde{\varepsilon}(k)}{1 + \boldsymbol{\phi}^T(k-d)\mathbf{F}(k)\boldsymbol{\phi}(k-d)} \qquad (9.99)$$

where

$$\tilde{\varepsilon}(k) = \Gamma(q^{-1})y(k) - \hat{\boldsymbol{\theta}}^T(k-1)\boldsymbol{\phi}(k-d) \qquad (9.100)$$

■ **REMARK 9.4.4** During the adaptation procedure we may have $\hat{b}_0(k) = 0$. To avoid division by zero in equation (9.86), if $|\hat{b}_0(k)| < \delta$ ($\delta > 0$) for a certain k, we repeat the evaluation of $\hat{\boldsymbol{\theta}}(k)$ from (9.94), using appropriate values for $\lambda_1(k-1)$ and $\lambda_2(k-1)$ in (9.96). These values must be chosen by trial and error so that $|\hat{b}_0(k)| \geq \delta$.

The control algorithm is summarized in Table 9.1. The control scheme for tracking and regulation with independently chosen dynamics, for the case of unknown parameters, is given in Figure 9.8.

■ **EXAMPLE 9.4.1** Consider the system

$$A(q^{-1})y(k) = q^{-1}B(q^{-1})u(k) \quad \text{with} \quad y(0) = 1$$

where $A(q^{-1}) = 1 + 2q^{-1} + q^{-2}$ and $B(q^{-1}) = 2 + q^{-1} + 0.5q^{-2}$ (asymptotically stable). It is desired to track the output of the reference model

$$y_m(k) = \left(\frac{q^{-1}B_m(q^{-1})}{A_m(q^{-1})}\right)\omega(k) = q^{-1}\left(\frac{1 + 0.3q^{-1}}{1 - q^{-1} + 0.25q^{-2}}\right)\omega(k) \quad \text{with} \quad y_m(0) = 2$$

The dynamics during regulation are characterized by the asymptotically stable polynomial $\Gamma(q^{-1}) = 1 + 0.5q^{-1}$.

Table 9.1 The model reference adaptive control algorithm.

$$\varphi^T(k) = [u(k) \mid u(k-1), \ldots, u(k-d-n_B+1), y(k), y(k-1), \ldots, y(k-n_R)]$$

$$= [u(k) \mid \varphi_0^T(k)] \tag{9.101}$$

$$\hat{\theta}^T(k) = [\hat{b}_0(k) \mid \hat{\theta}_0^T(k)] \tag{9.102}$$

$$u(k) = \frac{\Gamma(q^{-1})y_m(k+d) - \hat{\theta}_0^T(k)\varphi_0(k)}{\hat{b}_0(k)} \tag{9.103}$$

$$\bar{\varepsilon}(k) = \Gamma(q^{-1})y(k) - \hat{\theta}^T(k-1)\varphi_0(k-d) \tag{9.104}$$

$$\varepsilon(k) = \frac{\bar{\varepsilon}(k)}{1 + \varphi^T(k-d)F(k)\varphi(k-d)} \tag{9.105}$$

$$\hat{\theta}(k) = \hat{\theta}(k-1) + F(k)\varphi(k-d)\varepsilon(k) \tag{9.106}$$

$$F^{-1}(k+1) = \lambda_1(k)F^{-1}(k) + \lambda_2(k)\varphi(k-d)\varphi^T(k-d) \tag{9.107}$$

with initial conditions $y(0)$, $F(0) > 0$, $\hat{\theta}(0)$

(a) Determine a model reference control law which achieves the control objectives in the case of known parameters for $A(q^{-1})$ and $B(q^{-1})$.

(b) In the case of unknown plant parameters, determine a control law and appropriate adaptations (MRAC design) in order to satisfy the control objectives asymptotically.

▪▪▪ **SOLUTION**

(a) Here $d = 1$, $S(q^{-1}) = 1$ and we are looking for $R(q^{-1}) = r_0 + r_1 q^{-1}$ such that

$$\Gamma(q^{-1}) = 1 + 0.5q^{-1} = (1 + 2q^{-1} + q^{-2}) + q^{-1}(r_0 + r_1 q^{-1})$$
$$= A(q^{-1})S(q^{-1}) + q^{-d}R(q^{-1})$$

or equivalently

$$\begin{bmatrix} 1 & 0 & 0 \\ a_1 & 1 & 0 \\ a_2 & 0 & 1 \end{bmatrix}\begin{bmatrix} 1 \\ r_0 \\ r_1 \end{bmatrix} = \begin{bmatrix} 1 \\ \gamma_1 \\ 0 \end{bmatrix} \quad \text{or} \quad \begin{bmatrix} 1 & 0 & 0 \\ 2 & 1 & 0 \\ 1 & 0 & 1 \end{bmatrix}\begin{bmatrix} 1 \\ r_0 \\ r_1 \end{bmatrix} = \begin{bmatrix} 1 \\ 0.5 \\ 0 \end{bmatrix}$$

One easily obtains $r_0 = -1.5$ and $r_1 = -1$. Moreover,

$$\Psi(q^{-1}) = B(q^{-1})S(q^{-1}) = 2 + q^{-1} + 0.5q^{-2} \doteq 2 + q^{-1}(1 + 0.5q^{-1}) = b_0 + q^{-1}\Psi^*(q^{-1})$$

In the case of known parameters, the control law is

$$u(k) = \frac{\Gamma(q^{-1})y_m(k+1) - R(q^{-1})y(k) - \Psi^*(q^{-1})u(k-1)}{b_0}$$

$$= \frac{y_m(k+1) + 0.5y_m(k) + 1.5y(k) + y(k-1) - u(k-1) - 0.5u(k-2)}{2}$$

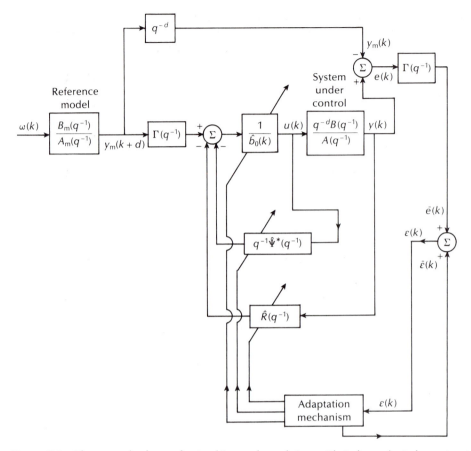

Figure 9.8 The control scheme for tracking and regulation with independent dynamics, in the case of unknown parameters.

(b) In the case of unknown plant parameters, the certainty equivalence control law is

$$u(k) = \frac{y_m(k+1) + 0.5y_m(k) - \hat{\theta}_1(k)u(k-1) - \hat{\theta}_2(k)u(k-2) - \hat{\theta}_3(k)y(k) - \hat{\theta}_4(k)y(k-1)}{\hat{\theta}_0(k)}$$

where

$$\hat{\boldsymbol{\theta}}^T(k) = [\hat{\theta}_0(k), \hat{\theta}_1(k), \hat{\theta}_2(k), \hat{\theta}_3(k), \hat{\theta}_4(k)]$$

The adaptation algorithm is given below. In this algorithm we let $\lambda_1(k) = 0.98$ and $\lambda_2(k) = 1$. This corresponds to a forgetting-factor algorithm, as explained in the third part of this section. The adaptation algorithm is

$$\mathbf{F}(0) = \frac{1}{10^{-3}}\,\mathbf{I}_5$$

$$\boldsymbol{\phi}^{\mathrm{T}}(k) = [u(k), u(k-1), u(k-2), y(k), y(k-1)] \quad \text{for} \quad k = 0, 1, 2, \ldots$$

$$\tilde{\varepsilon}(k) = y(k) + 0.5y(k-1) - \hat{\boldsymbol{\theta}}^{\mathrm{T}}(k-1)\boldsymbol{\phi}(k-1) \quad \text{for} \quad k = 1, 2, \ldots$$

$$\varepsilon(k) = \frac{\tilde{\varepsilon}(k)}{1 + \boldsymbol{\phi}^{\mathrm{T}}(k-1)\mathbf{F}(k)\boldsymbol{\phi}(k-1)} \quad \text{for} \quad k = 1, 2, \ldots$$

$$\hat{\boldsymbol{\theta}}(k) = \hat{\boldsymbol{\theta}}(k-1) + \mathbf{F}(k)\boldsymbol{\phi}(k-1)\varepsilon(k) \quad \text{for} \quad k = 1, 2, \ldots$$

$$\mathbf{F}^{-1}(k+1) = 0.98\mathbf{F}^{-1}(k) + \boldsymbol{\phi}(k-1)\boldsymbol{\phi}^{\mathrm{T}}(k-1) \quad \text{for} \quad k = 0, 1, 2, \ldots$$

We can initialize $\hat{\boldsymbol{\theta}}(k)$ with $\hat{\boldsymbol{\theta}}^{\mathrm{T}}(0) = [1, 0, 0, -1, -1]$ for convenience.

2. The hyperstability approach

The particular choice of the adaptation algorithm was guided by the objective of global asymptotic stability for the whole system, i.e. asymptotic stability for any finite initial parameter error and plant–model error. Moreover, the adaptation mechanism should ensure that the error between the plant output and the output of the reference model tends to zero asymptotically. The approach applied to satisfy the objectives relies on the fact that the model reference adaptive system can be viewed equivalently as a connection of a linear time-invariant feedforward block and a non-linear time-varying feedback block, as shown in Figure 9.9. The stability characteristics of this equivalent feedback configuration for the MRAC system are analyzed using hyperstability theory. Global asymptotic stability is guaranteed if the two blocks represent positive systems, as defined below. This analysis also leads to a general structure for the parameter adaptation algorithm to be used. Before the main theorem is stated, some preliminaries on positive systems and hyperstability theory will be given.

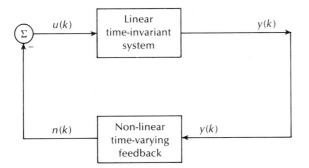

Figure 9.9 Definition of the hyperstability concept.

Consider the linear, time-invariant, single-input single-output, discrete-time system

$$\mathbf{x}(k+1) = \mathbf{A}\mathbf{x}(k) + \mathbf{b}u(k) \tag{9.108}$$

$$y(k) = \mathbf{c}^T\mathbf{x}(k) + du(k) \tag{9.109}$$

The transfer function of this state-space model is

$$G(z) = \mathbf{c}^T(z\mathbf{I} - \mathbf{A})^{-1}\mathbf{b} + d \tag{9.110}$$

It is assumed that the pair $(\mathbf{A}, \mathbf{b})$ is controllable and the pair $(\mathbf{A}, \mathbf{c}^T)$ is observable.

■ **DEFINITION 9.4.1** The transfer function $G(z)$ is **positive real** (PR) if:

1. $G(z)$ is analytic outside the unit circle (i.e. $G(z)$ has no poles for $|z| > 1$) and the poles of $G(z)$ which are on the unit circle $|z| = 1$ are simple with non-negative residues.
2. $G(e^{j\omega}) + G(e^{-j\omega}) \geq 0$ for all real ω for which $G(e^{j\omega})$ exists (i.e. for all z on the unit circle $|z| = 1$ which are not poles of $G(z)$).

The transfer function $G(z)$ is **strictly positive real** (SPR) if $G(\rho z)$ is positive real for some positive $\rho < 1$.

■ **LEMMA 9.4.1 (The Discrete Positive Real Lemma)** The transfer function $G(z)$ is SPR if and only if there exist a real positive definite matrix $\mathbf{P}$, a real matrix $\mathbf{L}$ and constants w and γ such that

$$\mathbf{A}^T\mathbf{P}\mathbf{A} - \mathbf{P} = -\mathbf{L}\mathbf{L}^T - \gamma^2\mathbf{I} \tag{9.111}$$

$$\mathbf{b}^T\mathbf{P}\mathbf{A} = \mathbf{c}^T + w\mathbf{L}^T \tag{9.112}$$

$$w^2 = 2d - \mathbf{b}^T\mathbf{P}\mathbf{b} \tag{9.113}$$

Consider now the closed-loop system of Figure 9.9 consisting of the linear controllable and observable system (9.108) and (9.109) in the feedforward path and a non-linear, time-varying, output feedback block of the form $n(k) = f(y(k), k, l)$, $l \leq k$. It is assumed that the feedback block satisfies the following inequality, known as the **Popov inequality**:

$$\sum_{k=k_0}^{k_1} n(k)y(k) \geq -\gamma_0^2 \quad \forall\, k_1 \geq k_0 \tag{9.114}$$

with γ_0^2 a positive constant. Actually, (9.114) is a sign preservation property. Note that this sign preservation property holds for discrete PR systems.

■ **DEFINITION 9.4.2**   The feedback connection of the two systems given in Figure 9.9 is **hyperstable**, if there exist positive constants $\delta > 0$ and $\gamma_0 > 0$ such that for every solution $\mathbf{x}(k)$ of (9.108) initialized at $\mathbf{x}(0)$, the inequality

$$\| \mathbf{x}(k) \| \leqslant \delta[\| \mathbf{x}(0) \| + \gamma_0] \quad \forall\, k > 0 \tag{9.115}$$

is satisfied for every non-linear feedback block $n(k) = f(y(k), k, l)$, $l \leqslant k$, satisfying the Popov inequality (9.114).

The feedback connection of Figure 9.9 is **asymptotically hyperstable**, if it is hyperstable and, moreover, $\lim_{k \to +\infty} \mathbf{x}(k) = \mathbf{0}$ for every feedback block $n(k) = f(y(k), k, l)$, $l \leqslant k$, satisfying the Popov inequality (9.114).

The following proposition relates asymptotically hyperstable and positive systems.

■ **PROPOSITION 9.4.1**   The necessary and sufficient condition for the feedback connection of Figure 9.9 to be asymptotically hyperstable for every feedback block satisfying the Popov inequality is that the transfer function $G(z)$ associated with the linear block should be strictly positive real.

Most adaptive control schemes, after an adequate analysis, lead to an equation of the form

$$v(k) = H(q^{-1})[\boldsymbol{\theta} - \hat{\boldsymbol{\theta}}(k)]^{\mathsf{T}} \boldsymbol{\phi}(k - d) \tag{9.116}$$

where $\boldsymbol{\theta}$ is an unknown parameter vector, $\hat{\boldsymbol{\theta}}(k)$ is the estimate of $\boldsymbol{\theta}$ resulting from an appropriate parameter adaptation algorithm, $\boldsymbol{\phi}(k - d)$ is a measurable regressor vector, $H(q^{-1})$ is a rational discrete transfer function of the form

$$H(z^{-1}) = \frac{1 + h_1' z^{-1} + \cdots + h_\alpha' z^{-\alpha}}{1 + h_1 z^{-1} + \cdots + h_\beta z^{-\beta}} \tag{9.117}$$

and the measurable quantity $v(k)$ is the so-called **processed augmented error**. A particular case of (9.116) is given by (9.93), where $H(q^{-d}) = 1$ and the *a posteriori* filtered plant–model error $\varepsilon(k)$ takes the place of $v(k)$. Then, the following stability theorem provides the appropriate adaptation mechanism sought.

■ **THEOREM 9.4.1** [21]   Assume that the relation between the measurable regressor vector $\boldsymbol{\phi}(k - d)$ and the processed augmented error $v(k)$ is given by the following output error type equation:

$$v(k) = H(q^{-1})[\boldsymbol{\theta} - \hat{\boldsymbol{\theta}}(k)]^{\mathsf{T}} \boldsymbol{\phi}(k - d) \tag{9.118}$$

Also assume that the following adaptation algorithm, which makes use of $v(k)$ as the basis of the parameter update law for $\hat{\boldsymbol{\theta}}(k)$, is used:

$$\hat{\boldsymbol{\theta}}(k) = \hat{\boldsymbol{\theta}}(k-1) + \mathbf{F}(k)\boldsymbol{\phi}(k-d)v(k) \tag{9.119}$$

$$\mathbf{F}^{-1}(k+1) = \lambda_1(k)\mathbf{F}^{-1}(k) + \lambda_2(k)\boldsymbol{\phi}(k-d)\boldsymbol{\phi}^{\mathrm{T}}(k-d) \quad \text{where} \quad \mathbf{F}(0) > 0 \tag{9.120}$$

with

$$0 < \lambda_1(k) \leqslant 1, \quad 0 \leqslant \lambda_2(k) < 2, \quad \forall k \tag{9.121}$$

Then, if the transfer function

$$\bar{H}(z^{-1}) = H(z^{-1}) - \frac{\lambda}{2} \tag{9.122}$$

is strictly positive real, where

$$2 > \lambda \geqslant \max_{0 \leqslant k < \infty} [\lambda_2(k)] \tag{9.123}$$

the following properties hold for all bounded initial conditions $v(0)$ and $\hat{\boldsymbol{\theta}}(0)$:

1. The global asymptotic stability of the processed augmented error

$$\lim_{k \to +\infty} v(k) = 0 \tag{9.124}$$

2. The global asymptotic stability of $\mathbf{x}(k)$, where $\mathbf{x}(k)$ is the state vector of any state realization of $H(z^{-1})$

$$\lim_{k \to +\infty} \mathbf{x}(k) = \mathbf{0} \tag{9.125}$$

3. The asymptotic orthogonality of the parametric error $\hat{\boldsymbol{\theta}}(k) - \boldsymbol{\theta}$ and the regressor vector $\boldsymbol{\phi}(k-d)$

$$\lim_{k \to +\infty} \boldsymbol{\phi}^{\mathrm{T}}(k-d)[\hat{\boldsymbol{\theta}}(k) - \boldsymbol{\theta}] = 0 \tag{9.126}$$

4. The asymptotic stability of a weighted norm of the parametric error

$$\lim_{k \to +\infty} [1 - \lambda_1(k)] \| \hat{\boldsymbol{\theta}}(k) - \boldsymbol{\theta} \|_{\mathbf{F}^{-1}(k)} = \lim_{k \to +\infty} [1 - \lambda_1(k)][\hat{\boldsymbol{\theta}}(k) - \boldsymbol{\theta}]^{\mathrm{T}}\mathbf{F}^{-1}(k)[\hat{\boldsymbol{\theta}}(k) - \boldsymbol{\theta}] = 0 \tag{9.127}$$

5. The asymptotic stability of a norm of the parameter variation. That is, letting

$$\Delta\hat{\boldsymbol{\theta}}(k) = \hat{\boldsymbol{\theta}}(k) - \hat{\boldsymbol{\theta}}(k-1) \tag{9.128}$$

we have that

$$\lim_{k \to +\infty} \Delta\hat{\boldsymbol{\theta}}^{\mathrm{T}}(k)\mathbf{F}^{-1}(k)\Delta\hat{\boldsymbol{\theta}}(k) = \lim_{k \to +\infty} \| \boldsymbol{\phi}(k-d)v(k) \|_{\mathbf{F}(k)} = 0 \tag{9.129}$$

6. The uniformly bounded character of a norm of the parametric error

$$[\hat{\boldsymbol{\theta}}(k-1) - \boldsymbol{\theta}]^{\mathrm{T}}\mathbf{F}^{-1}(k)[\hat{\boldsymbol{\theta}}(k-1) - \boldsymbol{\theta}] < M_1 < \infty \quad \forall k \tag{9.130}$$

7. The convergence of the above norm of the parametric error to a constant

$$\lim_{k \to +\infty} [\hat{\boldsymbol{\theta}}(k-1) - \boldsymbol{\theta}]^{\mathrm{T}} \mathbf{F}^{-1}(k)[\hat{\boldsymbol{\theta}}(k-1) - \boldsymbol{\theta}] = \text{constant} \tag{9.131}$$

Moreover, if $\mathbf{F}^{-1}(k) > \varepsilon\mathbf{F}^{-1}(0)$, $\mathbf{F}(0) > 0$ and $\varepsilon > 0$, $\forall\, k \geq 0$ and $\mathbf{F}^{-1}(k)$ is non-decreasing for $k \geq k_0$, then

8.

$$\lim_{k \to +\infty} \Delta\hat{\boldsymbol{\theta}}(k) = \lim_{k \to +\infty} \mathbf{F}(k)\boldsymbol{\phi}(k-d)v(k) = 0 \tag{9.132}$$

9.

$$\|\hat{\boldsymbol{\theta}}(k)\| \leq M_2 < \infty \quad \text{for} \quad k \geq 0 \tag{9.133}$$

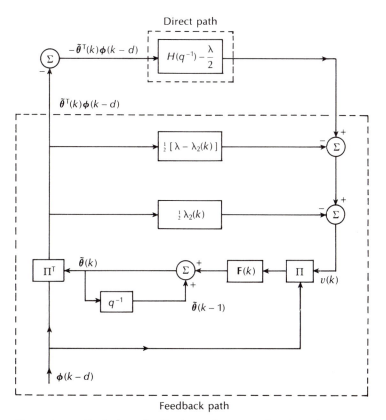

Figure 9.10 Equivalent feedback configuration for the adaptation mechanism, where Π denotes multiplication.

The proof of this stability theorem is based on hyperstability concepts and will be presented briefly. Equation (9.116), i.e. equation

$$v(k) = -H(q^{-1})\tilde{\boldsymbol{\theta}}^{\mathrm{T}}(k)\boldsymbol{\phi}(k-d) \quad \text{where} \quad \tilde{\boldsymbol{\theta}}(k) = \hat{\boldsymbol{\theta}}(k) - \boldsymbol{\theta} \tag{9.134}$$

combined with equation (9.119), which may be written as

$$\tilde{\boldsymbol{\theta}}(k) = \tilde{\boldsymbol{\theta}}(k-1) + \mathbf{F}(k)\boldsymbol{\phi}(k-d)v(k) \tag{9.135}$$

define a feedback system. In Figure 9.10 (p. 269), an equivalent configuration for this feedback system is given, involving the quantities λ and $\lambda_2(k)$. This feedback system has a linear block in the direct path and a time-varying non-linear block in the feedback path.

It can be proved that the feedback path, which realizes the adaptation algorithm (9.119), (9.120) and (9.121), satisfies the Popov inequality (9.114) provided that

$$0 < \lambda_1(k) \leqslant 1 \quad \forall k, \qquad 0 \leqslant \lambda_2(k) < 2 \quad \forall k, \qquad 2 > \lambda \geqslant \max_{0 \leqslant k < \infty} [\lambda_2(k)] \tag{9.136}$$

If there exists a λ such that $\bar{H}(z^{-1}) = H(z^{-1}) - \lambda/2$ is strictly positive real, then the equivalent feedback configuration of Figure 9.10 satisfies the conditions of Proposition 9.4.1 (hyperstability theorem). Consequently, $v(k) \to 0$ as $k \to +\infty$.

To prove that the plant–model error $e(k)$ converges to zero and the boundedness of the input and output signals is a cumbersome task and this additional analysis is omitted here.

3. Discussion of the parameter adaptation algorithm

Expression (9.120) defines a general law for the determination of the adaptation gain matrix $\mathbf{F}(k)$ and is repeated here for convenience:

$$\mathbf{F}^{-1}(k+1) = \lambda_1(k)\mathbf{F}^{-1}(k) + \lambda_2(k)\boldsymbol{\phi}(k-d)\,\boldsymbol{\phi}^{\mathrm{T}}(k-d) \quad \text{with} \quad \mathbf{F}(0) > \mathbf{0} \tag{9.137}$$

where $0 < \lambda_1(k) \leqslant 1$ and $0 \leqslant \lambda_2(k) < 2$. Using the matrix inversion lemma (8.26), the above equation may be written equivalently as follows:

$$\mathbf{F}(k+1) = \frac{1}{\lambda_1(k)} \left(\mathbf{F}(k) - \frac{\mathbf{F}(k)\boldsymbol{\phi}(k-d)\boldsymbol{\phi}^{\mathrm{T}}(k-d)\mathbf{F}(k)}{\lambda_1(k)\lambda_2^{-1}(k) + \boldsymbol{\phi}^{\mathrm{T}}(k-d)\mathbf{F}(k)\boldsymbol{\phi}(k-d)} \right) \tag{9.138}$$

We note that in general $\lambda_1(k)$ and $\lambda_2(k)$ have opposite effects on the adaptation gain. That is, as $\lambda_1(k) \leqslant 1$ increases the gain, $\lambda_2(k)$ does the opposite, i.e. it decreases the gain.

Different types of adaptation algorithms are obtained by appropriate choices of $\lambda_1(k)$ and $\lambda_2(k)$, $0 < \lambda_1(k) \leqslant 1$, $0 \leqslant \lambda_2(k) < 2$. We distinguish the following choices:

1. $\lambda_1(k) \equiv 1$ and $\lambda_2(k) \equiv 0$. In this case $\mathbf{F}(k) = \mathbf{F}(0)$. This choice corresponds to an algorithm with a constant gain. It is the simplest to implement, but also the least efficient. It is convenient for the estimation of unknown constant parameters, but not for time-varying parameters.

2. $\lambda_1(k) = \lambda_2(k) \equiv 1$. This choice corresponds to a recursive least-squares algorithm with decreasing gain.

3. $\lambda_1(k) \equiv \lambda_1 < 1$ (usually $0.95 \le \lambda_1 \le 0.99$) and $\lambda_2(k) \equiv 1$. This choice corresponds to an algorithm with a constant forgetting factor λ_1 (it 'forgets' old measurements exponentially).

4. $\lambda_1(k) < 1$ and $\lambda_2(k) \equiv 1$. This choice corresponds to a variable forgetting factor type of algorithm. Usually, $0.95 \le \lambda_1(k) \le 0.99$ or $\lambda_1(k+1) = \lambda_0\lambda_1(k) + 1 - \lambda_0$, with $0.95 \le \lambda_0 \le 0.99$ and $0.95 \le \lambda_1(0) \le 0.99$. In this last case, $\lim_{k \to +\infty} \lambda_1(k) = 1$ holds true.

5. When both $\lambda_1(k)$ and $\lambda_2(k)$ are time varying, we have extra freedom in choosing the gain profiles. For example, by choosing $\lambda_1(k)/\lambda_2(k) = \alpha(k)$ we have the following expression for the trace of $\mathbf{F}(k)$:

$$\text{tr } \mathbf{F}(k+1) = \frac{1}{\lambda_1(k)} \text{tr}\left(\mathbf{F}(k) - \frac{\mathbf{F}(k)\boldsymbol{\phi}(k-d)\boldsymbol{\phi}^{\mathrm{T}}(k-d)\mathbf{F}(k)}{\alpha(k) + \boldsymbol{\phi}^{\mathrm{T}}(k-d)\mathbf{F}(k)\boldsymbol{\phi}(k-d)}\right) \tag{9.139}$$

At each step we can choose $\alpha(k) = \lambda_1(k)/\lambda_2(k)$ and then specify $\lambda_1(k)$ from (9.139) such that the trace of $\mathbf{F}(k)$ has a prespecified value (constant or time varying) at each step.

■ **REMARK 9.4.5** We note that when $\boldsymbol{\phi}(k-d) = \mathbf{0}$ for a long period of time (this may happen in the steady state or in the absence of any signal in the input), using choices 3 or 4 may lead to an undesirable increase in the adaptation gain. In this case there is no change in the parameter estimates and $\mathbf{F}(k)$ will grow exponentially if $\lambda_1(k) < 1$, since in this case we have that $\mathbf{F}(k+1) = [1/\lambda_1(k)]\mathbf{F}(k)$. A new change in the set point can then lead to large changes in the parameter estimates and the plant output.

■ **REMARK 9.4.6** In practice we initialize at $\mathbf{F}(0) = (1/\delta)\mathbf{I}$, $0 < \delta \ll 1$.

9.5 Self-tuning regulators

9.5.1 *Introduction*

Another important class of adaptive systems with many industrial applications is that of self-tuning regulators. The block diagram of a self-tuning regulator (STR) is shown in Figure 9.11.

The STR is based on the idea of separating the estimation of the unknown parameters of the system under control from the design of the controller. The control scheme consists of two loops: the inner loop, which involves the plant with unknown parameters and a linear feedback controller with adjustable parameters; and the outer loop, which is used in the case of unknown plant parameters and is composed of a recursive parameter estimator and a block named 'controller design'.

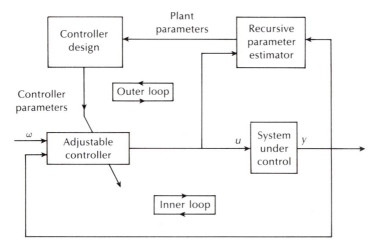

Figure 9.11 Block diagram of a self-tuning regulator.

In the case of known plant parameters, the design of the controller (i.e. the determination of its parameters as functions of the plant parameters) is carried out OFF-LINE. This controller satisfies a specific control design problem, such as minimum variance, pole placement, model following, etc. This control problem, in the context of the STRs, is called the **underlying control problem**.

When the plant parameters are uncertain, the recursive parameter estimator provides ON-LINE estimates of the unknown plant parameters. On the basis of these estimates, the solution of the control design problem (i.e. the determination of the controller parameters as functions of the plant parameters) is achieved ON-LINE in each step by the controller design block. The controller parameter estimates thus obtained are used to recalculate the control law at each step. Apart from the fact that the controller parameters are substituted by estimates obtained by the ON-LINE solution of the control design problem, the controller structure is kept the same as in the case of known plant parameters. This is the **certainty equivalence principle** (see also Section 9.3).

For the estimation of the plant parameters, various schemes can be used: least squares, recursive least squares, maximum likelihood, extended Kalman filtering, etc. Different combinations of appropriate parameter estimation methods and suitable control strategies lead to different adaptive controllers. For example, an adaptive controller based on least-squares estimation and deadbeat control was first described by Kalman in 1958, while the original STR design by Åström and Wittenmark [15] was based on least-squares estimation and the minimum variance control problem.

The control procedure discussed above leads to an **explicit STR**, where the term 'explicit' is due to the fact that the plant parameters are estimated explicitly. Such explicit STRs need to solve, at each step, the tedious controller design problem. It is sometimes possible, in order to eliminate the design calculations, to reparametrize the plant model so that it can be expressed in terms of the controller parameters which are then updated directly by the estimator. This leads to **implicit STRs**. These avoid controller design calculations and are based on estimates of an implicit plant model. Explicit STRs correspond to indirect adaptive control, while implicit STRs correspond to direct adaptive control.

A close relation has been established between STRs and MRAC systems, in spite of differences in their origin. Indeed, MRAC design was based on the deterministic servoproblem, while STR design was based on the stochastic regulation problem. Although the design methods of the inner loop and the parameter adjustments in the outer loop are different, direct MRAC systems are closely related to implicit STRs, while indirect MRAC systems are related to explicit STRs.

Implicit STRs are discussed in Section 9.5.2. In particular, a minimum variance control problem (combining both regulation and the servoproblem) is considered. The relation between the implicit STRs and the MRAC design of Section 9.4 is established. Explicit STRs, using the pole-placement technique, are treated in Section 9.5.3. Furthermore, it is explained how implicit pole-placement designs can also be derived.

9.5.2   *Regulation and tracking with minimum variance control*

Even though the original STR of Åström and Wittenmark [15] was designed for regulation purposes, here we also consider the servoproblem. Consider the system described by the single-input single-output model

$$A(q^{-1})y(k+d) = B(q^{-1})u(k) + \Delta(q^{-1})w(k+d) \tag{9.140}$$

where $u(k)$ is the input, $y(k)$ the output and $w(k)$ a sequence of independent, equally distributed Gaussian variables with zero mean value and variance σ^2. We are thus placed in a stochastic environment, where the disturbances can be characterized as filtered white noise. Furthermore, we have that

$$A(q^{-1}) = 1 + a_1 q^{-1} + \cdots + a_{n_A} q^{-n_A} \tag{9.141}$$

$$B(q^{-1}) = b_0 + b_1 q^{-1} + \cdots + b_{n_B} q^{-n_B} = b_0 + q^{-1}B^*(q^{-1}) \quad \text{with} \quad b_0 \neq 0 \tag{9.142}$$

$$\Delta(q^{-1}) = 1 + \delta_1 q^{-1} + \cdots + \delta_{n_A} q^{-n_A} = 1 + q^{-1}\Delta^*(q^{-1}) \tag{9.143}$$

It is assumed that the delay d of the system is known (so that $b_0 \neq 0$) and that $\deg A(q^{-1}) = \deg \Delta(q^{-1}) = n_A$. It is also necessary that the polynomials $B(q^{-1})$ and $\Delta(q^{-1})$ have all zeros inside the unit circle, thus restricting the approach of the present section to minimum-phase systems.

Assume that $y_m(k)$ is a reference sequence stored in memory or generated by the reference model

$$A_m(q^{-1})y_m(k) = q^{-d}B_m(q^{-1})\omega(k) \tag{9.144}$$

with

$$A_m(q^{-1}) = 1 + a_1^m q^{-1} + \cdots + a_{n_{A_m}}^m q^{-n_{A_m}} \tag{9.145}$$

$$B_m(q^{-1}) = b_0^m + b_1^m q^{-1} + \cdots + b_{n_{B_m}}^m q^{-n_{B_m}} \tag{9.146}$$

and $\omega(k)$ a bounded sequence. The control objectives in a stochastic environment will be related to the minimization of the variance of the error $y(k) - y_m(k)$ while tracking a trajectory, or to the minimization of the variance of the output in the case of regulation. That is, in the case of tracking a trajectory, it is desired that

$$E\{[y(k+d) - y_m(k+d)]^2\} \tag{9.147}$$

is minimized. In the case of regulation $(y_m(k+d) \equiv 0)$, it is desired that

$$E\{y^2(k+d)\} \tag{9.148}$$

is minimized. The control objectives will be satisfied in the case of known, as well as unknown, parameters in $A(q^{-1})$, $B(q^{-1})$ and $\Delta(q^{-1})$.

1. Minimum variance control in the case of known parameters

In order to obtain a causal controller satisfying the control objectives in the case of known parameters, the following polynomial (or Bezout) identity will be used:

$$\Delta(q^{-1}) = S(q^{-1})A(q^{-1}) + q^{-d}R(q^{-1}) \tag{9.149}$$

where

$$S(q^{-1}) = 1 + s_1 q^{-1} + \cdots + s_{n_S} q^{-n_S} \tag{9.150}$$

$$R(q^{-1}) = r_0 + r_1 q^{-1} + \cdots + r_{n_R} q^{-n_R} \tag{9.151}$$

This identity has a unique solution for $S(q^{-1})$ and $R(q^{-1})$ when $\deg S(q^{-1}) = n_S = d - 1$ and $\deg R(q^{-1}) = n_R = \max(n_A - 1, n_A - d) = n_A - 1$. Multiplying (9.140) by $S(q^{-1})$ yields

$$S(q^{-1})A(q^{-1})y(k+d) = S(q^{-1})B(q^{-1})u(k) + S(q^{-1})\Delta(q^{-1})w(k+d) \tag{9.152}$$

Using (9.149), relation (9.152) becomes

$$\Delta(q^{-1})y(k+d) - R(q^{-1})y(k) = \Psi(q^{-1})u(k) + S(q^{-1})\Delta(q^{-1})w(k+d) \tag{9.153}$$

where

$$\Psi(q^{-1}) = B(q^{-1})S(q^{-1}) = b_0 + q^{-1}\Psi^*(q^{-1}) = b_0 + \psi_1 q^{-1} + \cdots + \psi_{d+n_B-1}q^{-(d+n_B-1)} \tag{9.154}$$

Finally, (9.153) may be written as

$$y(k+d) = \left(\frac{R(q^{-1})}{\Delta(q^{-1})}\right)y(k) + \left(\frac{\Psi(q^{-1})}{\Delta(q^{-1})}\right)u(k) + S(q^{-1})w(k+d) \tag{9.155}$$

Here, we are searching for $u(k)$ which minimizes the cost function

$$E\{[y(k+d) - y_m(k+d)]^2\} =$$

$$E\left\{\left[\frac{R(q^{-1})}{\Delta(q^{-1})}y(k) + \frac{\Psi(q^{-1})}{\Delta(q^{-1})}u(k) - y_m(k+d) + S(q^{-1})w(k+d)\right]^2\right\} \tag{9.156}$$

Since $w(k)$ is white noise, the term $S(q^{-1})w(k+d)$, which is a moving average containing $w(k+d)$, $w(k+d-1)$, ..., $w(k+1)$, is not correlated to $y(k)$, $y_m(k+d)$ and $u(k)$. Hence, the expected values of terms containing the products $y(k)w(k+d)$, $y_m(k+d)w(k+d)$ and $u(k)w(k+d)$ are zero. We then have

$$E\{[y(k+d) - y_m(k+d)]^2\} =$$

$$E\left\{\left(\frac{R(q^{-1})}{\Delta(q^{-1})}y(k) + \frac{\Psi(q^{-1})}{\Delta(q^{-1})}u(k) - y_m(k+d)\right)^2\right\} + E\{[S(q^{-1})w(k+d)]^2\} \quad (9.157)$$

We note that the second term in the cost function is not affected by the control signal. Therefore, in order to minimize the variance of the error we must have

$$R(q^{-1})y(k) + \Psi(q^{-1})u(k) - \Delta(q^{-1})y_m(k+d)$$
$$= R(q^{-1})y(k) + b_0 u(k) + \Psi^*(q^{-1})u(k-1) - \Delta(q^{-1})y_m(k+d) = 0 \quad (9.158)$$

The minimum variance control is then given by

$$u(k) = \frac{\Delta(q^{-1})y_m(k+d) - \Psi^*(q^{-1})u(k-1) - R(q^{-1})y(k)}{b_0} \quad (9.159)$$

From (9.155) and (9.159) it is easy to see that the tracking error is a moving average, i.e.

$$y(k+d) - y_m(k+d) = S(q^{-1})w(k+d) \quad (9.160)$$

Equation (9.159) can be rewritten as

$$\Delta(q^{-1})y_m(k+d) = \theta^T\phi(k) \quad (9.161)$$

where

$$\theta^T = [b_0, \Psi_1, ..., \Psi_{d+n_B-1}, r_0, r_1, ..., r_{n_R}] \quad (9.162)$$

$$\phi^T(k) = [u(k), ..., u(k-1), u(k-d-n_B+1), y(k), y(k-1), ..., y(k-n_R)] \quad (9.163)$$

Note that the controller of Section 9.4, which achieves tracking and regulation with the two dynamics defined independently and the minimum variance controller derived above, have exactly the same structure. These two structures become identical if we choose $\Gamma(q^{-1}) = \Delta(q^{-1})$. In this case equations (9.161) and (9.73) are also identical. This establishes the connection between the two techniques and justifies the reason why we must secure the dynamics for tracking and regulation independently. Indeed, it is clear that the dynamics $\Gamma(q^{-1})$ during regulation must be chosen with respect to the characteristics of the disturbance model $\Delta(q^{-1})$ and not with respect to the desired dynamics during tracking.

2. Minimum variance control in the case of unknown parameters

In the case of uncertain parameters in the system model (i.e. in $A(q^{-1})$, $B(q^{-1})$ and $\Delta(q^{-1})$), an adjustable controller is used. The structure of the controller (9.159) is kept the same. The polynomials $\Delta^*(q^{-1})$, $\Psi^*(q^{-1})$ and $R(q^{-1})$ which involve unknown parameters and the constant b_0 are replaced by the estimates $\hat{\Delta}^*(q^{-1})$, $\hat{\Psi}^*(q^{-1})$, $\hat{R}(q^{-1})$ and $\hat{b}_0(k)$. Hence, the control law now becomes

$$u(k) = \frac{y_m(k+d) + \hat{\Delta}^*(q^{-1})y_m(k+d-1) - \hat{\Psi}^*(q^{-1})u(k-1) - \hat{R}(q^{-1})y(k)}{\hat{b}_0(k)} \quad (9.164)$$

where use was made of the relation $\Delta(q^{-1}) = 1 + q^{-1}\Delta^*(q^{-1})$. Expression (9.164) may be written compactly as

$$\Delta(q^{-1})y_m(k+d) = \hat{\boldsymbol{\theta}}_e^T(k)\boldsymbol{\phi}_e(k) \quad (9.165)$$

where

$$\hat{\boldsymbol{\theta}}_e^T(k) = [\hat{b}_0(k), \hat{\psi}_1(k), \dots, \hat{\psi}_{d+n_B-1}(k), \hat{r}_0(k), \dots, \hat{r}_{n_R}(k), \hat{\delta}_1(k), \dots, \hat{\delta}_{n_A}(k)] \quad (9.166)$$

and

$$\boldsymbol{\phi}_e^T(k) = [u(k), u(k-1), \dots, u(k-d-n_B+1), y(k), \dots,$$
$$y(k-n_R), -y_m(k+d-1), \dots, -y_m(k+d-n_A)] \quad (9.167)$$

In the adaptive case, the error $y(k+d) - y_m(k+d)$ cannot be identically equal to a moving average, as for the case of known parameters. The control objective is now to have the error converging asymptotically to a moving average, with probability equal to 1, i.e.

$$\text{Prob}\left\{ \lim_{k\to+\infty} [y(k+d) - y_m(k+d)] = S(q^{-1})w(k+d) \right\} = 1 \quad (9.168)$$

The control law (9.164) must be completed with adequate adaptation mechanisms for the parameters in $\hat{\boldsymbol{\theta}}_e^T(k)$. Let us reparametrize the system model in order to express this model in terms of the controller parameters. This will lead to an implicit self-tuning algorithm. Starting from (9.153), we have

$$y(k+d) = -q^{-1}\Delta^*(q^{-1})y(k+d) + q^{-d}R(q^{-1})y(k+d)$$
$$+ \Psi(q^{-1})u(k) + S(q^{-1})\Delta(q^{-1})w(k+d)$$

or

$$y(k) = -\Delta^*(q^{-1})y(k-1) + R(q^{-1})y(k-d) + \Psi(q^{-1})u(k-d) + S(q^{-1})\Delta(q^{-1})w(k)$$

or

$$y(k) = -\Delta^*(q^{-1})y_m(k-1) + R(q^{-1})y(k-d) + \Psi(q^{-1})u(k-d) + S(q^{-1})\Delta(q^{-1})w(k)$$
$$- \Delta^*(q^{-1})[y(k-1) - y_m(k-1)]$$

or

$$y(k) = \boldsymbol{\theta}_e^T\boldsymbol{\phi}_e(k-d) - \Delta^*(q^{-1})[y(k-1) - y_m(k-1)] + S(q^{-1})\Delta(q^{-1})w(k) \quad (9.169)$$

where

$$\boldsymbol{\theta}_e^T = [b_0, \psi_1, \dots, \psi_{d+n_B-1}, r_0, \dots, r_{n_R}, \delta_1, \dots, \delta_{n_A}] \quad (9.170)$$

Defining the **a priori error** $\bar{e}(k) = y(k) - y_\mathrm{m}(k)$, which is a measurable quantity, and using (9.165) and (9.169), we have

$$\bar{e}(k) = y(k) - y_\mathrm{m}(k) =$$
$$[\boldsymbol{\theta}_\mathrm{e} - \hat{\boldsymbol{\theta}}_\mathrm{e}(k-d)]^\mathrm{T} \boldsymbol{\phi}_\mathrm{e}(k-d) - q^{-1}\Delta^*(q^{-1})\bar{e}(k) + S(q^{-1})\Delta(q^{-1})w(k)$$

or

$$\Delta(q^{-1})\bar{e}(k) = [\boldsymbol{\theta}_\mathrm{e} - \hat{\boldsymbol{\theta}}_\mathrm{e}(k-d)]^\mathrm{T} \boldsymbol{\phi}_\mathrm{e}(k-d) + S(q^{-1})\Delta(q^{-1})w(k)$$

or

$$\bar{e}(k) = \frac{1}{\Delta(q^{-1})} [\boldsymbol{\theta}_\mathrm{e} - \hat{\boldsymbol{\theta}}_\mathrm{e}(k-d)]^\mathrm{T}\boldsymbol{\phi}_\mathrm{e}(k-d) + S(q^{-1})w(k) \tag{9.171}$$

Now, we can define the **auxiliary error**

$$\bar{\varepsilon}(k) = \frac{1}{\Delta(q^{-1})} [\hat{\boldsymbol{\theta}}_\mathrm{e}(k-d) - \hat{\boldsymbol{\theta}}_\mathrm{e}(k)]^\mathrm{T}\boldsymbol{\phi}_\mathrm{e}(k-d) \tag{9.172}$$

and the **a posteriori error**

$$\varepsilon(k) = \bar{e}(k) + \bar{\varepsilon}(k) = \frac{1}{\Delta(q^{-1})} [\boldsymbol{\theta}_\mathrm{e} - \hat{\boldsymbol{\theta}}_\mathrm{e}(k)]^\mathrm{T}\boldsymbol{\phi}_\mathrm{e}(k-d) + S(q^{-1})w(k) \tag{9.173}$$

In view of the first two parts of Section 9.4.4, the special form of the above expression (9.173) suggests that the following adaptation law should be used:

$$\hat{\boldsymbol{\theta}}_\mathrm{e}(k) = \hat{\boldsymbol{\theta}}_\mathrm{e}(k-1) + \frac{\mathbf{F}(k)\boldsymbol{\phi}_\mathrm{e}(k-d)\bar{e}(k)}{1 + \boldsymbol{\phi}_\mathrm{e}^\mathrm{T}(k-d)\mathbf{F}(k)\boldsymbol{\phi}_\mathrm{e}(k-d)} \tag{9.174}$$

$$\mathbf{F}^{-1}(k+1) = \lambda_1(k)\mathbf{F}^{-1}(k) + \lambda_2(k)\boldsymbol{\phi}_\mathrm{e}(k-d)\boldsymbol{\phi}_\mathrm{e}^\mathrm{T}(k-d) \quad \text{with} \quad \mathbf{F}(0) > 0 \tag{9.175}$$

In the stochastic case, in order to assure convergence to constant parameters, the adaptation gain should converge to zero. That is, we use $\lambda_1 \equiv 1$ (or $\lim_{k \to +\infty} \lambda_1(k) = 1$) and $0 \leqslant \lambda_2(k) < 2$, $\forall\, k$. Moreover, from (9.173) and referring to the second part of Section 9.4.4, we conclude that the transfer function

$$\frac{1}{\Delta(q^{-1})} - \frac{\lambda}{2} \quad \text{for} \quad 2 > \lambda \geqslant \max_{0 \leqslant k < \infty} [\lambda_2(k)] \tag{9.176}$$

must be strictly positive real. That is, the convergence of the minimum variance self-tuning controller depends on the characteristics of the noise. In order to relax this very strict condition, it has been proposed in the literature to filter the input and output data.

We can arrive at the same results by using the following considerations which reveal the separation between identification and control: starting from equation (9.169) we note that $S(q^{-1})w(k)$ is a moving average and, moreover, we expect that the error $\bar{e}(k-1) = y(k-1) - y_m(k-1)$ will converge asymptotically to a moving average with probability equal to 1. Therefore, the following equation can be used for the identification of the controller parameters (implicit self-tuning):

$$y(k) = \boldsymbol{\theta}_e^T \boldsymbol{\phi}_e(k-d) \tag{9.177}$$

Thus, the prediction error to be used in the adaptation algorithm is as follows:

$$\hat{e}(k) = y(k) - \hat{y}(k) = y(k) - \hat{\boldsymbol{\theta}}_e^T(k-d)\boldsymbol{\phi}_e(k-d) = y(k) - y_m(k) = \bar{e}(k) \tag{9.178}$$

where use was made of (9.165). Hence we arrive at the same adaptation law given by (9.174) and (9.175).

In their original work on implicit self-tuning, Åström and Wittenmark [15] considered the regulation problem $(y_m(k) \equiv 0)$. A control law similar to (9.159) where $y_m(k) = 0$ was used, in order to achieve minimum variance control. Furthermore, an expression similar to (9.178) was used in order to provide the prediction error $\hat{e}(k)$ for a least squares adaptation scheme.

The procedure outlined in the present section is not appropriate for non-minimum-phase systems. For such systems, the pole-placement design is appropriate and is considered in the following section.

9.5.3 *Pole-placement self-tuning regulators*

1. Pole-placement design with known parameters

Non-minimum-phase plants is one case where minimum variance control is not appropriate. For such plants, a pole-placement design (chosen as the underlying control problem) can be applied. The procedure consists of finding a feedback law for which the closed-loop poles have desired locations. Both explicit and implicit schemes may be formulated. Explicit schemes are based on estimates of parameters in an explicit system model, while implicit schemes are based on estimates of parameters in a modified implicit system model. Similarities between MRAC and STRs will emerge.

The discussion is limited to single-input single-output systems described by

$$A(q^{-1})y(k) = q^{-d}B(q^{-1})u(k) \tag{9.179}$$

where

$$A(q^{-1}) = 1 + a_1 q^{-1} + \cdots + a_{n_A} q^{-n_A} \tag{9.180}$$

$$B(q^{-1}) = b_0 + b_1 q^{-1} + \cdots + b_{n_B} q^{-n_B} \tag{9.181}$$

The polynomial $A(q^{-1})$ is thus monic, $A(q^{-1})$ and $B(q^{-1})$ are relatively prime (have no common factors) and the system delay is $d \geq 1$. It is desired to find a controller for which the relation from the command signal $\omega(k)$ to the output $y(k)$ becomes

$$A_m(q^{-1})y(k) = q^{-d}B_m(q^{-1})\omega(k) \tag{9.182}$$

where $A_m(q^{-1})$ is a stable monic polynomial and $A_m(q^{-1})$ and $B_m(q^{-1})$ are relatively prime. Restrictions on $B_m(q^{-1})$ will appear in what follows.

A general structure (R–S–T canonical structure) for the controller is presented in Figure 9.12. The controller is described by

$$R(q^{-1})u(k) = T(q^{-1})\omega(k) - S(q^{-1})y(k) \tag{9.183}$$

This controller offers a negative feedback with transfer function $-S(q^{-1})/R(q^{-1})$ and feedforward with transfer function $T(q^{-1})/R(q^{-1})$. Multiplying (9.183) by $q^{-d}B(q^{-1})$, one obtains

$$q^{-d}B(q^{-1})R(q^{-1})u(k) = q^{-d}T(q^{-1})B(q^{-1})\omega(k) - q^{-d}S(q^{-1})B(q^{-1})y(k)$$

or

$$[A(q^{-1})R(q^{-1}) + q^{-d}B(q^{-1})S(q^{-1})]y(k) = q^{-d}T(q^{-1})B(q^{-1})\omega(k) \tag{9.184}$$

Hence, the relation between $y(k)$ and $\omega(k)$ is given by

$$\frac{y(k)}{\omega(k)} = \frac{q^{-d}T(q^{-1})B(q^{-1})}{A(q^{-1})R(q^{-1}) + q^{-d}B(q^{-1})S(q^{-1})} \tag{9.185}$$

Relation (9.182), which represents the desired behaviour, may be written as

$$\frac{y(k)}{\omega(k)} = \frac{q^{-d}B_m(q^{-1})}{A_m(q^{-1})} \tag{9.186}$$

Thus, the design problem is equivalent to the algebraic problem of finding $R(q^{-1})$, $S(q^{-1})$ and $T(q^{-1})$ for which the following equation holds true:

$$\frac{q^{-d}T(q^{-1})B(q^{-1})}{A(q^{-1})R(q^{-1}) + q^{-d}B(q^{-1})S(q^{-1})} = \frac{q^{-d}B_m(q^{-1})}{A_m(q^{-1})} \tag{9.187}$$

From the left-hand side of (9.187), it is evident that the system zeros ($z^{n_B}B(z^{-1}) = 0$) will also be closed-loop zeros, unless they are cancelled out by corresponding closed-loop poles. But, unstable (or poorly damped) zeros should not

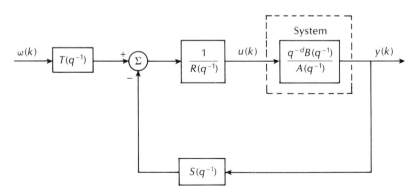

Figure 9.12　The R–S–T canonical structure used for the controller.

be cancelled out by the controller, since they would lead to instability. Thus, let us factor out $B(q^{-1})$ as follows:

$$B(q^{-1}) = B^+(q^{-1})B^-(q^{-1}) \tag{9.188}$$

where $B^+(q^{-1})$ contains the well-damped zeros (which can be cancelled out) and $B^-(q^{-1})$ contains the unstable and poorly damped zeros (which are not cancelled out). In order to obtain a unique factorization we also require that $B^+(q^{-1})$ is a monic polynomial. From (9.187), it follows that the characteristic polynomial of the closed-loop system is

$$A(q^{-1})R(q^{-1}) + q^{-d}B(q^{-1})S(q^{-1}) \tag{9.189}$$

The factors of this polynomial should be the desired reference model poles, i.e. the roots of $A_m(q^{-1})$, and the system zeros which can be cancelled out, i.e. the roots of $B^+(q^{-1})$. Moreover, since in general the order of the reference model, deg $A_m(q^{-1})$, is less than the order of the closed-loop system, $\deg[A(q^{-1})R(q^{-1}) + q^{-d}B(q^{-1})S(q^{-1})]$, there are factors in the left-hand side of (9.187) which cancel out. These factors correspond to a polynomial $A_o(q^{-1})$. The polynomial $A_o(q^{-1})$ is called the observer polynomial and is chosen to have well-damped roots. The appearance of this polynomial is more evident when a state-space solution to this problem is considered. In this case, the solution is a combination of state feedback and an observer. Hence, the characteristic polynomial of the closed-loop system assumes the following form:

$$A(q^{-1})R(q^{-1}) + q^{-d}B(q^{-1})S(q^{-1}) = B^+(q^{-1})A_m(q^{-1})A_o(q^{-1}) \tag{9.190}$$

Now, since $B^+(q^{-1})$ is a dividend of $B(q^{-1})$ and the polynomials $A(q^{-1})$ and $B(q^{-1})$ are relatively prime, it is clear from (9.190) that $B^+(q^{-1})$ should also be a dividend of the polynomial $R(q^{-1})$, i.e.

$$R(q^{-1}) = B^+(q^{-1})R_1(q^{-1}) \tag{9.191}$$

Equation (9.190) may then be rewritten as

$$A(q^{-1})R_1(q^{-1}) + q^{-d}B^-(q^{-1})S(q^{-1}) = A_m(q^{-1})A_o(q^{-1}) \tag{9.192}$$

Hence, equation (9.187) is then equivalent to the equation

$$\frac{q^{-d}B^+(q^{-1})B^-(q^{-1})T(q^{-1})}{B^+(q^{-1})A_m(q^{-1})A_o(q^{-1})} = \frac{q^{-d}B_m(q^{-1})}{A_m(q^{-1})} \tag{9.193}$$

In order that the foregoing equation holds true and since $B^-(q^{-1})$ cannot be cancelled out, it is clear that $B^-(q^{-1})$ must be a factor of $B_m(q^{-1})$, i.e.

$$B_m(q^{-1}) = B^-(q^{-1})B_m^+(q^{-1}) \tag{9.194}$$

and also that

$$T(q^{-1}) = A_o(q^{-1})B_m^+(q^{-1}) \tag{9.195}$$

It is thus evident that we are not absolutely free to choose $B_m(q^{-1})$ which corresponds to the specifications for the closed-loop zeros. We can choose freely the part $B_m^+(q^{-1})$ of $B_m(q^{-1})$, while equation (9.194) should be valid, otherwise there is no solution to the design problem.

It is necessary to establish conditions under which a solution for the polynomials $R_1(q^{-1})$ and $S(q^{-1})$ in equation (9.192) is guaranteed. This equation, linear in the polynomials $R_1(q^{-1})$ and $S(q^{-1})$, is a special case of the Diophantine equation (or Bezout identity), which has the general form (see also Remark 9.4.1)

$$\bar{A}(q^{-1})\bar{R}(q^{-1}) + \bar{B}(q^{-1})\bar{S}(q^{-1}) = \bar{C}(q^{-1}) \tag{9.196}$$

It can be proved that the Diophantine equation (9.196) always has a solution for $\bar{R}(q^{-1})$ and $\bar{S}(q^{-1})$, if the greatest common factor of $\bar{A}(q^{-1})$ and $\bar{B}(q^{-1})$ is a dividend of $\bar{C}(q^{-1})$. Therefore, (9.192) will always have a solution for $R_1(q^{-1})$ and $S(q^{-1})$, since we have assumed that $A(q^{-1})$ and $B(q^{-1})$ are co-prime, and consequently $A(q^{-1})$ and $q^{-d}B^-(q^{-1})$ are also co-prime.

Note that if a solution exists, then (9.196) in general has infinitely many solutions. Indeed, if $R^0(q^{-1})$ and $S^0(q^{-1})$ are solutions of (9.196), then it can be easily verified that $R^0(q^{-1}) + \bar{B}(q^{-1})Q(q^{-1})$ and $S^0(q^{-1}) - \bar{A}(q^{-1})Q(q^{-1})$, with $Q(q^{-1})$ an arbitrary polynomial, are also solutions of (9.196). Particular solutions can be specified in several ways. Different solutions give systems with different noise rejection properties.

It can be proved that there are unique solutions to (9.196) if in addition we impose the following restriction for the solution sought:

$$\deg \bar{R}(q^{-1}) < \deg \bar{B}(q^{-1}) \tag{9.197}$$

or

$$\deg \bar{S}(q^{-1}) < \deg \bar{A}(q^{-1}) \tag{9.198}$$

Moreover, for the pole-placement control problem, we seek particular solutions which lead to causal control laws (i.e. $\deg S(q^{-1}) \le \deg R(q^{-1})$ and $\deg T(q^{-1}) \le \deg R(q^{-1})$). Note also that it is often advantageous to keep $\deg S(q^{-1}) = \deg T(q^{-1}) = \deg R(q^{-1})$, in order to avoid an unnecessary delay in the controller.

Note from (9.190) that we must select either

$$\deg R(q^{-1}) = \deg A_m(q^{-1}) + \deg A_o(q^{-1}) + \deg B^+(q^{-1}) - \deg A(q^{-1}) \tag{9.199}$$

or

$$\deg S(q^{-1}) = \deg A_m(q^{-1}) + \deg A_o(q^{-1}) - \deg B^-(q^{-1}) - d \tag{9.200}$$

The degrees of $R(q^{-1})$ or $S(q^{-1})$ are thus imposed by the structure of the system and the structure of the desired closed-loop transfer function. In order to assure unique solutions, if we choose to satisfy (9.199) we must have $\deg S(q^{-1}) \le \deg A(q^{-1}) - 1$ (this results from (9.198)) and if we choose to satisfy (9.200) we must have $\deg R(q^{-1}) \le \deg B(q^{-1}) + d - 1$ (this results from (9.197)).

By selecting (9.199) or (9.200), possible choices for the degrees of $R_1(q^{-1})$ and $S(q^{-1})$ in (9.192), corresponding to unique solutions and minimum-order polynomials, are consequently given below:

$$\deg R_1(q^{-1}) = \deg A_m(q^{-1}) + \deg A_o(q^{-1}) - \deg A(q^{-1}) \tag{9.201}$$

$$\deg S(q^{-1}) = \deg A(q^{-1}) - 1 \tag{9.202}$$

or

$$\deg R_1(q^{-1}) = \deg B^-(q^{-1}) + d - 1 \tag{9.203}$$

$$\deg S(q^{-1}) = \deg A_m(q^{-1}) + \deg A_o(q^{-1}) - \deg B^-(q^{-1}) - d \tag{9.204}$$

By selecting (9.201) and (9.202) and in order to have causal control laws (i.e. $\deg S(q^{-1}) = \deg A(q^{-1}) - 1 \leqslant \deg R(q^{-1})$), relation (9.199) leads to

$$\deg A_o(q^{-1}) \geqslant 2 \deg A(q^{-1}) - \deg A_m(q^{-1}) - \deg B^+(q^{-1}) - 1 \tag{9.205}$$

Relation (9.205) is a restriction on the degree of the observer polynomial $A_o(q^{-1})$. Moreover, requiring that $\deg T(q^{-1}) \leqslant \deg R(q^{-1})$ and using (9.195) and (9.199) we obtain

$$\deg A_o(q^{-1}) + \deg B_m^+(q^{-1}) = \deg T(q^{-1}) \leqslant \deg R(q^{-1})$$
$$= \deg A_m(q^{-1}) + \deg A_o(q^{-1}) + \deg B^+(q^{-1}) - \deg A(q^{-1})$$

or

$$\deg A(q^{-1}) - \deg B(q^{-1}) \leqslant \deg A_m(q^{-1}) - \deg B_m(q^{-1}) \tag{9.206}$$

The pole excess of the system should thus be less than the pole excess of the reference model. Condition (9.205) in combination with (9.206) guarantees that the feedback will be causal when (9.201) and (9.202) are chosen. This in turn implies that the transfer functions S/R and T/R will be causal.

The control algorithm, in the case of known parameters, is summarized in Table 9.2.

2. Pole placement design in the case of unknown parameters

In the case of uncertain system model parameters, an STR is used on the basis of the following separation principle. Here, the unknown system parameters are estimated recursively. Based on the certainty equivalence principle, the controller is recomputed at each step using the estimated system parameters. The controller design problem (Diophantine equation) therefore needs to be solved at each step.

The parameter estimator is based on the system model

$$A(q^{-1})y(k) = B(q^{-1})u(k - d) \tag{9.207}$$

or explicitly

$$y(k) + a_1 y(k - 1) + \cdots + a_{n_A} y(k - n_A) =$$
$$b_0 u(k - d) + b_1 u(k - d - 1) + \cdots + b_{n_B} u(k - d - n_B) \tag{9.208}$$

Table 9.2 The pole-placement control algorithm for the case of known parameters.

Given $A(q^{-1})$ and $B(q^{-1})$ with $A(q^{-1})$ monic and $A(q^{-1})$ and $B(q^{-1})$ co-prime:

Step 1   Factor $B(q^{-1}) = B^+(q^{-1})B^-(q^{-1})$ with $B^+(q^{-1})$ monic
Choose $A_m(q^{-1})$, $B_m(q^{-1}) = B^-(q^{-1})B_m^+(q^{-1})$ and $A_o(q^{-1})$ such that
(9.205) and (9.206) are satisfied

Step 2   Select the degrees of $R_1(q^{-1})$ and $S(q^{-1})$ in order to satisfy (9.201) and
(9.202) or (9.203) and (9.204)
Solve $A(q^{-1})R_1(q^{-1}) + q^{-d}B^-(q^{-1})S(q^{-1}) = A_m(q^{-1})A_o(q^{-1})$ for $R_1(q^{-1})$
and $S(q^{-1})$

Step 3   Compute $R(q^{-1}) = B^+(q^{-1})R_1(q^{-1})$ and $T(q^{-1}) = A_o(q^{-1})B_m^+(q^{-1})$

The foregoing steps are executed once OFF-LINE

Step 4   Apply the control law

$$u(k) = \left(\frac{T(q^{-1})}{R(q^{-1})}\right)\omega(k) - \left(\frac{S(q^{-1})}{R(q^{-1})}\right)y(k)$$

at each step.

Introducing the parameter vector

$$\boldsymbol{\theta}^{\mathrm{T}} = [a_1, \ldots, a_{n_A}, b_0, \ldots, b_{n_B}] \tag{9.209}$$

and the regression vector

$$\boldsymbol{\phi}^{\mathrm{T}}(k) = [-y(k-1), \ldots, -y(k-n_A), u(k-d), \ldots, u(k-d-n_B)] \tag{9.210}$$

equation (9.208) is expressed compactly as

$$y(k) = \boldsymbol{\theta}^{\mathrm{T}}\boldsymbol{\phi}(k) \tag{9.211}$$

Based on the prediction model (9.211), the recursive least-squares estimator is described by the recursive equation

$$\hat{\boldsymbol{\theta}}(k) = \hat{\boldsymbol{\theta}}(k-1) + \mathbf{F}(k)\boldsymbol{\phi}(k)\varepsilon(k) \tag{9.212}$$

with the prediction error

$$\varepsilon(k) = y(k) - \boldsymbol{\phi}^{\mathrm{T}}(k)\hat{\boldsymbol{\theta}}(k-1) \tag{9.213}$$

The gain matrix $\mathbf{F}(k)$ can be deduced recursively using the expression

$$\mathbf{F}(k+1) = \frac{1}{\lambda}\left(\mathbf{F}(k) - \frac{\mathbf{F}(k)\boldsymbol{\phi}(k)\boldsymbol{\phi}^{\mathrm{T}}(k)\mathbf{F}(k)}{1 + \boldsymbol{\phi}^{\mathrm{T}}(k)\mathbf{F}(k)\boldsymbol{\phi}(k)}\right) \quad \text{with} \quad \mathbf{F}(0) > 0 \tag{9.214}$$

where $0 < \lambda \leq 1$ is a **forgetting factor**. The restrictions of Remark 9.4.5 hold for (9.214). Since it is assumed that there is no noise in the system model (9.179), there will be no bias error using recursive least squares.

In self-tuning, the convergence of the parameter estimates to the true values is of great importance. In order to obtain good estimates using (9.212), it is necessary that the process input be sufficiently rich in frequencies, or persistently exciting. The concept of **persistent excitation** was first introduced in identification problems. This states that we cannot identify all the parameters of a model unless we have enough distinct frequencies in the spectrum of the input signal. In general, when the input to a system is the result of feedback and it is therefore a dependent variable within the adaptive loop, the input signal is not persistently exciting.

In the explicit STR based on the pole-placement design discussed above, the estimated parameters are the parameters of the system model. This explicit adaptive pole-placement algorithm is summarized in Table 9.3.

An implicit STR design procedure based on pole placement may also be considered. To this end, we reparametrize the system model such that the controller parameters appear. These controller parameters can be updated directly. The proper system model structure sought is obtained by multiplying (9.192) by $y(k)$ to yield

$$
\begin{aligned}
A_m(q^{-1})A_o(q^{-1})y(k) &= A(q^{-1})R_1(q^{-1})y(k) + q^{-d}B^-(q^{-1})S(q^{-1})y(k) \\
&= q^{-d}B(q^{-1})R_1(q^{-1})u(k) + q^{-d}B^-(q^{-1})S(q^{-1})y(k) \\
&= q^{-d}B^-(q^{-1})[R(q^{-1})u(k) + S(q^{-1})y(k)]
\end{aligned}
\tag{9.215}
$$

The reparametrization (9.215), which is an implicit system model, is redundant, since it has more parameters than (9.207). It is also bilinear in the parameters of $B^-(q^{-1})$, $R(q^{-1})$ and $S(q^{-1})$ This leads to a non-trivial bilinear estimation problem. We can obtain the regulator parameters by estimating $B^-(q^{-1})$, $R(q^{-1})$ and $S(q^{-1})$ in (9.215) directly, avoiding at each step the control design problem, i.e. the solution of the Diophantine equation. This leads to a less time-consuming algorithm, in the sense that the design calculations become trivial. The implicit STR is summarized in Table 9.4.

Table 9.3 The explicit adaptive pole-placement algorithm for the case of unknown parameters.

Step 1   Estimate the model parameters in $A(q^{-1})$ and $B(q^{-1})$ using (9.212), (9.213) and (9.214), recursively, at each step
It is assumed that $A(q^{-1})$ and $B(q^{-1})$ have no common factors

Step 2   Factorize the polynomial $B(q^{-1})$ so that the decomposition
$B^+(q^{-1})B^-(q^{-1})$ can be made ON-LINE at each step
Solve the controller design problem with the estimates obtained in step 1, i.e. solve (9.192) for $R_1(q^{-1})$ and $S(q^{-1})$ using $A(q^{-1})$ and $B^-(q^{-1})$ calculated on the basis of the estimation at step 1
Calculate $R(q^{-1})$ and $T(q^{-1})$ from (9.191) and (9.195), respectively

Step 3   Compute the control law using (9.183)

Step 4   Repeat steps 1 to 3 at each sampling period

Table 9.4 The implicit pole-placement STR for the case of unknown parameters.

Step 1 Estimate the coefficients in $R(q^{-1})$, $B^-(q^{-1})$ and $S(q^{-1})$ recursively based on the reparametrized model (bilinear estimation problem)

$$A_m(q^{-1})A_o(q^{-1})y(k) = q^{-d}B^-(q^{-1})[R(q^{-1})u(k) + S(q^{-1})y(k)] \qquad (9.216)$$

Step 2 Compute the control law using the relations

$$T(q^{-1}) = A_o(q^{-1})B_m^+(q^{-1}) \qquad (9.217)$$

$$u(k) = \left(\frac{T(q^{-1})}{R(q^{-1})}\right)\omega(k) - \left(\frac{S(q^{-1})}{R(q^{-1})}\right)y(k) \qquad (9.218)$$

Step 3 Repeat steps 1 and 2 at each sampling period

In order to avoid non-linear parametrization, (9.215) is rewritten equivalently as

$$A_m(q^{-1})A_o(q^{-1})y(k) = q^{-d}\bar{R}(q^{-1})u(k) + q^{-d}\bar{S}(q^{-1})y(k) \qquad (9.219)$$

where

$$\bar{R}(q^{-1}) = B^-(q^{-1})R(q^{-1}) \qquad (9.220)$$

and

$$\bar{S}(q^{-1}) = B^-(q^{-1})S(q^{-1}) \qquad (9.221)$$

Based on the linear model (9.219), it is possible to estimate the coefficients of the polynomials $\bar{R}(q^{-1})$ and $\bar{S}(q^{-1})$. However, it should be noted that in general this is not a minimal parametrization since the coefficients of the polynomial $B^-(q^{-1})$ are estimated twice. Moreover, possible common factors in $\bar{R}(q^{-1})$ and $\bar{S}(q^{-1})$ (corresponding to $B^-(q^{-1})$) should be cancelled out in order to avoid cancellation of

Table 9.5 An alternative implicit pole-placement STR for the case of unknown parameters.

Step 1 Using the model (9.219) and least squares, estimate the coefficients of the polynomials $\bar{R}(q^{-1})$ and $\bar{S}(q^{-1})$

Step 2 Cancel out possible common factors in $\bar{R}(q^{-1})$ and $\bar{S}(q^{-1})$ in order to obtain $R(q^{-1})$ and $S(q^{-1})$

Step 3 Compute the control law using the relations

$$T(q^{-1}) = A_o(q^{-1})B_m^+(q^{-1})$$

$$u(k) = \left(\frac{T(q^{-1})}{R(q^{-1})}\right)\omega(k) - \left(\frac{S(q^{-1})}{R(q^{-1})}\right)y(k)$$

Step 4 Repeat steps 1 to 3 at each sampling period

unstable modes in the control law. The algorithm thus obtained is summarized in Table 9.5 (p. 285).

■ **EXAMPLE 9.5.1**   Consider the system

$$A(q^{-1})y(k) = q^{-1}B(q^{-1})u(k) \quad \text{with} \quad y(0) = 1$$

where $A(q^{-1}) = 1 + 2q^{-1} + q^{-2}$ and $B(q^{-1}) = 2 + q^{-1} + q^{-2}$. The polynomial $B(q^{-1})$ can be factored as

$$B(q^{-1}) = 2(1 + 0.5q^{-1} + 0.5q^{-2}) = B^-(q^{-1})B^+(q^{-1})$$

with $B^+(q^{-1})$ monic. The desired closed-loop behaviour is given by

$$A_m(q^{-1})y(k) = q^{-1}B_m(q^{-1})\omega(k)$$

with $A_m(q^{-1}) = 1 - q^{-1} + 0.25q^{-2}$ and $B_m(q^{-1}) = 1 + 0.3q^{-1} = 2(0.5 + 0.15q^{-1}) = B^-(q^{-1})B_m^+(q^{-1})$.

(a) In the case of known plant parameters, calculate the pole-placement control law.
(b) In the case of unknown plant parameters, define an explicit adaptive pole-placement control scheme.
(c) Repeat part (b) for an implicit adaptive pole-placement control scheme.

■ **SOLUTION**

   (a) In the case of known parameters and in order to satisfy (9.201), (9.202) and (9.205) we select deg $A_o(q^{-1}) = 0$, deg $R_1(q^{-1}) = 0$ and deg $S(q^{-1}) = 1$. Hence, $A_o(q^{-1}) = 1$, $R_1(q^{-1}) = r_0$ and $S(q^{-1}) = s_0 + s_1q^{-1}$. We now solve the following Diophantine equation for r_0, s_0 and s_1:

$$A(q^{-1})R_1(q^{-1}) + q^{-1}B^-(q^{-1})S(q^{-1}) = A_m(q^{-1})A_o(q^{-1})$$

or

$$(1 + 2q^{-1} + q^{-2})r_0 + 2q^{-1}(s_0 + s_1q^{-1}) = 1 - q^{-1} + 0.25q^{-2}$$

One easily obtains $r_0 = 1$, $s_0 = -1.5$ and $s_1 = -0.375$, and consequently $R_1(q^{-1}) = 1$ and $S(q^{-1}) = -1.5 - 0.375q^{-1}$. Note from (9.192) that when $A_o(q^{-1})$ is a monic polynomial, then $R_1(q^{-1})$ and $R(q^{-1})$ are restricted to be monic polynomials also. Now, $R(q^{-1}) = B^+(q^{-1})R_1(q^{-1}) = 1 + 0.5q^{-1} + 0.5q^{-2}$ and $T(q^{-1}) = B_m^+(q^{-1})A_o(q^{-1}) = 0.5 + 0.15q^{-1}$. The control law is given by

$$u(k) = \left(\frac{T(q^{-1})}{R(q^{-1})}\right)\omega(k) - \left(\frac{S(q^{-1})}{R(q^{-1})}\right)y(k)$$

or

$$u(k) = \left(\frac{0.5 + 0.15q^{-1}}{1 + 0.5q^{-1} + 0.5q^{-2}}\right)\omega(k) + \left(\frac{1.5 + 0.375q^{-1}}{1 + 0.5q^{-1} + 0.5q^{-2}}\right)y(k)$$

(b) The system model belongs to the following class of models:

$$(1 + a_1 q^{-1} + a_2 q^{-2}) y(k) = (b_0 + b_1 q^{-1} + b_2 q^{-2}) u(k-1)$$

which may be rewritten as

$$y(k) = \boldsymbol{\theta}^{\mathrm{T}} \boldsymbol{\phi}(k)$$

where

$$\boldsymbol{\theta}^{\mathrm{T}} = [a_1, a_2, b_0, b_1, b_2]$$

$$\boldsymbol{\phi}^{\mathrm{T}}(k) = [-y(k-1), -y(k-2), u(k-1), u(k-2), u(k-3)]$$

The parameters a_i, b_i can be estimated ON-LINE using the following algorithm:

$$\boldsymbol{\phi}^{\mathrm{T}}(k) = [-y(k-1), -y(k-2), u(k-1), u(k-2), u(k-3)]$$

$$\mathbf{F}(k+1) = \frac{1}{0.99} \left(\mathbf{F}(k) - \frac{\mathbf{F}(k)\boldsymbol{\phi}(k)\boldsymbol{\phi}^{\mathrm{T}}(k)\mathbf{F}(k)}{1 + \boldsymbol{\phi}^{\mathrm{T}}(k)\mathbf{F}(k)\boldsymbol{\phi}(k)} \right) \quad \text{with} \quad \mathbf{F}(0) = \frac{1}{10^{-3}} \mathbf{I}_5$$

$$\varepsilon(k) = y(k) - \boldsymbol{\phi}^{\mathrm{T}}(k)\hat{\boldsymbol{\theta}}(k-1)$$
$$\hat{\boldsymbol{\theta}}(k) = \hat{\boldsymbol{\theta}}(k-1) + \mathbf{F}(k)\boldsymbol{\phi}(k)\varepsilon(k)$$
$$\hat{\boldsymbol{\theta}}^{\mathrm{T}}(k) = [\hat{a}_1(k), \hat{a}_2(k), \hat{b}_0(k), \hat{b}_1(k), \hat{b}_2(k)]$$

initialized for example at $\hat{\boldsymbol{\theta}}^{\mathrm{T}}(0) = [1, 0, 1, 0, 0]$. At each step, $\hat{B}(q^{-1}) = \hat{b}_0(k) + \hat{b}_1(k)q^{-1} + \hat{b}_2(k)q^{-2}$ is factored as

$$\hat{B}_0(q^{-1}) = \hat{B}^-(q^{-1})\hat{B}^+(q^{-1})$$

where $\hat{B}^+(q^{-1})$ is chosen to be monic. Moreover, at each step, the following Diophantine equation is solved for $\hat{r}_0(k)$, $\hat{s}_0(k)$ and $\hat{s}_1(k)$:

$$\hat{A}(q^{-1})\hat{r}_0(k) + q^{-1}\hat{B}^-(q^{-1})[\hat{s}_0(k) + \hat{s}_1(k)q^{-1}] = 1 - q^{-1} + 0.25 q^{-2}$$

where $\hat{A}(q^{-1}) = 1 + \hat{a}_1(k)q^{-1} + \hat{a}_2(k)q^{-2}$ and $\hat{S}(q^{-1}) = \hat{s}_0(k) + \hat{s}_1(k)q^{-1}$. Then, the following computations are made:

$$\hat{R}(q^{-1}) = \hat{R}_1(q^{-1})\hat{B}^+(q^{-1}) = \hat{r}_0(k)\hat{B}^+(q^{-1})$$
$$T(q^{-1}) = B_m^+(q^{-1})$$

where

$$B_m(q^{-1}) = \hat{B}^-(q^{-1})B_m^+(q^{-1})$$

Here, $B_m^+(q^{-1})$ can be any polynomial of our choice. The control law to be applied to the system at each step is

$$u(k) = \left(\frac{T(q^{-1})}{\hat{R}(q^{-1})} \right) \omega(k) - \left(\frac{\hat{S}(q^{-1})}{\hat{R}(q^{-1})} \right) y(k)$$

(c) In the case of an implicit adaptive pole-placement design, the parameters of the following implicit system model are estimated using recursive least squares:

$$A_o(q^{-1})A_m(q^{-1})y(k) = q^{-1}\bar{R}(q^{-1})u(k) + q^{-1}\bar{S}(q^{-1})y(k)$$

or

$$(1 - q^{-1} + 0.25q^{-2})y(k) = [\bar{r}_0(k) + \bar{r}_1(k)q^{-1} + \bar{r}_2(k)q^{-2}]u(k-1)$$
$$+ [\bar{s}_0(k) + \bar{s}_1(k)q^{-1}]y(k-1)$$

or

$$y(k) - y(k-1) + 0.25y(k-2) = [\bar{r}_0(k), \quad \bar{r}_1(k), \quad \bar{r}_2(k), \quad \bar{s}_0(k), \quad \bar{s}_1(k)]\begin{bmatrix} u(k-1) \\ u(k-2) \\ u(k-3) \\ y(k-1) \\ y(k-2) \end{bmatrix}$$

with $\bar{R}(q^{-1}) = B^-(q^{-1})R(q^{-1})$ and $\bar{S}(q^{-1}) = B^-(q^{-1})S(q^{-1})$. Next, any common factors in $\bar{R}(q^{-1})$ and $\bar{S}(q^{-1})$ (corresponding to $B^-(q^{-1})$) are cancelled out to obtain $R(q^{-1})$ and $S(q^{-1})$. One then has

$$T(q^{-1}) = B_m^+(q^{-1})$$

and the control law is given by

$$u(k) = \left(\frac{T(q^{-1})}{R(q^{-1})}\right)\omega(k) - \left(\frac{S(q^{-1})}{R(q^{-1})}\right)y(k)$$

The procedure described above is repeated at each sampling period.

9.6 Problems

1. Consider a system described by

$$y(k) = \theta_0 G(q^{-1})u(k)$$

where θ_0 is an unknown parameter and $G(q^{-1})$ a known rational function of q^{-1}. The reference model is described by

$$y_m(k) = \theta_m G(q^{-1})\omega(k)$$

where θ_m is a known parameter. The controller is of the form

$$u(k) = \theta\omega(k)$$

Find an adaptation mechanism for the feedforward gain θ, by using the MIT rule.

2. Consider a system described by

$$y(k) = \left(\frac{q^{-1}(0.36 + 0.28q^{-1})}{1 - 1.36q^{-1} + 0.36q^{-2}} \right) u(k)$$

and a reference model given by

$$y_m(k) = \left(\frac{q^{-1}(0.38 + 0.24q^{-1})}{1 - 0.78q^{-1} + 0.37q^{-2}} \right) \omega(k)$$

In the case of known system parameters, determine a pole placement control law in order to follow the reference model. In the case of unknown system parameters, determine an adaptive controller in order to achieve model following by using the MIT rule.

3. Consider a system described by

$$y(k) = \left(\frac{q^{-2}\beta_0}{1 + a_1q^{-1} + a_2q^{-2}} \right) u(k)$$

where β_0, a_1 and a_2 are free parameters. The desired input–output behaviour is given by

$$y_m(k) = q^{-2} \left(\frac{b_0}{1 + a_1q^{-1} + a_2q^{-2}} \right) \omega(k)$$

where $1 + a_1q^{-1} + a_2q^{-2}$ is an asymptotically stable polynomial.

(a) In the case where β_0, a_1 and a_2 are assumed to be known, determine a pole-placement control law.

(b) Suppose that β_0, a_1 and a_2 are unknown. Determine an adaptive pole-placement control law by using the MIT rule.

(c) Design an explicit adaptive pole-placement algorithm in the case where β_0, a_1 and a_2 are unknown.

4. In a plastics extruder, the raw polymer is fed in solid form (see also Problem 7 in Chapter 5). Then, it is pushed forward by a screw through different temperature zones where it is melted and it is finally extruded from a die in order to be processed further. The transfer function from the screw speed (which is the main controlling variable) to the temperature of the polymer at the output is given by

$$G(s) = \frac{Ke^{-s\tau}}{\beta s + 1}$$

where K is the static gain, β is the time constant and τ the system delay. The system delay τ is such that

$$\tau = (d-1)T + L \quad \text{where} \quad 0 < L < T$$

where T is the sampling period and d is a positive integer.

(a) Verify that the equivalent discrete-time transfer function, using a zero-order hold circuit, is given by

$$G(z) = \frac{z^{-d}(b_0 + b_1 z^{-1})}{1 + a_1 z^{-1}}$$

with

$$a_1 = -e^{-T/\beta}$$
$$b_0 = K(1 - e^{(L-T)/\beta})$$
$$b_1 = Ke^{-T/\beta}(e^{L/\beta} - 1)$$

The discretized system is thus equivalently described by the difference equation

$$y(k+d) = -a_1 y(k+d-1) + b_0 u(k) + b_1 u(k-1)$$

It is clear that d is the discretized (sampled-data) system delay.

(b) Choose $d = 1$ and a sampling period T such that the sampled-data system has a stable zero (i.e. choose $-b_1/b_0$ to be inside the unit circle). The minimum-phase system thus obtained is described by

$$y(k+1) = -a_1 y(k) + b_0 u(k) + b_1 u(k-1)$$

For this system model determine a model reference control law which, in the case of known model parameters, satisfies the control objective

$$\Gamma(q^{-1})[y(k+1) - y_m(k+1)] \equiv 0 \quad \text{for} \quad k \geq 0$$

where $\Gamma(q^{-1}) = 1 + \gamma_1 q^{-1}$. The polynomial $\Gamma(q^{-1})$ is assumed to be asymptotically stable. The asymptotically stable reference model is given by

$$y_m(k) = \left(\frac{q^{-1}(b_0^m + b_1^m q^{-1})}{1 + a_1^m q^{-1}}\right) \omega(k)$$

(c) In the case of unknown plant parameters, determine an MRAC design in order to satisfy the control objective of part (b) asymptotically.

5.   A system is described by the model (initially known to the designer) [21]

$$y(k) = \left(\frac{q^{-2}(1 + 0.4q^{-1})}{(1 - 0.5q^{-1})[1 - (0.8 + 0.3j)q^{-1}][1 - (0.8 - 0.3j)q^{-1}]}\right) u(k)$$

It is desired to follow the reference model

$$y_m(k) = \left(\frac{q^{-2}(0.28 + 0.22q^{-1})}{(1 - 0.5q^{-1})[1 - (0.7 + 0.2j)q^{-1}][1 - (0.7 - 0.2j)q^{-1}]} \right) \omega(k)$$

(a) By choosing $\Gamma(q^{-1}) = 1$, or $\Gamma(q^{-1}) = (1 - 0.4q^{-1})^3$, design a model reference control law, assuming the model parameters known.

(b) Suppose now that parameter changes occur in the system model. The system, after the parameter changes, is described by (the model is assumed unknown to the designer)

$$y(k) = \left(\frac{q^{-2}(0.9 + 0.5q^{-1})}{(1 - 0.5q^{-1})[1 - (0.9 + 0.5j)q^{-1}][1 - (0.9 - 0.5j)q^{-1}]} \right) u(k)$$

When the control objective is tracking, the changes occur at $t = 25$ s, and when the control objective is regulation, the changes occur at $t = 0$ s. In the case of unknown parameters determine an adaptive MRAC scheme. Simulate the behaviour of the closed-loop system during tracking and during regulation, with the choices $\Gamma(q^{-1}) = 1$ or $\Gamma(q^{-1}) = (1 - 0.4q^{-1})^3$. Use as initial parameter values for the adaptive controller those obtained from the design in the non-adaptive case. Assume that a constant trace algorithm is used for the adaptation gain, with $\lambda_1(k)/\lambda_2(k) = 1$ and tr $\mathbf{F}(k) = $ tr $\mathbf{F}(0)$, with $\mathbf{F}(0) = 10\mathbf{I}$.

6. Consider the first-order system

$$y(k) + ay(k-1) = bu(k-1) + w(k) + cw(k-1)$$

Determine an adaptive minimum variance control law when a, b and c are unknown. The control objective is either regulation or tracking of a sequence $y_m(k)$ stored in memory.

7. For the system of Problem 4, it is desired to design an explicit STR, considering the model reference control as the 'underlying control problem'. Define a least squares algorithm in order to estimate the plant parameters. Reparametrize the control law of Problem 4, in order that the plant parameters appear explicitly. Solve the controller design problem and define the STR. Simulate the behaviour of the system.

8. For the system of Problem 4, design an explicit pole-placement STR. Distinguish the cases of stable and unstable system zero.

9. Determine explicit and implicit self-tuners for the plant

$$y(k) = 0.86y(k-1) + 0.08u(k-1) + 0.06u(k-2)$$

The desired characteristic polynomial is chosen as follows:

$$A_m(q^{-1}) = 1 - 1.5q^{-1} + 0.6q^{-2}$$

Also, choose $B_m^+(q^{-1}) = 1$. Use a forgetting factor adaptation algorithm with $\lambda = 0.95$. Moreover, use $\mathbf{F}(0) = 100\mathbf{I}$. Simulate the behaviour of the system.

9.7 References

Books

[1] B.D.O. Anderson, R.R. Bitmead, C.R. Johnson Jr, P.V. Kokotovic, R.L. Kosut, I.M.Y. Mareels, L. Praly and B.D. Riedle, *Stability of Adaptive Systems: Passivity and Averaging Analysis*, MIT Press, Cambridge, Massachusetts, 1986.

[2] K.J. Åström and B. Wittenmark, *Computer Controlled Systems*, Prentice Hall, Englewood Cliffs, New Jersey, 1984.

[3] K.J. Åström and B. Wittenmark, *Adaptive Control*, Addison-Wesley, New York, 1989.

[4] V.V. Chalam, *Adaptive Control Systems: Techniques and Applications*, Marcel Dekker, New York, 1987.

[5] B. Egardt, *Stability of Adaptive Controllers*, Springer Verlag, Berlin, 1989.

[6] G. Goodwin and K. Sin, *Adaptive Filtering, Prediction and Control*, Prentice Hall, Englewood Cliffs, New Jersey, 1984.

[7] C.J. Harris and S.A. Billings (eds), *Self Tuning and Adaptive Control: Theory and Applications*, Peter Peregrinus, Stevenage, 1981.

[8] I.D. Landau, *Adaptive Control: The Model Reference Approach*, Marcel Dekker, New York, 1979.

[9] E. Mishkin and L. Braun (eds), *Adaptive Control Systems*, McGraw-Hill, New York, 1961.

[10] K.S. Narendra and A. Annaswamy, *Stable Adaptive Systems*, Prentice Hall, Englewood Cliffs, New Jersey, 1989.

[11] K.S. Narendra and R.V. Monopoli (eds), *Applications of Adaptive Control*, Academic Press, New York, 1980.

[12] S.S. Sastry and M. Bodson, *Adaptive Control: Stability Convergence and Robustness*, Prentice Hall, Englewood Cliffs, New Jersey, 1988.

[13] H. Ubenhauen (ed.), *Methods and Applications in Adaptive Control*, Springer Verlag, Berlin, 1980.

Papers

[14] K.J. Åström, 'Theory and Applications of Adaptive Control – A Survey', *Automatica*, Vol. 19, No. 5, pp. 471–486, 1983.

[15] K.J. Åström and B. Wittenmark, 'On Self Tuning Regulators', *Automatica*, Vol. 9, pp. 185–199, 1973.

[16] K.J. Åström and B. Wittenmark, 'Self-Tuning Controllers Based on Pole-Zero Placement', *IEE Proceedings*, Vol. 127, Part D, No. 3, pp. 120–130, May 1980.

[17] K.J. Åström, U. Gorisson, L. Ljung and B. Wittenmark, 'Theory and Applications of Self-Tuning Regulators', *Automatica*, Vol. 13, pp. 457–476, 1977.

[18] T. Ionescu and R. Monopoli, 'Discrete Model Reference Adaptive Control with an Augmented Error Signal', *Automatica*, Vol. 13, pp. 507–517, 1977.

[19] I.D. Landau, 'A Survey of Model Reference Adaptive Techniques – Theory and Applications', *Automatica*, Vol. 10, pp. 353–379, 1974.

[20] I.D. Landau, 'An Extension of a Stability Theorem Applicable to Adaptive Control', *IEEE Transactions on Automatic Control*, Vol. AC–25, No. 4, pp. 814–817, 1980.

[21] I.D. Landau and R. Lozano, 'Unification of Discrete Time Explicit Model Reference Adaptive Control Designs', *Automatica*, Vol. 17, No. 4, pp. 593–611, 1981.

[22] I.D. Landau and H.M. Silveiva, 'A Stability Theorem with Applications to Adaptive Control', *IEEE Transactions on Automatic Control*, Vol. AC–24, No. 2, pp. 305–312, 1979.

[23] R. Lozano and I.D. Landau, 'Redesign of Explicit and Implicit Discrete-Time Model Reference Adaptive Control Schemes', *International Journal of Control*, Vol. 33, No. 2, pp. 247–268, 1981.

[24] P. Parks, 'Liapunov Redesign of Model Reference Adaptive Control Systems', *IEEE Transactions on Automatic Control*, Vol. AC–11, No. 3, pp. 362–367, 1966.

[25] B. Wittenmark, 'Stochastic Adaptive Control Methods – A Survey', *International Journal of Control*, Vol. 21, No. 5, pp. 705–730, 1975.

[26] B. Wittenmark and K.J. Åström, 'Practical Issues in the Implementation of Self-Tuning Control', *Automatica*, Vol. 20, No. 5, pp. 595–605, 1984.

10

Realization of discrete-time controllers and quantization errors

10.1 Introduction

The previous chapters referred to the description, analysis and design of discrete-time systems. Most attention was given to design, covering both classical and modern controller design techniques.

Assume that for a given system under control we have determined an appropriate controller $G_c(z)$ or a feedback control law of the form $\mathbf{u}(t) = \mathbf{F}\mathbf{x}(t) + \mathbf{G}\boldsymbol{\omega}(t)$. In the present chapter, we will show how this controller or control law is realized practically. The realization may be done using either hardware or software. Both realizations have theoretical as well as practical interest. The practical interest is by far the greater since practical issues like inexpensive circuits, memory saving, quantization of signals and parameters, round-off errors, selection of appropriate microprocessors, etc., must be carefully considered.

10.2 Hardware controller realization

Consider a controller described by the transfer function

$$G_c(z) = \frac{U(z)}{E(z)} = \frac{b(z)}{a(z)} = \frac{b_0 + b_1 z^{-1} + b_2 z^{-2} + \cdots + b_n z^{-n}}{1 + a_1 z^{-1} + a_2 z^{-2} + \cdots + a_n z^{-n}} \tag{10.1}$$

or equivalently by the difference equation

$$u(k) + a_1 u(k-1) + \cdots + a_n u(k-n) = b_0 e(k) + b_1 e(k-1) + \cdots + b_n e(k-n) \tag{10.2}$$

The term **hardware realization** for a discrete-time controller implies a realization with discrete network elements, such as hardware delay units, multipliers and summers. The term **software realization** implies a realization using a digital computer program. There are several forms of hardware realization for a controller.

In the following sections, the three most popular forms will be described, namely direct, cascade and parallel realization.

10.2.1 *Direct realization*

A direct hardware realization of the transfer function (10.1) can be implemented using the least possible number of delay units, i.e. with n delay units, and is shown in Figure 10.1.

Another direct hardware realization of the transfer function (10.1) is based on the difference equation (10.2). This equation is an nth-order difference equation which can easily be written in state form as follows:

$$\mathbf{x}(k+1) = \mathbf{A}\mathbf{x}(k) + \mathbf{b}e(k) \tag{10.3a}$$

$$u(k) = \mathbf{c}^{\mathrm{T}}\mathbf{x}(k) + de(k) \tag{10.3b}$$

where the matrix $\mathbf{A}$ and the vectors $\mathbf{b}$ and $\mathbf{c}$ have the following forms:

$$\mathbf{A} = \begin{bmatrix} 0 & 1 & 0 & \cdots & 0 \\ 0 & 0 & 1 & \cdots & 0 \\ \vdots & \vdots & \vdots & \ddots & \vdots \\ 0 & 0 & 0 & \cdots & 1 \\ -a_n & -a_{n-1} & -a_{n-2} & \cdots & -a_1 \end{bmatrix}, \quad \mathbf{b} = \begin{bmatrix} 0 \\ 0 \\ \vdots \\ 0 \\ 1 \end{bmatrix} \quad \text{and} \quad \mathbf{c} = \begin{bmatrix} b_n - a_n b_0 \\ b_{n-1} - a_{n-1} b_0 \\ \vdots \\ b_2 - a_2 b_0 \\ b_1 - a_1 b_0 \end{bmatrix} \tag{10.3c}$$

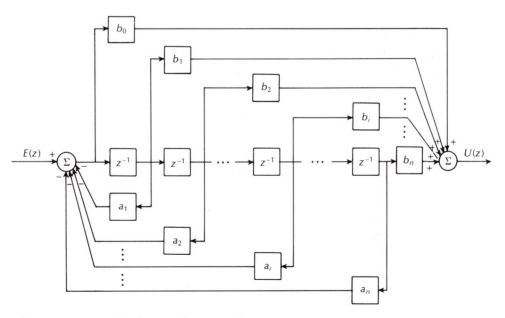

Figure 10.1   Direct hardware realization of the transfer function (10.1).

and the constant $d = b_0$. The state equations (10.3) are in phase canonical form (known as the controllable form) which was studied in Section 4.4.6. The realization of equations (10.3) is given in Figure 10.2.

Another realization of $G_c(z)$ in state space is

$$\tilde{\mathbf{x}}(k+1) = \tilde{\mathbf{A}}\tilde{\mathbf{x}}(k) + \tilde{\mathbf{b}}e(k) \tag{10.4a}$$

$$u(k) = \tilde{\mathbf{c}}^T\tilde{\mathbf{x}}(k) + \tilde{d}e(k) \tag{10.4b}$$

where the matrix $\tilde{\mathbf{A}}$ and the vectors $\tilde{\mathbf{b}}$ and $\tilde{\mathbf{c}}$ have the following forms:

$$\tilde{\mathbf{A}} = \mathbf{A}^T, \quad \tilde{\mathbf{b}} = \mathbf{c}, \quad \tilde{\mathbf{c}} = \mathbf{b} \quad \text{and} \quad \tilde{d} = d = b_0 \tag{10.4c}$$

where $\mathbf{A}$, $\mathbf{b}$ and $\mathbf{c}$ are defined in (10.3c). The state equations (10.4) are in phase canonical form (known as the observable form). Figure 10.3 shows the hardware realization for the system (10.4).

To verify the foregoing results, we calculate the transfer function $G(z) = \mathbf{c}^T(z\mathbf{I} - \mathbf{A})^{-1}\mathbf{b} + d$ of system (10.3) and $\tilde{G}(z) = \tilde{\mathbf{c}}^T(z\mathbf{I} - \tilde{\mathbf{A}})^{-1}\tilde{\mathbf{b}} + \tilde{d}$ of system (10.4). After some algebraic calculations we obtain that $G(z) = \tilde{G}(z) = G_c(z)$, where $G_c(z)$ is given by (10.1). Therefore, the descriptions (10.1), (10.3) and (10.4) are equivalent, as expected.

■ **REMARK 10.2.1** Every software realization of the transfer function $G_c(z)$, which is defined by equation (10.1), presents certain errors. The sources of error are basically the following three: quantization of the input signal, rounding off of numbers in calculations, and quantization of the values of the controller parameters a_i and b_i, $i = 1, 2, \ldots, n$ of $G_c(z)$. These errors appear because of practical limitations in the number of bits which present signals or numbers. Note that the third type of error can be reduced significantly in cases where the initial nth-order transfer function can be broken down into subsystems of order significantly smaller than n (namely, first

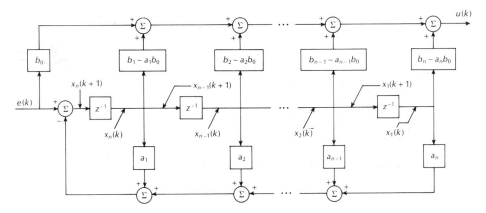

Figure 10.2 Direct hardware realization of state equations (10.3).

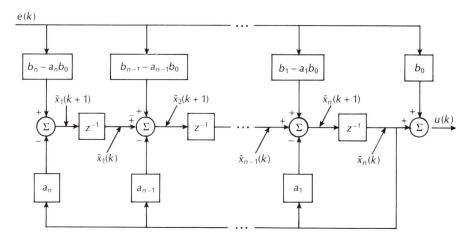

Figure 10.3 Direct hardware realization of the state equations (10.4).

or second order). In this way the realization becomes less sensitive to parameter errors. For more on quantization errors see Section 10.4.

The foregoing remark has led to the development of various hardware realizations of $G_c(z)$, which reduce the error produced by the quantization of the parameters a_i and b_i of $G_c(z)$. From these forms, the two most popular, namely the cascade and the parallel realizations, are presented below.

10.2.2 Cascade realization

In the cascade realization (also known as series realization), it is assumed that the transfer function $G_c(z)$ can be factored out into a product of terms of the form

$$G_c(z) = \beta G_1(z)G_2(z) \dots G_q(z) \quad \text{where} \quad 1 \leq q \leq n \tag{10.5}$$

where β is a constant and $G_1(z), \dots, G_q(z)$ are either first- or second-order transfer functions. The block diagram of (10.5) is given in Figure 10.4.

The circuit realization of $G_1(z), \dots, G_q(z)$ can be constructed as follows.

1. First-order terms

The first-order terms are of the form

$$G_c(z) = \frac{U(z)}{E(z)} = \frac{1 + bz^{-1}}{1 + az^{-1}} \tag{10.6}$$

Figure 10.4   Cascade block diagram of (10.5).

The circuit realization of (10.6) is shown in Figure 10.5. To verify the realization of Figure 10.5, we proceed as follows: we have that $X(z) = z^{-1}R(z)$, $R(z) = zX(z)$, $U(z) = bX(z) + R(z)$, $U(z) = (b + z)X(z)$, $R(z) = E(z) - aX(z)$ and $E(z) = (a + z)X(z)$. Therefore

$$G_c(z) = \frac{U(z)}{E(z)} = \frac{(b + z)X(z)}{(a + z)X(z)} = \frac{b + z}{a + z} = \frac{1 + bz^{-1}}{1 + az^{-1}}$$

which is in agreement with (10.6).

2. Second-order terms

The second-order terms are of the form

$$G_c(z) = \frac{1 + b_1 z^{-1} + b_2 z^{-2}}{1 + a_1 z^{-1} + a_2 z^{-2}} \tag{10.7}$$

The circuit realization of (10.7) is given in Figure 10.6.

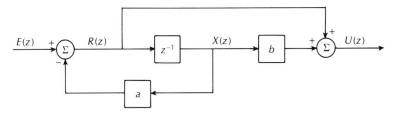

Figure 10.5   Realization of the first-order term (10.6).

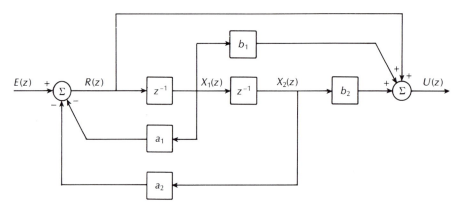

Figure 10.6   Realization of the second-order term (10.7).

From Figure 10.6 we obtain: $X_1(z) = z^{-1}R(z)$ and $X_2(z) = z^{-1}X_1(z) = z^{-2}R(z)$. Moreover, we have: $R(z) = E(z) - a_1X_1(z) - a_2X_2(z)$ or $R(z) = E(z) - a_1z^{-1}R(z) - a_2z^{-2}R(z)$. Therefore

$$E(z) = (1 + a_1z^{-1} + a_2z^{-2})R(z)$$

Also we have $U(z) = R(z) + b_1X_1(z) + b_2X_2(z)$ or $U(z) = R(z) + b_1z^{-1}R(z) + b_2z^{-2}R(z)$. Hence

$$U(z) = (1 + b_1z^{-1} + b_2z^{-2})R(z)$$

Finally, dividing the respective sides of the above two equations we obtain:

$$G_c(z) = \frac{U(z)}{E(z)} = \frac{1 + b_1z^{-1} + b_2z^{-2}}{1 + a_1z^{-1} + a_2z^{-2}}$$

which is in agreement with (10.7).

The general case is examined below.

3. Transfer function with more than one term

Figure 10.7 shows the cascade realization of a transfer function with several first- and second-order terms. The case where only one second-order term appears is examined below. The cascade realization, using the following difference equations derived from the structure of Figure 10.7, leads to the state equations:

$$x_1(k + 1) = -b_1x_1(k) + \beta e(k)$$
$$x_2(k + 1) = (a_1 - b_1)x_1(k) - b_2x_2(k) - b_3x_3(k) + \beta e(k)$$
$$x_3(k + 1) = x_2(k)$$
$$x_4(k + 1) = (a_1 - b_1)x_1(k) + (a_2 - b_2)x_2(k) + (a_3 - b_3)x_3(k) - b_4x_4(k) + \beta e(k)$$
$$\vdots$$
$$x_n(k + 1) = (a_1 - b_1)x_1(k) + (a_2 - b_2)x_2(k) + \cdots + (a_{n-1} - b_{n-1})x_{n-1}(k)$$
$$- b_nx_n(k) + \beta e(k)$$

The above equations can be written compactly as follows:

$$\mathbf{x}(k + 1) = \mathbf{A}\mathbf{x}(k) + \mathbf{b}e(k) \tag{10.8a}$$
$$u(k) = \mathbf{c}^T\mathbf{x}(k) + de(k) \tag{10.8b}$$

where the matrix $\mathbf{A}$ and the vectors $\mathbf{b}$ and $\mathbf{c}$ have the following forms:

$$\mathbf{A} = \begin{bmatrix} -b_1 & 0 & 0 & \cdots & 0 \\ a_1 - b_1 & -b_2 & -b_3 & \cdots & 0 \\ 0 & 1 & 0 & \cdots & 0 \\ a_1 - b_1 & a_2 - b_2 & a_3 - b_3 & \cdots & 0 \\ \vdots & \vdots & \vdots & \ddots & \vdots \\ a_1 - b_1 & a_2 - b_2 & a_3 - b_3 & \cdots & -b_n \end{bmatrix}, \quad \mathbf{b} = \begin{bmatrix} \beta \\ \beta \\ 0 \\ \beta \\ \vdots \\ \beta \end{bmatrix} \quad \text{and} \quad \mathbf{c} = \begin{bmatrix} a_1 - b_1 \\ a_2 - b_2 \\ a_3 - b_3 \\ a_4 - b_4 \\ \vdots \\ a_n - b_n \end{bmatrix} \tag{10.8c}$$

and the constant $d = \beta$.

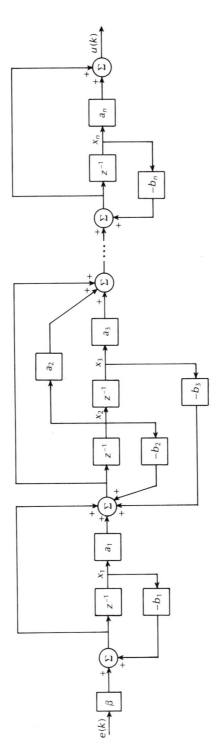

Figure 10.7 Cascade realization (general case).

10.2.3 *Parallel realization*

In the parallel realization it is assumed that the transfer function $G_c(z)$ may be expanded as a sum of terms of the form

$$G_c(z) = \frac{U(z)}{E(z)} = \gamma + G_1(z) + G_2(z) + \ldots + G_q(z) \quad \text{where} \quad 1 \leq q \leq n \tag{10.9}$$

where γ is a constant and $G_1(z), \ldots, G_q(z)$ are transfer functions of either first or second order. The block diagram realization of (10.9) is given in Figure 10.8.

1. First-order terms

The first-order terms are of the form

$$G(z) = \frac{b_0}{1 + a_1 z^{-1}} \tag{10.10}$$

The circuit realization of (10.10) is shown in Figure 10.9. To verify this realization we note that $X(z) = E(z) - a_1 z^{-1} X(z)$, or $E(z) = (1 + a_1 z^{-1}) X(z)$. Moreover, $U(z) = b_0 X(z)$. Hence, dividing the respective sides of the two equations we obtain

$$G(z) = \frac{U(z)}{E(z)} = \frac{b_0}{1 + a_1 z^{-1}}$$

which is in agreement with equation (10.10).

2. Second-order terms

The second-order terms are of the form

$$G(z) = \frac{b_0 + b_1 z^{-1}}{1 + a_1 z^{-1} + a_2 z^{-2}} \tag{10.11}$$

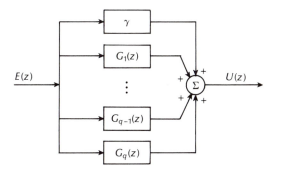

Figure 10.8 Parallel realization block diagram of (10.9).

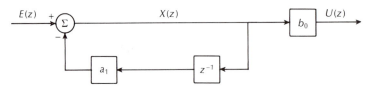

Figure 10.9   Realization of the first-order term (10.10).

The circuit realization of (10.11) is shown in Figure 10.10. To verify this realization we observe that $X_1(z) = z^{-1}R(z)$ and $X_2(z) = z^{-1}X_1(z) = z^{-2}R(z)$. Moreover, $R(z) = E(z) - a_1X_1(z) - a_2X_2(z)$ or $R(z) = E(z) - a_1z^{-1}R(z) - a_2z^{-2}R(z)$. Hence

$$E(z) = (1 + a_1z^{-1} + a_2z^{-2})R(z)$$

We also have $U(z) = b_0R(z) + b_1X_1(z)$ or $U(z) = b_0R(z) + b_1z^{-1}R(z)$. Thus

$$U(z) = (b_0 + b_1z^{-1})R(z)$$

Finally, dividing the respective sides of the expressions for $E(z)$ and $U(z)$, we obtain

$$G(z) = \frac{U(z)}{E(z)} = \frac{b_0 + b_1z^{-1}}{1 + a_1z^{-1} + a_2z^{-2}}$$

which is in agreement with (10.11).

3. Transfer function with more than one term

The general case (equation (10.9)) can be realized if the first- and second-order terms, shown in Figures 10.9 and 10.10, respectively, are interconnected as in Figure 10.11. The parallel realization (assuming, for simplicity, that there exists only one second-order term) can be constructed on the basis of Figure 10.11. Here the

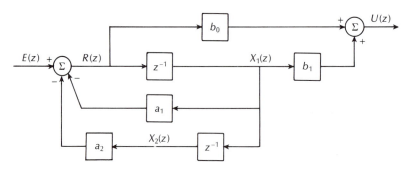

Figure 10.10   Realization of the second-order term (10.11).

difference equations are

$$x_1(k+1) = -a_1 x_1(k) + e(k)$$
$$x_2(k+1) = -a_3 x_3(k) - a_2 x_2(k) + e(k)$$
$$x_3(k+1) = x_2(k)$$
$$\vdots$$
$$x_n(k+1) = -a_n x_n(k) + e(k)$$

These equations can be written compactly as follows:

$$\mathbf{x}(k+1) = \mathbf{A}\mathbf{x}(k) + \mathbf{b}e(k) \tag{10.12a}$$

$$u(k) = \mathbf{c}^T \mathbf{x}(k) + de(k) \tag{10.12b}$$

where

$$\mathbf{A} = \begin{bmatrix} -a_1 & 0 & 0 & 0 & \cdots & 0 \\ 0 & -a_2 & -a_3 & 0 & \cdots & 0 \\ 0 & 0 & 1 & 0 & \cdots & 0 \\ 0 & 0 & 0 & -a_4 & \cdots & 0 \\ \vdots & \vdots & \vdots & \vdots & \ddots & \vdots \\ 0 & 0 & 0 & 0 & \cdots & -a_n \end{bmatrix}, \quad \mathbf{b} = \begin{bmatrix} 1 \\ 1 \\ 0 \\ 1 \\ \vdots \\ 1 \end{bmatrix} \quad \text{and} \quad \mathbf{c} = \begin{bmatrix} -a_1 b_1 \\ -a_2 b_2 + b_3 \\ -a_3 b_3 - b_2 a_3 \\ -a_4 b_4 \\ \vdots \\ -a_n b_n \end{bmatrix} \tag{10.12c}$$

and the constant $d = \gamma + b_1 + b_2 + b_4 + \cdots + b_n$.

■ **EXAMPLE 10.2.1** Consider a controller with transfer function

$$G_c(z) = \frac{U(z)}{E(z)} = \frac{4 + 2z^{-1} - 2z^{-2}}{1 + 0.1z^{-1} - 0.12z^{-2}}$$

Determine the three forms of realization of $G_c(z)$.

■ **SOLUTION** We have

1. Direct realization: according to the first realization procedure, the difference equation is

$$u(k) + 0.1u(k-1) - 0.12u(k-2) = 4e(k) + 2e(k-1) - 2e(k-2)$$

Furthermore, the state equations are

$$\begin{bmatrix} x_1(k+1) \\ x_2(k+1) \end{bmatrix} = \begin{bmatrix} 0 & 1 \\ 0.12 & -0.1 \end{bmatrix} \begin{bmatrix} x_1(k) \\ x_2(k) \end{bmatrix} + \begin{bmatrix} 0 \\ 1 \end{bmatrix} e(k)$$

$$u(k) = \begin{bmatrix} -1.52 & 1.6 \end{bmatrix} \begin{bmatrix} x_1(k) \\ x_2(k) \end{bmatrix} + 4e(k)$$

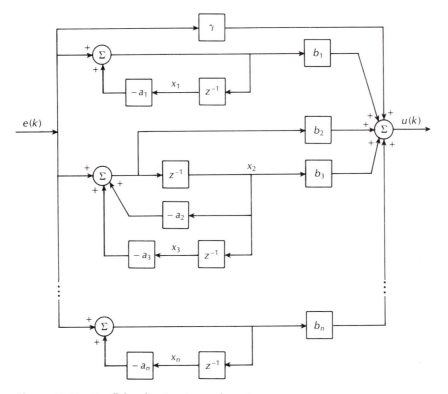

Figure 10.11 Parallel realization (general case).

According to the second realization procedure the difference equation remains the same, i.e.

$$u(k) + 0.1u(k-1) - 0.12u(k-2) = 4e(k) + 2e(k-1) - 2e(k-2)$$

whose state equations are

$$\begin{bmatrix} \tilde{x}_1(k+1) \\ \tilde{x}_2(k+1) \end{bmatrix} = \begin{bmatrix} 0 & 0.12 \\ 1 & -0.1 \end{bmatrix} \begin{bmatrix} \tilde{x}_1(k) \\ \tilde{x}_2(k) \end{bmatrix} + \begin{bmatrix} -1.52 \\ 1.6 \end{bmatrix} e(k)$$

$$u(k) = \begin{bmatrix} 0 & 1 \end{bmatrix} \begin{bmatrix} \tilde{x}_1(k) \\ \tilde{x}_2(k) \end{bmatrix} + 4e(k)$$

2. Cascade realization: the transfer function $G_c(z)$ is written as

$$G_c(z) = \frac{E(z)}{U(z)} = 4 \frac{(1 + z^{-1})(1 - 0.5z^{-1})}{(1 + 0.4z^{-1})(1 - 0.3z^{-1})} = 4 \frac{(z + 1)(z - 0.5)}{(z + 0.4)(z - 0.3)}$$

The state-space equations are

$$\begin{bmatrix} x_1(k+1) \\ x_2(k+1) \end{bmatrix} = \begin{bmatrix} -0.4 & 0 \\ 0.6 & 0.3 \end{bmatrix} \begin{bmatrix} x_1(k) \\ x_2(k) \end{bmatrix} + 4 \begin{bmatrix} 1 \\ 1 \end{bmatrix} e(k)$$

$$u(k) = [0.6 \quad -0.2] \begin{bmatrix} x_1(k) \\ x_2(k) \end{bmatrix} + 4e(k)$$

3. Parallel realization: expanding $G_c(z)$ in partial fractions we have

$$G_c(z) = 4 \left(\frac{(1 + z^{-1})(1 - 0.5z^{-1})}{(1 + 0.4z^{-1})(1 - 0.3z^{-1})} \right) = A + \frac{B}{1 + 0.4z^{-1}} + \frac{\Gamma}{1 - 0.3z^{-1}}$$

After several calculations we find that $A = 50/3$, $B = -54/7$ and $\Gamma = 104/21$. Hence

$$G_c(z) = \frac{50}{3} - \frac{54/7}{1 + 0.4z^{-1}} - \frac{104/21}{1 - 0.3z^{-1}}$$

The state equations are

$$\begin{bmatrix} x_1(k+1) \\ x_2(k+1) \end{bmatrix} = \begin{bmatrix} -0.4 & 0 \\ 0 & 0.3 \end{bmatrix} \begin{bmatrix} x_1(k) \\ x_2(k) \end{bmatrix} + \begin{bmatrix} 1 \\ 1 \end{bmatrix} e(k)$$

■ **EXAMPLE 10.2.2** Consider a discrete-time PID controller with transfer function

$$G_c(z) = K_p + K_p \frac{T}{T_i} \left(\frac{z}{z-1} \right) + K_p \frac{T_d}{T_i} \left(\frac{z-1}{z} \right) = K \left(\frac{z^2 - az + b}{z(z-1)} \right)$$

where all the parameters K_p, T, T_i, T_d, K, a and b are defined in Section 5.5.4. Determine the direct realization (both forms) and the cascade realization.

■ **SOLUTION** We have:

1. Direct realization: for the first realization procedure we write $G_c(z)$ as follows:

$$G_c(z) = K \left(\frac{z^2 - az + b}{z(z-1)} \right) = K \left(\frac{1 - az^{-1} + bz^{-2}}{1 - z^{-1}} \right)$$

The state equations are

$$\begin{bmatrix} x_1(k+1) \\ x_2(k+1) \end{bmatrix} = \begin{bmatrix} 0 & 1 \\ 0 & 1 \end{bmatrix} \begin{bmatrix} x_1(k) \\ x_2(k) \end{bmatrix} + \begin{bmatrix} 0 \\ 1 \end{bmatrix} e(k)$$

$$u(k) = K[b \quad -a + 1] \begin{bmatrix} x_1(k) \\ x_2(k) \end{bmatrix} + Ke(k)$$

For the second realization procedure, the state equations are

$$
\begin{bmatrix} \tilde{x}_1(k+1) \\ \tilde{x}_2(k+1) \end{bmatrix} = \begin{bmatrix} 0 & 0 \\ 1 & 1 \end{bmatrix} \begin{bmatrix} \tilde{x}_1(k) \\ \tilde{x}_2(k) \end{bmatrix} + K \begin{bmatrix} b \\ -a+1 \end{bmatrix} e(k)
$$

$$
u(k) = \begin{bmatrix} 0 & 1 \end{bmatrix} \begin{bmatrix} \tilde{x}_1(k) \\ \tilde{x}_2(k) \end{bmatrix} + Ke(k)
$$

2. Cascade realization: the state equations are

$$
\begin{bmatrix} x_1(k+1) \\ x_2(k+1) \end{bmatrix} = \begin{bmatrix} 1 & 0 \\ 1 & 0 \end{bmatrix} \begin{bmatrix} x_1(k) \\ x_2(k) \end{bmatrix} + K \begin{bmatrix} 1 \\ 0 \end{bmatrix} e(k)
$$

$$
u(k) = \begin{bmatrix} 1-a & b \end{bmatrix} \begin{bmatrix} x_1(k) \\ x_2(k) \end{bmatrix} + Ke(k)
$$

10.3 Software controller realization

The hardware implementation of digital control algorithms presented in the previous section uses electrical and electronic devices (delay units, multiplicators, etc.) and therefore has some limitations. In particular, these hardware realizations cannot be very reliable and accurate, since the characteristics of the elements used may change in time, owing to temperature or other variations. Moreover, a change in the structure of the control algorithm results in major changes in the circuit layout. Furthermore, it is very difficult, when using hardware implementation, to realize sophisticated controllers, as for example the observers of Chapter 6, the identification schemes of Chapter 8 and the adaptive control schemes of Chapter 9.

Owing to the rapid development of inexpensive, fast computer hardware and software systems, the use of microprocessors for real-time (ON-LINE) control of real systems has now become standard practice. The computer is in this case an integral part of the control loop and the microprocessor implements the control algorithm (software realization). This leads to the computer-controlled system of Figure 10.12. Another configuration of the computer-controlled system is given in Figure 1.1 (Chapter 1).

Since the computer uses digital data representation, the analog output signal $y(t)$ in Figure 10.12 is approximated by a sequence of numbers. These numbers are the values of $y(t)$ at the time instants kT, $k = 0, 1, 2, \ldots$, where T is the sampling period. The samples are then encoded into a binary word using a finite number of bits. This procedure, called quantization, is carried out by analog-to-digital (A/D) converters.

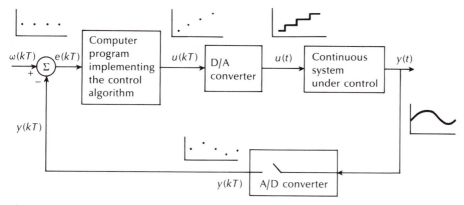

Figure 10.12 Computer-controlled system.

The computer reads the A/D converter output every T seconds. The reference signal $\omega(kT)$ is stored as a sequence of binary numbers in the computer memory. The error $e(kT)$ is formed and is then processed by the computer, i.e. the computer program performs the calculations needed to realize the control algorithm. Finally, the computer feeds its output to the system under control via a digital-to-analog (D/A) converter (or hold circuit). The D/A converter holds constant the numerical value given by the computer until the next sample arrives. This procedure is repeated for each sampling period T, provided that the computation time (i.e. the time needed to perform all calculations for the realization of the control algorithm) is less than the sampling period T.

The advantages of controller realization using microprocessors are the low hardware cost, reliability and accuracy. Moreover, the controller designer has the flexibility to use the same hardware equipment to realize and test different control algorithms. This can be done because only the program which realizes the control algorithm has to be changed for every controller realization. However, the performance of a digital control system is strongly dependent on many factors, among which are the sampling period, possible computational, and the word length in the microprocessor, and the interface devices (A/D and D/A converters) used. Large sampling periods, T, give the necessary computation time required to realize the control algorithm, but the bandwidth needed for the closed-loop system may require small values for T. Calculations must be performed within T, otherwise computational delays may even lead to instability of the closed-loop system. Finally, the coefficients appearing in the control algorithms and the values of the reference input must be represented by a finite number of bits (finite word length). All these operations, together with quantization, introduce inaccuracies and trade-offs in the control loop.

The various aspects of software realization will not be investigated further in this book. The interested reader may see [1–11]. In the following section some introductory material is presented with regard to the important issue of quantization errors.

10.4 Quantization errors

10.4.1 *Fixed point and floating point representations*

Microprocessors use a binary representation of numbers. Numbers are stored as sequences of bits, each bit taking the value 0 or 1. We distinguish between the **fixed point** and **floating point** representations.

In the fixed point representation, we assume that a number is represented by a sequence of n bits as $c_{n-1}c_{n-2} \ldots c_2c_1c_0$, where the c_is take on the value 0 or 1. The radix point (separating the integer from the fractional part) is fixed at a certain position in the sequence. Integers are represented in the case where the radix point is fixed at the rightmost position, while, if the radix point is somewhere between the bits c_{n-1} and c_0, fractional parts are also assumed. Suppose that there are $l < n$ bits to the right of the radix point (representing the fractional part) and $n - l$ bits to the left of the radix point (representing the integer part).

One can distinguish three forms of fixed point representation, depending on the way negative numbers are represented. In the **sign–magnitude** fixed point representation, the most significant bit c_{n-1} represents the sign (usually 0 for a positive number and 1 for a negative number). The remaining sequence $c_{n-2} \ldots c_2c_1c_0$ represents the magnitude, where each bit represents a power of 2 according to its position in the sequence. For example, the sequence 01001.101 represents the decimal number +9.625, as is shown below:

$$01001.101_2 = +[1 \times 2^3 + 0 \times 2^2 + 0 \times 2^1 + 1 \times 2^0 + 1 \times 2^{-1} + 0 \times 2^{-2} + 1 \times 2^{-3}]$$
$$= +[8 + 0 + 0 + 1 + 0.5 + 0 + 0.125] = +9.625_{10}$$

With l fractional bits, the magnitude $c_{n-2}c_{n-1} \ldots c_2c_1c_0$ is calculated to be equal to

$$2^{-l}\left(\sum_{i=0}^{n-2} c_i 2^i\right)$$

Then, the least significant bit c_0 represents the magnitude 2^{-l}. The **resolution** or **quantization level**, which represents the smallest difference that can be distinguished, is thus equal to $q = 2^{-l}$. Obviously, for integers (radix point to the rightmost position) it is $q = 1$. In general, when n bits are used for the sign–magnitude representation, the least number which can be represented has a magnitude of $q = 2^{-(n-1)}$. This corresponds to the case where the radix point is just after the sign bit c_{n-1} ($l = n - 1$). Note that in the sign–magnitude form, there are two representations for zero, namely $00 \ldots 00$ and $10 \ldots 00$.

In the **one's complement** form, the sign–magnitude representation is used for positive numbers, while the negative of a number is obtained by complementing all bits (changing the zeros to ones and vice versa). The most significant bit, c_{n-1}, still represents the sign. If we still assume l bits to the right of the radix point, then the bit sequence $c_{n-1}c_{n-2} \ldots c_2c_1c_0$ represents the number

$$2^{-l}\left(-c_{n-1}(2^{n-1} - 1) + \sum_{i=0}^{n-2} c_i 2^i\right)$$

For example, when using $n = 4$ bits with no fractional part, then 0101 represents the positive integer $+5$, while 1010 represents -5. We have two representations for zero in this form also, namely all bits being equal to zero or one.

In the **two's complement** form, the sign–magnitude representation is still used for positive numbers, while the negative of a number is obtained by complementing all bits and adding 1. For example, when using $n = 4$ bits with no fractional part, then 0101 represents $+5$ as in the sign–magnitude and one's complement forms, and $1010 + 1 = 1011$ represents -5. Generalizing, the sequence $c_{n-1}c_{n-2} \ldots c_2 c_1 c_0$ represents the decimal

$$2^{-l}\left(-c_{n-1}2^{n-1} + \sum_{i=0}^{n-2} c_i 2^i\right)$$

(c_{n-1} still representing the sign). In Table 10.1, the equivalent decimal representation of binary numbers, when using $n = 4$ bits with no fractional part and the above forms of fixed point arithmetic, is given. Note that in the two's complement form one extra level is allowed in the negative direction (here, it corresponds to -8), while $+8$ cannot be represented.

In the floating point representation, a number is represented as $m \times b^c$, where m is called the mantissa, c the characteristic and b the radix or base (usually $b = 2$). Fixed point arithmetic is used for representing m and c.

Table 10.1 Equivalent decimal representation of a 4 bit integer, using the sign–magnitude, one's and two's complement forms.

Binary representation $n = 4$ bits with no fractional part	Decimal equivalents		
	Sign–magnitude	One's complement	Two's complement
0000	0	0	0
0001	+1	+1	+1
0010	+2	+2	+2
0011	+3	+3	+3
0100	+4	+4	+4
0101	+5	+5	+5
0110	+6	+6	+6
0111	+7	+7	+7
1000	0	−7	−8
1001	−1	−6	−7
1010	−2	−5	−6
1011	−3	−4	−5
1100	−4	−3	−4
1101	−5	−2	−3
1110	−6	−1	−2
1111	−7	0	−1

10.4.2 Truncation and rounding

Because of the finite word length used for representing numbers by microprocessors (e.g. registers or memory locations of finite length), accuracy problems occur when a number having m bits needs to be represented by a number having $n < m$ bits. This process, called **quantization**, usually appears in A/D conversions, after multiplications or divisions, when inserting constants, etc. It is of interest to calculate the errors due to quantization and how these errors propagate through transfer functions. This is shown in the rest of this section.

Quantization can be achieved either by using a process called **truncation** or by using a process called **rounding**. When truncating a number of m bits to a number with $n < m$ bits, the $m - n$ least significant bits are ignored. If $l < n$ is the length of the fractional part of the n bit number, then the part ignored is at most as large as $q = 2^{-l}$. The exact relation between the values of a continuous variable x and the values of its quantized version x_{qT} using truncation are given in Figure 10.13 for the case of two's complement representation and in Figure 10.14 for the cases of one's complement or sign–magnitude representations.

Denote by $\varepsilon_T = x_{qT} - x$ the error during truncation in the case of two's complement representation and by $\varepsilon'_T = x'_{qT} - x$ the error in the case of one's complement or sign–magnitude representations. It can be easily seen from Figures 10.13 and 10.14 that

$$-q < \varepsilon_T \leqslant 0 \tag{10.13a}$$

and

$$0 \leqslant \varepsilon'_T < q \text{ for } x < 0 \quad \text{and} \quad -q < \varepsilon'_T \leqslant 0 \text{ for } x > 0 \tag{10.13b}$$

When the procedure followed for quantization is rounding, the quantized value is the same as in truncation if the first lost bit is equal to 0, otherwise one adds the term 2^{-l} if the first lost bit equals 1. The respective relation between the continuous signal x

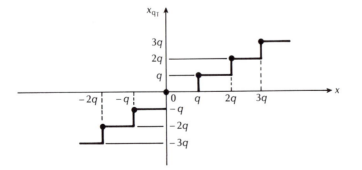

Figure 10.13 Truncation, two's complement. Continuous signal x and truncated values x_{qT} with quantization level q.

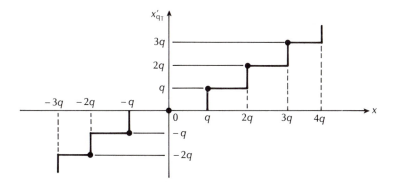

Figure 10.14 Truncation, one's complement or sign–magnitude. Continuous signal x and truncated values x'_{qT} with quantization level q.

and the result x_{qR} after rounding is shown in Figure 10.15. One easily obtains

$$-q/2 \leq \varepsilon_R = x_{qR} - x \leq q/2$$

This is valid for all cases of fixed point representation used.

■ **EXAMPLE 10.4.1**   Quantize a continuous signal x, by assuming an 8 bit word length and two's complement representation. Distinguish the cases of x taking values in the range -2 volts to 2 volts and -3 volts to 3 volts, respectively.

■ **SOLUTION**   Consider first that x takes values in the range -2 to 2 volts. Since we wish to represent both positive and negative values, 1 bit will be used for the sign. Notice that in this case 1 bit is enough for the representation of the integer part. The remaining $l = 6$ bits will be used to represent the fractional part. Therefore, the least

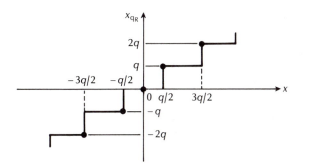

Figure 10.15 Rounding, all fixed point representations. Continuous signal x and values x_{qR} after rounding with quantization level q.

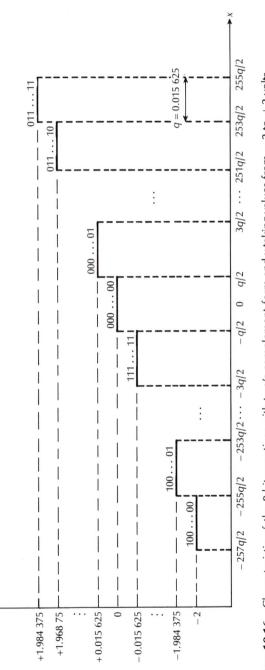

Figure 10.16 Characteristic of the 8 bit quantizer with two's complement form and x taking values from −2 to +2 volts.

number which can be represented (the resolution) has a magnitude of $q = 2^{-l} = 2^{-6} = 0.015\,625$. The resolution (interval between successive levels) can also be found by following another, more general, reasoning. Since 8 bits are used for the representation, we can achieve $2^8 = 256$ different levels. The full dynamic range of the signal equals 2 volts $-$ (-2 volts) $= 4$ volts. Consequently, each level has a width (resolution) of $4/256$ volts $= 0.015\,625$ volts. The relation between the binary and the corresponding decimal representations, when using the two's complement form, is given in Table 10.2. In this case, the radix point is between the second and third bit from the left. The first bit from the left represents the sign. The characteristic of the quantizer is given in Figure 10.16.

When x takes values in the range -3 to 3 volts and since then the full signal range equals $3 - (-3) = 6$ volts, the quantization level is $q = 6/256 = 0.023\,437\,5$ volts. In this case, a one-to-one correspondence between the binary and a decimal representation can be established, making use of all the 256 different quantization levels. This is shown in Table 10.2.

10.4.3 A stochastic model for the quantization error

A stochastic model can be established for the quantization error. To this end, one assumes that the quantization error can take all the values in the ranges determined in the previous section, with equal probability. This assumption is completely justified

Table 10.2 Quantization of a continuous signal, with a word length of 8 bits and two's complement form.

Binary representation	Equivalent decimal representation $x \in [-2, 2]$	Respective decimal representation $x \in [-3, 3]$
000 ... 00	0	0
000 ... 01	$+0.015\,625$	$+0.023\,437\,5$
000 ... 10	$+0.031\,25$	$+0.046\,875$
000 ... 11	$+0.046\,875$	$+0.070\,312\,5$
⋮	⋮	⋮
011 ... 10	$+1.968\,75$	$+2.953\,125$
011 ... 11	$+1.984\,375$	$+2.976\,562\,5$
100 ... 00	-2	-3
100 ... 01	$-1.984\,375$	$-2.976\,562\,5$
100 ... 10	$-1.968\,75$	$-2.953\,125$
⋮	⋮	⋮
111 ... 10	$-0.031\,25$	$-0.046\,875$
111 ... 11	$-0.015\,625$	$-0.023\,437\,5$

in the case when x itself is a random variable. When x is deterministic, this assumption is reasonable for most practical cases and leads to a uniform probability density. In the case of truncation, using a two's complement scheme and assuming that ε_T takes all the values in the range $[-q, 0]$ with equal probability, it is obvious that the probability density should be

$$f_{\varepsilon_T}(\xi) = 1/q$$

as shown in Figure 10.17, in order to have

$$\int_{-\infty}^{+\infty} f_{\varepsilon_T}(\xi)\,\mathrm{d}\xi = \int_{-q}^{0} \frac{1}{q}\,\mathrm{d}\xi = 1$$

Then, we can easily compute the mean value and the variance of the truncation error as follows:

$$\mu_{\varepsilon_T} = E\{\varepsilon_T\} = \int_{-\infty}^{+\infty} \xi f_{\varepsilon_T}(\xi)\,\mathrm{d}\xi = \int_{-q}^{0} \xi\,\frac{1}{q}\,\mathrm{d}\xi = -\frac{q}{2}$$

$$\sigma_{\varepsilon_T}^2 = E\{(\varepsilon_T - \mu_{\varepsilon_T})^2\} = \int_{-\infty}^{+\infty} \left(\xi + \frac{q}{2}\right)^2 f_{\varepsilon_T}(\xi)\,\mathrm{d}\xi = \frac{q^2}{12}$$

In an analogous manner, when rounding is considered, we have the probability density of Figure 10.18, in which the mean value is given by

$$\mu_{\varepsilon_R} = E\{\varepsilon_R\} = \int_{-\infty}^{+\infty} \xi f_{\varepsilon_R}(\xi)\,\mathrm{d}\xi = \int_{-q/2}^{q/2} \xi\,\frac{1}{q}\,\mathrm{d}\xi = 0$$

and the variance is given by

$$\sigma_{\varepsilon_R}^2 = E\{(\varepsilon_R - \mu_{\varepsilon_R})^2\} = \int_{-\infty}^{+\infty} (\xi - 0)^2 f_{\varepsilon_R}(\xi)\,\mathrm{d}\xi$$

$$= \int_{-q/2}^{q/2} \xi^2\,\frac{1}{q}\,\mathrm{d}\xi = \frac{\xi^3}{3q}\bigg|_{-q/2}^{q/2} = \frac{1}{3q}\left(\frac{q^3}{8} + \frac{q^3}{8}\right) = \frac{q^2}{12}$$

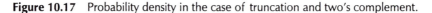

Figure 10.17 Probability density in the case of truncation and two's complement.

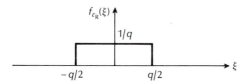

Figure 10.18 Probability density in the case of rounding.

10.4.4 *Propagation of the quantization error through transfer functions*

We are now in a position to investigate how quantization errors are transmitted through transfer functions. This will give us the opportunity to analyze the effect of quantization errors, when realizing control algorithms implemented by microprocessors. Three procedures can be followed, namely the worst case analysis to obtain error bounds, the steady state analysis, and statistical analysis.

1. Worst case and steady state analysis

Suppose that the input to a transfer function $H(z)$ is a quantization error $\varepsilon(k)$, which is assumed bounded, i.e. $|\varepsilon(k)| < a$. We search for an upper bound to the output signal $y(k)$. In the z-domain we have

$$Y(z) = H(z)E(z)$$

Equivalently, in the time domain one has the convolution sum

$$y(k) = \sum_{i=0}^{k} h(i)\varepsilon(k - i)$$

where $h(k) = \mathcal{Z}^{-1}[H(z)]$ is the system's impulse response. Then,

$$|y(k)| = \left| \sum_{i=0}^{k} h(i)\varepsilon(k - i) \right| \leq \sum_{i=0}^{k} |h(i)|\,|\varepsilon(k - i)| < a \sum_{i=0}^{k} |h(i)|$$

If the output signal $y(k)$ reaches a constant steady state value, then

$$\lim_{k \to \infty} |y(k)| = \lim_{k \to \infty} \left| \sum_{i=0}^{k} h(i)\varepsilon(k - i) \right| < a \lim_{k \to \infty} \sum_{i=0}^{k} |h(i)|$$

$$= a \left| \sum_{i=0}^{\infty} h(i) \right| = a\,|H(z = 1)|$$

An error bound and a bound for the steady state value of the error have thus been established.

2. Statistical analysis

A statistical model for the quantization error, with known statistics (mean value and variance), is available. In what follows we will compute the statistics of the output of a transfer function $H(z)$, having the quantization error as input. To this end, we define the autocorrelation function of a stochastic signal $w(k)$ as

$$R_w(j) = E\{w(k)w(k + j)\} \tag{10.14a}$$

and its spectrum $S_w(z)$ as the Z-transform of R_w, i.e.

$$S_w(z) = \mathcal{Z}[R_w(j)] \tag{10.14b}$$

Note that $R_w(0) = E\{w^2(k)\}$, i.e. $R_w(0)$ equals the variance of $w(k)$. We denote by $x(k)$ the output of the transfer function $H(z)$ to the signal $w(k)$. In the z-domain we have

$$X(z) = H(z)W(z)$$

Equivalently, in the time domain we have the convolution sum

$$x(k) = \sum_{i=-\infty}^{+\infty} h(i)w(k-i) \tag{10.15}$$

where $h(k) = \mathcal{L}^{-1}[H(z)]$ is the system's impulse response. Since $w(k)$ is a stochastic variable, then $x(k)$ is a stochastic variable as well. Taking the expected values in equation (10.15), we obtain

$$E\{x(k)\} = \sum_{i=-\infty}^{+\infty} h(i)E\{w(k-i)\} = E\{w(k)\} \sum_{i=-\infty}^{+\infty} h(i)$$

But

$$H(z) = \sum_{i=-\infty}^{+\infty} h(i)z^{-i}$$

and hence

$$\sum_{i=-\infty}^{+\infty} h(i) = H(z)|_{z=1}$$

Finally

$$E\{x(k)\} = H(z)|_{z=1} E\{w(k)\} \tag{10.16}$$

Equation (10.16) relates the mean value of $x(k)$ to the mean value of $w(k)$. Obviously, when $w(k)$ has a zero mean value, the output signal $x(k)$ also has a zero mean value, independently of $H(z)$.

In order to compute the variance of $x(k)$, we will compute its spectrum $S_x(z)$ and then its autocorrelation function R_x. It is $\sigma_x^2 = R_x(0)$. We thus need a relation between the spectra $S_w(z)$ and $S_x(z)$ of the input and output stochastic variables $w(k)$ and $x(k)$, respectively. Since the output signal $x(k)$ is given by

$$x(j) = \sum_{k=-\infty}^{+\infty} h(k)w(j-k)$$

its autocorrelation function, according to definition (10.14a), is given by

$$R_x(l) = E\{x(j+l)x(j)\} = E\left\{ \left(\sum_{k=-\infty}^{+\infty} h(k)w(j+l-k) \right) \left(\sum_{n=-\infty}^{+\infty} h(n)w(j-n) \right) \right\}$$

However, the system's impulse response is not a stochastic variable and consequently

$$R_x(l) = \sum_{k=-\infty}^{+\infty} h(k) \sum_{n=-\infty}^{+\infty} h(n) E\{w(j+l-k)w(j-n)\} \qquad (10.17)$$

Since $E\{w(j+l-k)w(j-n)\} = R_w(l-k+n)$, equation (10.17) may be rewritten as

$$R_x(l) = \sum_{k=-\infty}^{+\infty} h(k) \sum_{n=-\infty}^{+\infty} h(n) R_w(l-k+n)$$

Following the definition (10.14b), the expression for the spectrum $S_x(z)$ of $x(k)$ is derived as follows:

$$S_x(z) = \mathscr{Z}[R_x(l)] = \sum_{l=-\infty}^{+\infty} R_x(l) z^{-1}$$

$$= \sum_{l=-\infty}^{+\infty} \sum_{k=-\infty}^{+\infty} h(k) \sum_{n=-\infty}^{+\infty} h(n) R_w(l-k+n) z^{-1}$$

$$= \sum_{k=-\infty}^{+\infty} h(k) \sum_{n=-\infty}^{+\infty} h(n) \sum_{l=-\infty}^{+\infty} R_w(l-k+n) z^{-1}$$

$$= \sum_{k=-\infty}^{+\infty} h(k) \sum_{n=-\infty}^{+\infty} h(n) \sum_{m=-\infty}^{+\infty} R_w(m) z^{-(m+k-n)}$$

where a change in the order of summation has been applied and we have put $m = l - k + n$. Since $z^{-(m+k-n)} = z^{-m} z^{-k} z^n$, we have

$$S_x(z) = \sum_{k=-\infty}^{+\infty} h(k) z^{-k} \sum_{n=-\infty}^{+\infty} h(n) z^n \sum_{m=-\infty}^{+\infty} R_w(m) z^{-m}$$

Finally, by noting that

$$\sum_{k=-\infty}^{+\infty} h(k) z^{-k} = H(z) \quad \text{and} \quad \sum_{m=-\infty}^{+\infty} R_w(m) z^{-m} = \mathscr{Z}[R_w(m)] = S_w(z)$$

we conclude that the following equation relating the spectra of $w(k)$ and $x(k)$ holds true:

$$S_x(z) = H(z) H(z^{-1}) S_w(z)$$

The autocorrelation function of $x(k)$ is now rewritten as

$$R_x(l) = \mathscr{Z}^{-1}[S_x(z)] = \mathscr{Z}^{-1}[H(z) H(z^{-1}) S_w(z)]$$

$$= \frac{1}{2\pi j} \oint_{|z|=1} H(z) H(z^{-1}) S_w(z) z^{l-1} \, dz$$

Finally for $l = 0$ we have $\sigma_z^2 = R_x(0)$

$$= \frac{1}{2\pi j} \oint_{|z|=1} H(z) H(z-1) S_w(z) z^{-1} \, dz \qquad (10.18)$$

■ **REMARK 10.4.1** Regarding the calculation of the integral in (10.18), we mention the following. If a function $f(z)$ is analytic inside and on the unit circle $|z| = 1$, except at a finite number of poles which are inside or on the unit circle, then

$$\oint_{|z|=1} f(z)\,dz = 2\pi j\ (\Sigma \text{ residues computed at the poles of } f(z) \text{ inside the unit circle})$$

The residue at a pole a of $f(z)$ is computed as $\lim_{z \to a} f(z)(z - a)$.

■ **EXAMPLE 10.4.2** This example refers to rounding. For this case, we have already determined that the quantization error has a mean value $\mu_{\varepsilon R} = 0$ and a variance $\sigma_{\varepsilon R}^2 = q^2/12$. Since $\varepsilon_R(k)$ is assumed not to be correlated, we have

$$R_{\varepsilon_R}(j) = E\{\varepsilon_R(k)\varepsilon_R(k+j)\} = \begin{cases} q^2/12 & \text{if } j = 0 \\ 0 & \text{if } j \neq 0 \end{cases}$$

Determine the statistics of the output signal $y(k)$ when $\varepsilon_R(k)$ is the input to a transfer function $H(z)$.

■ **SOLUTION** We have

$$E\{y(k)\} = H(z)\big|_{z=1}\,\mu_{\varepsilon_R} = 0$$

Moreover, the spectrum of $\varepsilon_R(k)$ is

$$S_{\varepsilon_R}(z) = \mathcal{Z}[R_{\varepsilon_R}(j)] = \sum_{j=-\infty}^{+\infty} R_{\varepsilon_R}(j)z^{-j} = R_{\varepsilon_R}(0) = q^2/12$$

The variance of the output signal $y(k)$ to the quantization error $\varepsilon_R(k)$ is then given by

$$\sigma_y^2 = R_y(0) = \frac{1}{2\pi j}\oint_{|z|=1} H(z)H(z^{-1})S_{\varepsilon_R}(z)\,\frac{dz}{z} = \frac{1}{2\pi j}\oint_{|z|=1} H(z)H(z^{-1})\,\frac{dz}{z}\,\sigma_{\varepsilon_R}^2$$

$$= \frac{1}{2\pi j}\frac{q^2}{12}\oint_{|z|=1} H(z)H(z^{-1})\,\frac{dz}{z}$$

In the case for example where $H(z) = 1/(z - a)$ with $|a| < 1$ (stable transfer function), one has

$$\oint_{|z|=1}\left(\frac{1}{z-a}\right)\left(\frac{1}{z^{-1}-a}\right)\frac{dz}{z} = \frac{2\pi j}{z(z^{-1}-a)}\bigg|_{z=a} = 2\pi j\left(\frac{1}{1-a^2}\right)$$

and finally

$$\sigma_y^2 = \sigma_{\varepsilon_R}^2\left(\frac{1}{1-a^2}\right) = \frac{q^2}{12}\left(\frac{1}{1-a^2}\right)$$

■ **EXAMPLE 10.4.3**   In this example, we will examine the truncation errors after multiplication due to the use of registers with finite word length. Parallel and direct realizations of a transfer function will be considered. It will be clear that amplification or attenuation of the quantization noise depends strongly on the particular realization chosen.

Consider the transfer function

$$H(z) = \frac{Y(z)}{U(z)} = \frac{\beta_1}{1 - \alpha_1 z^{-1}} + \frac{\beta_2}{1 - \alpha_2 z^{-1}}$$

with $|\alpha_1| < 1$ and $|\alpha_2| < 1$. Compare the parallel and direct realizations with quantization errors after the multiplications needed. Distinguish different parallel realizations, according to the position of the multipliers β_1 and β_2. Compute the mean value and the variance of the output signal.

■ **SOLUTION**   We distinguish two different ways of parallel realization, shown in Figures 10.19 and 10.20, respectively, according to the position of the multipliers β_1 and β_2. We have the following truncation errors, inserted in Figures 10.19 and 10.20 at points where multiplications take place: ε_{m_1} (multiplication by α_1), ε_{m_2} (multiplication by α_2), ε_{m_3} (multiplication by β_1) and ε_{m_4} (multiplication by β_2). Consider the system of Figure 10.21, having the transfer function $H(z)$. Its input is the stochastic variable $w(k)$ with mean value $E\{w(k)\}$ and variance $\sigma^2_{w(k)}$. The conclusion of the analysis of Section 10.4.4 is that the statistics of the output signal $x(k)$ are the following:

$$E\{x(k)\} = H(z)|_{z=1} E\{w(k)\}$$

$$\sigma^2_{x(k)} = \sigma^2_{w(k)} \frac{1}{2\pi j} \oint_{|z|=1} H(z)H(z^{-1}) \frac{dz}{z}$$

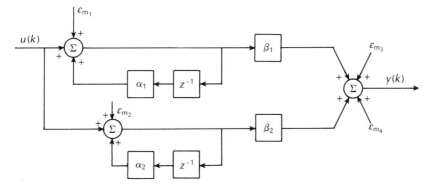

Figure 10.19   A parallel realization of $H(z)$.

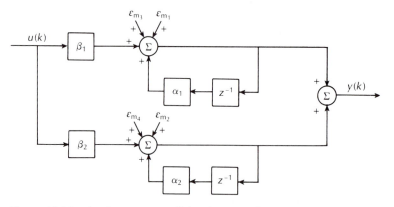

Figure 10.20   An alternative parallel realization of $H(z)$.

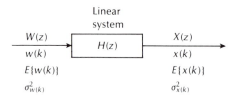

Figure 10.21   A linear system $H(z)$ with a stochastic input $w(k)$, having mean value $E\{w(k)\}$ and variance $\sigma^2_{w(k)}$.

In Figure 10.19 one easily finds the transfer functions needed as follows:

$$H_1(z) = \frac{Y_1(z)}{E_{m_1}(z)} = \frac{\beta_1}{1 - \alpha_1 z^{-1}}, \quad H_2(z) = \frac{Y_2(z)}{E_{m_2}(z)} = \frac{\beta_2}{1 - \alpha_2 z^{-1}}$$

$$H_3(z) = \frac{Y_3(z)}{E_{m_3}(z)} = 1 = \frac{Y_4(z)}{E_{m_4}(z)} = H_4(z)$$

where $y_i(k)$ is the output due to the input $\varepsilon_{m_i}(k)$ and $y(k) = \sum_{i=1}^{4} y_i(k)$ is the output due to the sum of all the truncation errors. Here

$$E\{\varepsilon_{m_i}(k)\} = -\frac{q}{2} \quad \text{and} \quad \sigma^2_{\varepsilon_{m_i}(k)} = \frac{q^2}{12} \quad \text{for} \quad i = 1, 2, 3, 4$$

with q the quantization level. Then

$$E\{y(k)\} = \sum_{i=1}^{4} (H_i(z)|_{z=1} E\{\varepsilon_{m_i}(k)\}) = \left(\frac{\beta_1}{1 - \alpha_1} + \frac{\beta_2}{1 - \alpha_2} + 2\right)\left(-\frac{q}{2}\right)$$

Moreover, because of the additivity of the variations, we have

$$\sigma^2_{y(k)} = \sum_{i=1}^{4} \sigma^2_{\epsilon_{m_i}(k)} \frac{1}{2\pi j} \oint_{|z|=1} H_i(z) H_i(z^{-1}) \frac{dz}{z}$$

$$= \left(\frac{q^2}{12}\right) \left(\frac{1}{2\pi j}\right) \sum_{i=1}^{4} \oint_{|z|=1} H_i(z) H_i(z^{-1}) \frac{dz}{z}$$

Furthermore,

$$\oint_{|z|=1} H_1(z) H_1(z^{-1}) \frac{dz}{z} = 2\pi j \left. \frac{\beta_1^2}{1 - \alpha_1 z} \right|_{z=\alpha_1} = 2\pi j \left(\frac{\beta_1^2}{1 - \alpha_1^2}\right)$$

$$\oint_{|z|=1} H_2(z) H_2(z^{-1}) \frac{dz}{z} = 2\pi j \left(\frac{\beta_2^2}{1 - \alpha_2^2}\right)$$

$$\oint_{|z|=1} H_3(z) H_3(z^{-1}) \frac{dz}{z} = \oint_{|z|=1} H_4(z) H_4(z^{-1}) \frac{dz}{z} = 2\pi j$$

Finally

$$\sigma^2_{y(k)} = \frac{q^2}{12} \left(\frac{\beta_1^2}{1 - \alpha_1^2} + \frac{\beta_2^2}{1 - \alpha_2^2} + 2\right)$$

The same procedure is repeated for the parallel realization of Figure 10.20. For this case

$$H_1(z) = \frac{Y_1(z)}{E_{m_1}(z)} = \frac{1}{1 - \alpha_1 z^{-1}} = \frac{Y_2(z)}{E_{m_2}(z)} = H_2(z)$$

$$H_3(z) = \frac{Y_3(z)}{E_{m_3}(z)} = \frac{1}{1 - \alpha_2 z^{-1}} = \frac{Y_4(z)}{E_{m_4}(z)} = H_4(z)$$

The statistics of the output signal are then evaluated as follows:

$$E\{y(k)\} = 2\left(\frac{1}{1 - \alpha_1} + \frac{1}{1 - \alpha_2}\right)\left(-\frac{q}{2}\right)$$

and

$$\sigma^2_{y(k)} = \left(\frac{q^2}{12}\right) \left(\frac{1}{2\pi j}\right) 2 \left(\oint_{|z|=1} H_1(z) H_1(z^{-1}) \frac{dz}{z} + \oint_{|z|=1} H_3(z) H_3(z^{-1}) \frac{dz}{z}\right)$$

However,

$$\oint_{|z|=1} H_1(z)H_1(z^{-1})\, \frac{dz}{z} = 2\pi j\, \frac{1}{1 - \alpha_1 z}\bigg|_{z=\alpha_1} = 2\pi j\left(\frac{1}{1 - \alpha_1^2}\right)$$

and

$$\oint_{|z|=1} H_3(z)H_3(z^{-1})\, \frac{dz}{z} = 2\pi j\left(\frac{1}{1 - \alpha_2^2}\right)$$

Consequently

$$\sigma_{y(k)}^2 = \left(\frac{q^2}{12}\right)2\left(\frac{1}{1 - \alpha_1^2} + \frac{1}{1 - \alpha_2^2}\right)$$

For large β_is, the second realization amplifies the quantization noise less.

A direct realization of $H(z)$ is given in Figure 10.22, according to

$$H(z) = \frac{\beta_1}{1 - \alpha_1 z^{-1}} + \frac{\beta_2}{1 - \alpha_2 z^{-1}} = \frac{b_1 + b_2 z^{-1}}{1 + a_1 z^{-1} + a_2 z^{-2}}$$

with $b_1 = \beta_1 + \beta_2$, $b_2 = -(\alpha_1\beta_2 + \alpha_2\beta_1)$, $a_1 = -(\alpha_1 + \alpha_2)$ and $a_2 = \alpha_1\alpha_2$. In this case $\varepsilon_{m_1}(k)$ is the truncation error after multiplication by b_1, $\varepsilon_{m_2}(k)$ the error after multiplication by b_2, $\varepsilon_{m_3}(k)$ the error after multiplication by a_1 and $\varepsilon_{m_4}(k)$ the error after multiplication by a_2. The transfer functions needed are the following:

$$H_1(z) = \frac{Y_1(z)}{E_{m_1}(z)} = \frac{1}{1 + a_1 z^{-1} + a_2 z^{-2}} = \frac{z^2}{(z - \alpha_1)(z - \alpha_2)}$$

$$H_2(z) = \frac{Y_2(z)}{E_{m_2}(z)} = \frac{Y_3(z)}{E_{m_3}(z)} = H_3(z) = \frac{z^{-1}}{1 + a_1 z^{-1} + a_2 z^{-2}} = \frac{z}{(z - \alpha_1)(z - \alpha_2)}$$

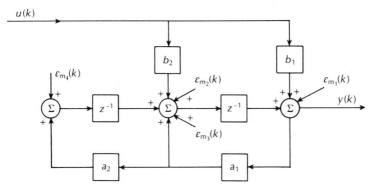

Figure 10.22   A direct realization of $H(z)$.

and

$$H_4(z) = \frac{Y_4(z)}{E_{m_4}(z)} = \frac{z^{-2}}{1 + a_1 z^{-1} + a_2 z^{-2}} = \frac{1}{(z - \alpha_1)(z - \alpha_2)}$$

Then

$$E\{y(k)\} = \sum_{i=1}^{4} H_i(z)|_{z=1} \, E\{\varepsilon_{m_i}(k)\} = \left(\frac{4}{(1 - \alpha_1)(1 - \alpha_2)} \right) \left(-\frac{q}{2} \right)$$

and

$$\sigma^2_{y(k)} =$$

$$\left(\frac{q^2}{12} \right) \left(\frac{1}{2\pi j} \right) \left(\oint_{|z|=1} H_1(z) H_1(z^{-1}) \frac{dz}{z} + 2 \oint_{|z|=1} H_2(z) H_2(z^{-1}) \frac{dz}{z} + \oint_{|z|=1} H_4(z) H_4(z^{-1}) \frac{dz}{z} \right)$$

Further

$$\oint_{|z|=1} H_1(z) H_1(z^{-1}) \frac{dz}{z}$$

$$= 2\pi j \left[\left(\frac{1}{z - \alpha_2} \right) \left(\frac{1}{1 - \alpha_1 z} \right) \left(\frac{1}{z^{-1} - \alpha_2} \right) \Big|_{z=\alpha_1} + \left(\frac{1}{z - \alpha_1} \right) \left(\frac{1}{1 - \alpha_2 z} \right) \left(\frac{1}{z^{-1} - \alpha_1} \right) \Big|_{z=\alpha_2} \right]$$

$$= 2\pi j \left[\left(\frac{1}{\alpha_1 - \alpha_2} \right) \left(\frac{1}{1 - \alpha_1^2} \right) \left(\frac{\alpha_1}{1 - \alpha_1 \alpha_2} \right) + \left(\frac{1}{\alpha_2 - \alpha_1} \right) \left(\frac{1}{1 - \alpha_2^2} \right) \left(\frac{\alpha_2}{1 - \alpha_1 \alpha_2} \right) \right]$$

$$= 2\pi j \left(\frac{1}{1 - \alpha_1 \alpha_2} \right) \left(\frac{1}{\alpha_1 - \alpha_2} \right) \left(\frac{\alpha_1}{1 - \alpha_1^2} - \frac{\alpha_2}{1 - \alpha_2^2} \right)$$

$$= \oint_{|z|=1} H_i(z) H_i(z^{-1}) \frac{dz}{z} \quad \text{for} \quad i = 2, 3, 4$$

and finally

$$\sigma^2_{y(k)} = \frac{q^2}{12} \left(\frac{4}{(\alpha_1 - \alpha_2)(1 - \alpha_1 \alpha_2)} \right) \left(\frac{\alpha_1}{1 - \alpha_1^2} - \frac{\alpha_2}{1 - \alpha_2^2} \right)$$

$$= 4 \, \frac{q^2}{12} \left(\frac{1 + \alpha_1 \alpha_2}{(1 - \alpha_1 \alpha_2)(1 - \alpha_1^2)(1 - \alpha_2^2)} \right)$$

10.4.5   *Loss of controllability and quantized closed-loop poles owing to quantization*

The quantization of gains or signals in the control loop may drastically influence the controllability or the pole positions of a closed-loop system. This is studied in this section using two examples.

■ **EXAMPLE 10.4 4**   A discrete-time linear system is described in state space by

$$\mathbf{x}(k+1) = \mathbf{Ax}(k) + \mathbf{b}u(k) \tag{10.19}$$

with

$$\mathbf{A} = \begin{bmatrix} 0 & 1 \\ -0.8 & -1.8 \end{bmatrix} \quad \text{and} \quad \mathbf{b} = \begin{bmatrix} 0 \\ 1 \end{bmatrix}$$

Examine the controllability of the closed-loop system if the input signal $u(k)$ is quantized, with quantization level equal to q (i.e. $u(k) = \gamma_k q$, $\gamma_k = 0, \pm1, \pm2, \ldots$).

■ **SOLUTION**   The state of the system at the Nth step (i.e. at time NT, with T being the sampling period) is given by (see (3.20))

$$\mathbf{x}(N) = \mathbf{A}^N \mathbf{x}(0) + \sum_{k=0}^{N-1} \mathbf{A}^{N-k-1} \mathbf{b}u(k) \tag{10.20}$$

with $\mathbf{x}(0)$ the initial state vector. We also have

$$\mathbf{A}^N = \begin{bmatrix} 0 & 1 \\ -0.8 & -1.8 \end{bmatrix}^N = \begin{bmatrix} 4(-1)^{N+1} + 5(-0.8)^N & 5[(-1)^{N+1} + (-0.8)^N] \\ 4(-1)^{N+2} + 5(-0.8)^{N+1} & 5[(-1)^{N+2} + (-0.8)^{N+1}] \end{bmatrix}$$

Since the pair $(\mathbf{A}, \mathbf{b})$ is controllable, then the vector $\mathbf{A}^{N-k-1}\mathbf{b}$ can be expressed as a linear combination of the vectors

$$\mathbf{b} = \begin{bmatrix} 0 \\ 1 \end{bmatrix} \quad \text{and} \quad \mathbf{Ab} = \begin{bmatrix} 1 \\ -1.8 \end{bmatrix}$$

More precisely, we have

$$\mathbf{A}^{N-k-1}\mathbf{b} = [4(-1)^{N-k} + 5(-0.8)^{N-k-1}]\mathbf{b} + 5[(-1)^{N-k} + (-0.8)^{N-k-1}]\mathbf{Ab}$$

Since $\mathbf{A}^{N-k-1}\mathbf{b}$ depends linearly on $\mathbf{b}$ and $\mathbf{Ab}$ and under the assumption that $u(k)$ is quantized with quantization level q, expression (10.20) takes on the form

$$\mathbf{x}(N) = \mathbf{A}^N\mathbf{x}(0) + n_1 q\mathbf{b} + n_2 q\mathbf{Ab} = \mathbf{A}^N\mathbf{x}(0) + n_1 q\begin{bmatrix} 0 \\ 1 \end{bmatrix} + n_2 q\begin{bmatrix} 1 \\ -1.8 \end{bmatrix}$$

with n_1 and n_2 appropriate integers. Consequently, even though the initial system is controllable, the values that $\mathbf{x}(N)$ can assume is a discrete set only. Hence the system loses controllability owing to quantization and finite word length. In Figure 10.23, we give some reachable states (marked with dots) from the initial state $\mathbf{x}^\mathsf{T}(0) = (0, 0)$. These possible states are at distances which are multiples of q, from $x_1(0) = 0$ and $x_2(0) = 0$, in the (x_1, x_2)-plane.

■ **EXAMPLE 10.4.5**   Consider the system

$$\mathbf{x}(k+1) = \mathbf{A}\mathbf{x}(k) + \mathbf{b}u(k)$$

with

$$\mathbf{A} = \begin{bmatrix} 0 & 1 \\ 0 & 0 \end{bmatrix} \quad \text{and} \quad \mathbf{b} = \begin{bmatrix} 0 \\ 1 \end{bmatrix}$$

Consider also the control law

$$u(k) = -\mathbf{f}^\mathsf{T}\mathbf{x}(k) \quad \text{where } \mathbf{f}^\mathsf{T} = [f_1, f_1 + f_2]$$

Examine the influence of the quantization of f_1 and f_2 with quantization level q, in the position of the closed-loop poles.

■ **SOLUTION**   In Section 6.2, it was shown that if a system is controllable, then the closed-loop poles can be placed arbitrarily at desired locations by means of state feedback. However, as will become clear, if the feedback gains assume values in a discrete set only (e.g. because of quantization), the closed-loop poles also assume values in a discrete set. The characteristic equation of the closed-loop system is

$$|z\mathbf{I} - \mathbf{A} + \mathbf{b}\mathbf{f}^\mathsf{T}| = 0 \quad \text{or} \quad z^2 + f_2 z + f_1 = 0 = (z - \rho_1)(z - \rho_2) = z^2 - (\rho_1 + \rho_2)z + \rho_1\rho_2$$

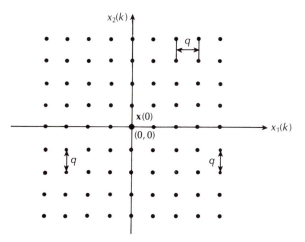

Figure 10.23   Reachable states $x_1(k)$, $x_2(k)$ due to quantization of the input signal $u(k)$ with quantization level q.

with ρ_1 and ρ_2 the eigenvalues of the closed-loop system. Obviously, we have

$$f_1 = \rho_1 \rho_2 \quad \text{and} \quad f_2 = -(\rho_1 + \rho_2)$$

By application of the Jury criterion of Chapter 4 and in order for the polynomial $z^2 + f_2 z + f_1$ to be stable, the following inequalities should hold:

$$-1 < f_1 < 1, \quad f_1 > -1 + f_2 \quad \text{and} \quad f_1 > -1 - f_2$$

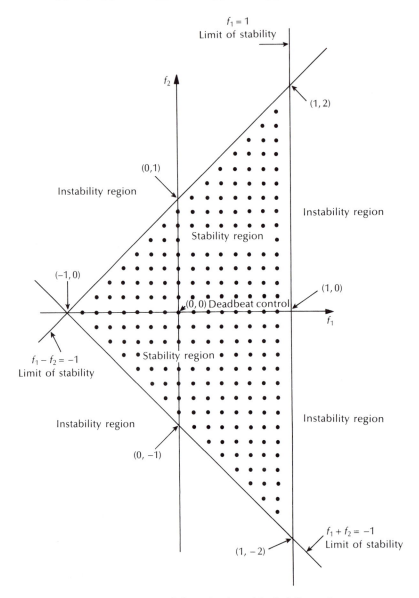

Figure 10.24 Quantization of the gains f_1 and f_2. Stability region.

The region of stability in the (f_1, f_2)-plane is given in Figure 10.24. In this figure, it was assumed that f_1 and f_2 are quantized with 4 and 5 bits respectively, thus guaranteeing the same quantization level $q = \frac{1}{8}$ for both gains. The acceptable values for f_1 and f_2 are marked with dots in the (f_1, f_2)-plane. Only values leading to a stable closed-loop system are shown. In polar form, the closed-loop poles are $z_{1,2} = \rho e^{\pm j\theta}$ and the characteristic equation is rewritten as

$$z^2 - 2\rho[\cos \theta]z + \rho^2 = 0$$

The magnitude of the poles is thus equal to

$$\rho = \sqrt{|f_1|}$$

and their real part is

$$\sigma = \rho \cos \theta = -f_2/2$$

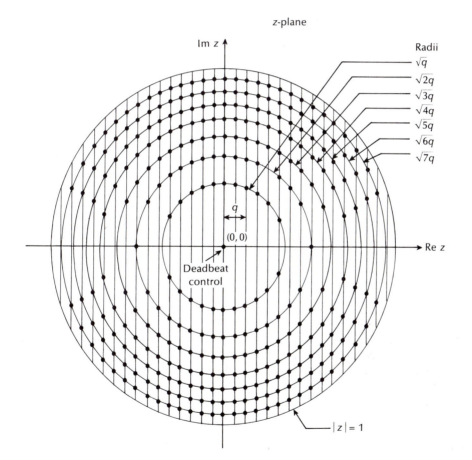

Figure 10.25   Possible closed-loop poles due to the quantization of the gains f_1 and f_2.

The quantization of f_1 with quantization level q restricts the roots of the characteristic equation to lie on circles in the z-plane. These circles are centred at $z = 0$ and have radii 0, $\sqrt{q}$, $\sqrt{2q}$, ..., $\sqrt{7q}$. The quantization of f_2 restricts the real part of the roots to take values only in a discrete set. The intersection of the circles with the vertical lines corresponding to possible values for the real part of the eigenvalues gives the possible values of the eigenvalues of the closed-loop system. They are shown by the dots in Figure 10.25.

10.5 Problems

1. The transfer function of a discrete-time controller is

$$G_c(z) = \frac{U(z)}{E(z)} = \frac{5(1 + 0.25z^{-1})}{(1 - 0.5z^{-1})(1 - 0.1z^{-1})}$$

Determine the direct, cascade and parallel realizations.

2. Repeat Problem 1 for the following transfer functions:

(a) $G_c(z) = \dfrac{8(1 + z^{-1})(1 + 0.4z^{-1})}{(1 + 0.1z^{-1})(1 + 0.8z^{-1})}$

(b) $G_c(z) = \dfrac{1 + z^{-1}}{z^{-1}(1 + 0.2z^{-1})(1 - 0.4z^{-1})}$

(c) $G_c(z) = \dfrac{10(1 + 0.3z^{-1})}{z^{-2}(1 + 0.4z^{-1})}$

(d) $G_c(z) = \dfrac{4(1 - 0.2z^{-1})}{z^{-1}(1 + z^{-1} + z^{-2})}$

(e) $G_c(z) = \dfrac{4(1 + 0.2z^{-1})(1 - 0.4z^{-1})(1 + 0.9z^{-1})}{(1 + 0.1z^{-1})(1 + 0.3z^{-1})(1 - 0.8z^{-1})}$

3. Examine the truncation errors after multiplication, due to finite word length, in the case of the transfer functions of Problem 2. Compare the cascade, parallel and direct realizations to quantization errors after the required multiplications.

4. Consider the system

$$\mathbf{x}(k+1) = \begin{bmatrix} 0 & 1 & 0 \\ 0 & 0 & 1 \\ -0.105 & -0.71 & -1.5 \end{bmatrix} \mathbf{x}(k) + \begin{bmatrix} 0 \\ 0 \\ 1 \end{bmatrix} u(k) = \mathbf{A}\mathbf{x}(k) + \mathbf{b}u(k)$$

Examine the closed-loop system's controllability, if the input signal $u(k)$ is quantized, with quantization level q.

5. Consider the system

$$\mathbf{x}(k+1) = \begin{bmatrix} 0 & 1 \\ -0.25 & -1 \end{bmatrix} \mathbf{x}(k) + \begin{bmatrix} 0 \\ 1 \end{bmatrix} u(k)$$

and the control law

$$u(k) = -\mathbf{f}^{\mathrm{T}}\mathbf{x}(k) = -[f_1 \; f_2]\mathbf{x}(k)$$

Examine the influence of the quantization of f_1 and f_2, with quantization level q, in the position of the closed-loop poles.

10.6 References

Books

[1] B. Brighouse and G. Loveday, *Microprocessors in Engineering Systems*, Pitman, London, 1987.

[2] D.A. Cassell, *Microcomputers and Modern Control Engineering*, Reston Publishing Co., Reston, Virginia, 1983.

[3] G.F. Franklin, J.D. Powell and M.L. Workman, *Digital Control of Dynamic Systems*, Addison-Wesley, London, 1990.

[4] C.H. Houpis and G.B. Lamont, *Digital Control Systems*, McGraw-Hill, New York, 1985.

[5] C.D. Johnson, *Microprocessor Based Process Control*, Prentice Hall, Englewood Cliffs, New Jersey, 1984.

[6] P. Katz, *Digital Control Using Microprocessors*, Prentice Hall, London, 1981.

[7] B.C. Kuo, *Digital Control Systems*, Holt–Saunders, Tokyo, 1980.

[8] S.A. Money, *Microprocessors in Instrumentation and Control*, Collins, London, 1985.

[9] G.A. Perdikaris, *Computer Controlled Systems*, Kluwer Academic, London, 1991.

Papers

[10] I. Shenberg, 'The Design and Implementation of Digital Compensation Networks', *IEEE Israel Tenth Conference*, Tel Aviv, 1977.

[11] J. Slaughter, 'Quantization Errors in Digital Control Systems', *IEEE Trans. Automatic Control*, Vol. 9, pp. 70–74, 1964.

11

An intelligent approach to control: fuzzy controllers

11.1 Introduction to intelligent control

In the last two decades, a new approach to control has gained considerable attention. This new approach is called **intelligent control** (to distinguish it from **conventional** or traditional control) [1]. The term 'conventional control' refers to theories and methods that are employed to control dynamic systems whose behaviour is primarily described by differential and difference equations. Thus, all the well-known classical and state-space techniques presented thus far in this book fall into this category.

The term 'intelligent control' has a more general meaning and it addresses more general control problems. That is, it may refer to systems which cannot be adequately described by a differential/difference equations framework but may require other mathematical models, as, for example, discrete event system models. More often, it treats control problems where a qualitative model is available and the control strategy is formulated and executed on the basis of a set of linguistic rules [2], [5], [9], [10], [12], [14], [32–35]. Overall, intelligent control techniques can be applied to systems whose complexity defies conventional control methods.

There are three basic approaches to intelligent control: **knowledge-based expert systems, fuzzy logic** and **neural networks**. All these approaches are interesting and very promising areas of research and development. In this book, we present only the fuzzy logic approach. For the interested reader we suggest references [6] and [8] for knowledge-based systems and neural networks.

The fuzzy control approach has been studied extensively over the last decade and many important theoretical, as well as practical, results have been reported. The fuzzy controller is based on fuzzy logic. Fuzzy logic was first introduced by Zadeh in 1965 [32], whereas the first practical fuzzy logic controller was implemented by Mamdani in 1974 [15]. Today, fuzzy control applications cover a variety of practical systems, such as the control of cement kilns [3], [18], train operation [31], parking control of a car [26], heat exchangers [4], robots [22] and in many other systems, such as home appliances, video cameras, elevators, aerospace, etc.

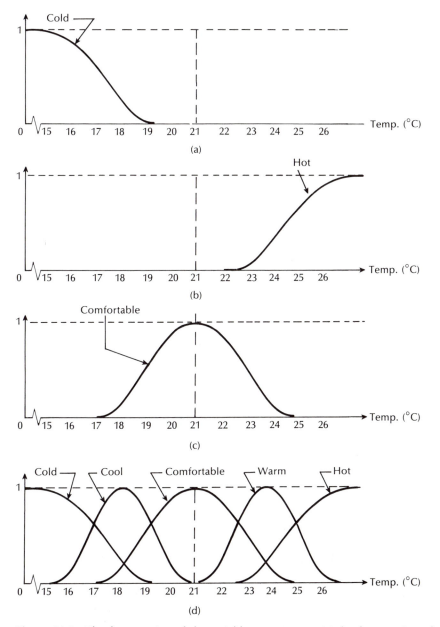

Figure 11.1 The fuzzy notion of the variable temperature: (a) the fuzzy notion of cold; (b) the fuzzy notion of hot; (c) the fuzzy notion of comfortable; (d) the whole domain.

In this chapter, a brief introduction to fuzzy controllers is presented. This material aims to give the reader the heuristics of this approach to control, which may be quite useful in many practical control problems, but treats the theoretical aspects in an introductory manner only. Furthermore, we hope that this material will inspire further investigation, not only in the area of fuzzy controllers, but also in the more general area of intelligent control. For further reading on the subject of fuzzy controllers see the books [1–14] and papers [15–35] cited in the References.

11.2   General remarks on fuzzy controllers

A principal characteristic of fuzzy control is that it works with linguistic rules (such as 'if the temperature is high then increase cooling') rather than with mathematical models and functional relationships. With conventional control, the decisions made by a controller are a rigid 'true' or 'false'. Fuzzy control uses fuzzy logic, which is much closer in spirit to human thinking and natural language than conventional control systems. Furthermore, fuzzy logic facilitates the computer implementation of imprecise (fuzzy) statements.

Fuzzy logic provides an effective means of capturing the approximate and inexact nature of the real world. To put it simply, the basic idea in fuzzy logic, instead of specifying a truth or falsehood, 0 or 1, etc., is to exert a gradual transition depending on the circumstances. For example, an air conditioning unit using conventional control recognizes room temperature only as warm when the temperature is greater than 21 °C, and cold when the temperature is less than 21 °C. Using fuzzy control, room temperature can be recognized as cold, cool, comfortable, warm or hot and, furthermore, if this temperature is increasing or decreasing. On the basis of these fuzzy variables, a fuzzy controller makes its decision on how to cool the room.

In Figure 11.1 (a–c) (see p. 331) the fuzzy notions of cold, hot and comfortable are presented in graphical form. The magnitude of these graphical representations lies between 0 and 1. The whole domain of fuzzy variables referring to the notion of temperature may be constructed by adding other variables such as cool, warm, etc., as shown in Figure 11.1 (d).

11.3   Fuzzy sets

A non-fuzzy set (or class) is any collection of items (or elements or members) which can be treated as a whole. Consider the following examples:

1. The set of all positive integers less than 11. This is a finite set of 10 members, i.e. the numbers 1, 2, 3, ..., 9, 10. This set is written as $\{1, 2, ..., 9, 10\}$.
2. The set of all positive integers greater than 4. This set has an infinite number of members and can be written as $x > 4$.
3. The set of all humans having four eyes. This set does not have any members and is called an empty (or null) set.

In contrast to non-fuzzy (or crisp) sets, in a fuzzy set there is no precise criterion for membership. Consider for example the set of **middle-aged people**. What are the members of this set? Of course, babies or 100 year old people are not middle-aged people! One may argue that people from 40 to 60 appear to be in the set of middle-aged people. This may not, however, hold true for all people, in all places and at all times. For example, centuries ago, in most countries, the mean life expectancy was around 50 (this is true today for certain developing countries). We may now ask: are the ages 32, 36, 38, 55, 58, 60 and 65 members of the set of middle-aged people? The answer is that the set 'middle-aged people' is a fuzzy set, where there is no precise criterion for membership and it depends on time, place, circumstances, subjective point of view, etc. Other examples of fuzzy sets are intelligent people, tall people, strong feelings, strong winds, bad weather, feeling ill etc.

To distinguish between members of a fuzzy set which are more probable than those which are less probable of belonging to the set, we use the **grade of membership**, denoted by μ which lies in the range from 0 to 1. Hence, the grade of membership μ is a measure of the confidence or certainty that a particular element belongs to a fuzzy set. If $\mu = 1$, then that particular element is certainly a member, and of course if $\mu = 0$, then it is certainly not a member.

The elements of a fuzzy set are taken from a **universe**. The universe contains all elements. Consider the examples:

1. The set of numbers from 1 to 1000. The elements are taken from the universe of all numbers.
2. The set of tall people. The elements are taken from the universe of all people.

If x is an element of a fuzzy set then the associated grade of x, with its fuzzy set, is described via a **membership function**, denoted by $\mu(x)$. There are two methods for defining fuzzy sets, depending on whether the universe of discourse is discrete or continuous. In the discrete case, the grade of membership function of a fuzzy set is represented as a vector whose dimension depends on the degree of discretization. An example of a discrete membership function, referring to the fuzzy set **middle**, is $(0.6/30, 0.8/40, 1/50, 0.8/60, 0.6/70)$, where the universe of discourse represents the ages $[0, 100]$. This is a bell-shaped membership function. In the continuous case, a functional definition expresses the membership function of a fuzzy set in a functional form. Some typical examples of continuous membership functions are given below:

$$\mu(x) = \exp\left(-\frac{(x - x_0)^2}{2\sigma^2}\right) \tag{11.1}$$

$$\mu(x) = \left[1 + \left(\frac{x - x_0}{\sigma}\right)^2\right]^{-1} \tag{11.2}$$

$$\mu(x) = 1 - \exp\left[-\left(\frac{\sigma}{x_0 - x}\right)^{\xi}\right] \tag{11.3}$$

where x_0 is the point where $\mu(x)$ is maximum (i.e. $\mu(x_0) = 1$) and σ is the standard deviation. Expression (11.1) is the well-known standard Gaussian curve. In expression (11.3) the exponent ξ shapes the gradient of the sloping sides.

To facilitate our understanding further, we refer to Figure 11.2, where the very simple case of the fuzzy and non-fuzzy interpretation of an old man is given. In the non-fuzzy or crisp case, everyone older than 70 is old, whereas in the fuzzy case the

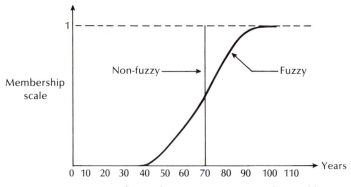

Figure 11.2 Fuzzy and non-fuzzy interpretation of an old man. In the non-fuzzy case anyone older than 70 is old; in the fuzzy case the transition is gradual.

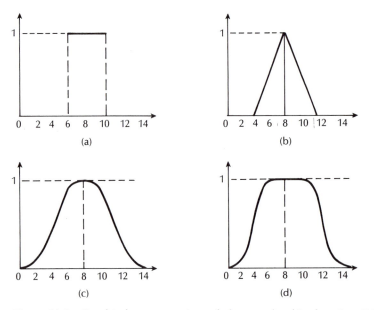

Figure 11.3 Graphical representation of the membership function (11.4). In (a) the crisp membership function is shown, while in (b), (c) and (d) three fuzzy membership functions, known as triangular, bell-shaped and trapezoidal (or flattened bell-shaped), are good candidates for the fuzzy presentation of (11.4).

transition is gradual. This graphical presentation reveals the distinct difference between fuzzy and non-fuzzy (or crisp) sets. Next, consider the membership functions given in Figure 11.3. Figure 11.3(a) presents the crisp membership function $\mu(x)$ defined as follows:

$$\mu(x) = \begin{cases} 1 & 6 \leqslant x \leqslant 10 \\ 0 & \text{otherwise} \end{cases} \tag{11.4}$$

The non-fuzzy (or crisp) membership function is unique. The corresponding fuzzy membership function may have several forms: triangular (11.3b), bell-shaped curve (11.3c), trapezoidal or flattened bell shaped (11.3d), etc.

In fuzzy sets, the variable x may be algebraic, as in relations (11.1) to (11.3), or it may be a **linguistic variable**. A linguistic variable takes on words or sentences as values. This type of value is called a **term set**. For example, let the variable x be the linguistic variable 'age'. Then, one may construct the following term: {very young, young, middle-aged, old, very old}. Note that each term in the set (e.g. young) is a fuzzy variable itself. In Figure 11.4 the three sets young (Y), middle-aged (M) and old (O) are shown. In Figure 11.5, the four sets young (Y) and very young (VY),

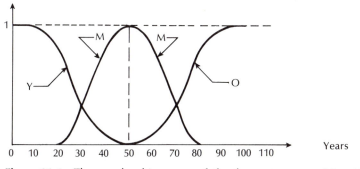

Figure 11.4 The membership curves of the three sets: young (Y), middle-aged (M) and old (O).

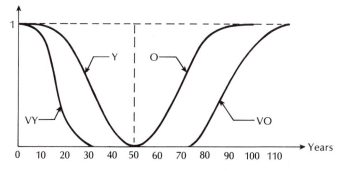

Figure 11.5 The membership curves of the four sets: very young (VY), young (Y), old (O) and very old (VO).

old (O) and very old (VO), are shown. We say that the sets 'young' and 'old' are **primary sets**, whereas the sets 'very young' and 'very old' are derived from them.

11.4 Fuzzy controllers

The non-fuzzy (crisp) PID controller was presented in Section 5.5, where it was pointed out that this type of controller has many practical merits and has become the most popular type of controller in industrial applications. The same arguments hold true for the fuzzy PID controller and for this reason, we will focus our attention on this controller. We will examine, in increasing order of complexity, the fuzzy proportional (FP), the fuzzy proportional–derivative (FPD) and the fuzzy proportional–derivative plus integral (FDP + I) controller.

1. The FP controller

Let $e(k)$ be the input and $u(k)$ be the output of the controller, respectively. The input $e(k)$ is the error

$$e(k) = \omega(k) - y(k) \tag{11.5}$$

where $\omega(k)$ is the reference signal and $y(k)$ is the output of the system (see any closed-loop figure in the book, or Figure 11.6 below). Then

$$u(k) = f(e(k)) \tag{11.6}$$

That is, the output of the controller is a non-linear function of $e(k)$. A simplified diagram of (11.5) is given in Figure 11.6. In comparison with Figure 5.24, where only one tuning parameter appears (the parameter P), for FP controllers we have two tuning parameters, namely the parameters G_e and G_u, which are the error and controller output gains, respectively. The block designated 'rule base' is the heart of the fuzzy controller whose function is explained in Sections 11.5 and 11.6 below.

2. The FPD controller

Let $ce(k)$ denote the change in the error (for continuous-time systems $ce(k)$ corresponds to the derivative of the error de/dt). An approximation to $ce(t)$ is given by

$$ce(k) = \frac{e(k) - e(k-1)}{T} \tag{11.7}$$

where T is the sampling time. The block diagram of an FPD controller is given in Figure 11.7. Here, the output of the controller is a non-linear function of two variables, namely the variables $e(k)$ and $ce(k)$, i.e.

$$u(k) = f(e(k), \, ce(k)) \tag{11.8}$$

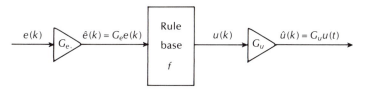

Figure 11.6   The fuzzy proportional (FP) controller.

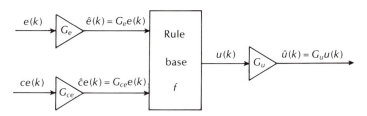

Figure 11.7   The fuzzy proportional–derivative (FPD) controller.

Note that here we have three tuning gains $(G_e$, G_{ce} and $G_u)$ as compared with the crisp PI controller which only has two (see Section 5.5).

3. The FPD + I controller

It has been shown that it is not straightforward to write rules regarding integral action. Furthermore, the rule base simultaneously involving three control actions (proportional, derivative and integral) becomes very large. To circumvent these difficulties, we separate the integral action from the other two actions, resulting in an FPD + I controller, as shown in Figure 11.8. In the present case the output of the controller $u(k)$ is a non-linear function of three variables, namely the variables $e(k)$, $ce(k)$ and $ie(k)$, i.e.

$$u(k) = f(e(k), ce(k), ie(k)) \tag{11.9}$$

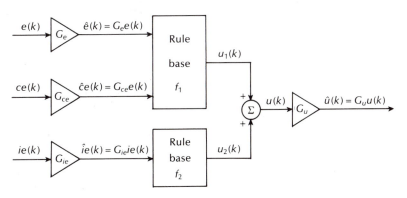

Figure 11.8   The fuzzy proportional–derivative and integral (FPD + I) controller.

where $ie(k)$ denotes the integral of the error. Since the integral action has been separated from the proportional and derivative actions, relation (11.9) breaks down to two terms:

$$u(k) = u_1(k) + u_2(k) = f_1(e(k), \, ce(k)) + f_2(ie(k)) \tag{11.10}$$

Note that here we have four tuning parameters (G_e, G_{ce}, G_{ie} and G_u) as compared with the crisp PID controller of Section 5.5 which only has three.

11.5   Elements of a fuzzy controller

A simplified block diagram of a fuzzy logic controller (FLC) incorporated in a closed-loop system is shown in Figure 11.9. The fuzzy controller involves four basic operations, namely the **fuzzification interface**, the **rule base**, the **inference engine** and the **defuzzification interface**. A brief explanation of these four elements is given below. A more detailed explanation is given in the following four sections.

1. Fuzzification interface

Here, the crisp error signal $e(k)$ is converted into a suitable linguistic fuzzy set.

2. Rule base

The rule base is the heart of a fuzzy controller, since the control strategy used to control the closed-loop system is stored as a collection of **control rules**. For example, consider a controller with three inputs e_1, e_2 and e_3 and output u. Then, a typical control rule has the form

if e_1 is A, e_2 is B and e_3 is C, then u is D \hfill (11.11)

where A, B, C and D are linguistic terms, such as very low, very high, medium, etc. The control rule (11.11) is composed of two parts: the 'if' part and the 'then' part. The 'if' part is the input to the controller and the 'then' part is the output of the

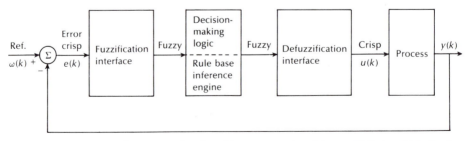

Figure 11.9   Basic configuration of a closed-loop system involving an FLC, in which the fuzzy and crisp data flow is identified.

controller. The 'if' part is called the **premise** (or **antecedent** or **condition**) and the 'then' part is called the **consequence** (or **action**).

3. Inference engine

The basic operation of the inference engine is that it 'infers', i.e. it deduces (from evidence or data) a logical conclusion. Consider the following example described by the logical rule known as *modus ponens*:

Premise 1: If an animal is a cat, then it has four legs
Premise 2: My pet is a cat

Conclusion: My pet has four legs

Here, premise 1 is the base rule, premise 2 is the fact (or the evidence or the data) and the conclusion is the consequence.

The inference engine is a program which uses the rule base and the input data of the controller to draw the conclusion, very much in the manner shown by the above *modus ponens* rule. The conclusion of the inference engine is the fuzzy output of the controller, which subsequently becomes the input to the defuzzification interface.

4. Defuzzification interface

In this last operation, the fuzzy conclusion of the inference engine is defuzzified, i.e. it is converted into a crisp signal. This last signal is the final product of the FLC which is, of course, the crisp control signal to the process.

11.6 Fuzzification

The fuzzification procedure consists of finding appropriate membership functions to describe crisp data. For example, let speed be a linguistic variable. Then the set $T(\text{speed})$ could be

$$T(\text{speed}) = \{\text{slow, medium, fast}\} \tag{11.12}$$

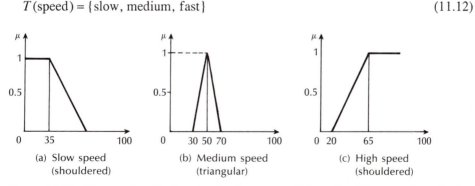

(a) Slow speed
(shouldered)

(b) Medium speed
(triangular)

(c) High speed
(shouldered)

Figure 11.10   The membership function for the term set T (speed) = {slow, medium, fast}.

On a scale from 0 to 100, slow speed may be up to 35, medium speed could be from 30 to 70, and high speed could be from 65 to 100. The membership functions for the three fuzzy variables may have several shapes. Figure 11.10 (see p. 339) shows some membership functions for each of the three fuzzy variables.

Other examples of fuzzification of crisp data have already been presented in Figures 11.1, 11.2, 11.3 (b–d), 11.4 and 11.5.

11.7 The rule base

The most usual source for constructing linguistic control rules is human experts. We start by questioning experts or operators using a carefully prepared questionnaire. Using their answers, a collection of if–then rules is established. These rules contain all the information regarding the control of the process. Note that there are other types of sources for constructing the rule base, such as control engineering knowledge, fuzzy models, etc. [19].

The linguistic control rules are usually presented to the end-user in different formats. One such format has the verbal form of Table 11.1 which refers to the two-input one-output controller of Figure 11.7 for the control of the temperature of a room. Here, the controller inputs e and ce refer to the error and change in error, respectively, whereas the variable u refers to the output of the controller. This format involves the following five fuzzy sets: zero (Z), small positive (SP), large positive (LP), small negative (SN) and large negative (LN). Clearly, the set of if–then rules presented in Table 11.1 is an example of a linguistic control strategy applied by the controller in order to maintain the room temperature close to the desired optimum value of 21 °C.

In Figure 11.11 the **graphical representation** of the five fuzzy sets Z, SP, LP, SN and LN is given. Using Figure 11.11, the **graphical forms** of the nine rules of Table 11.1 are presented in Figure 11.12.

Table 11.1 Verbal format of if–then rules.

RULE 1: if Z(e) and Z(ce), then Z(u)
RULE 2: if SP(e) and Z(ce), then SN(u)
RULE 3: if LP(e) and Z(ce), then LN(u)
RULE 4: if SN(e) and Z(ce), then SP(u)
RULE 5: if LN(e) and Z(ce), then LP(u)
RULE 6: if SP(e) and SN(ce), then Z(u)
RULE 7: if SN(e) and SP(ce), then Z(u)
RULE 8: if SP(e) and SP(ce), then LN(u)
RULE 9: if LP(e) and LP(ce), then LN(u)

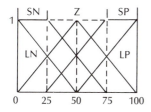

Figure 11.11 Graphical representation of the five fuzzy sets: zero (Z), small positive (SP), large positive (LP), small negative (SN) and large negative (LN).

11.8 The inference engine

The task of the inference engine is to deduce a logical conclusion, using the rule base. To illustrate how this is performed, we present three examples, in somewhat increasing order of complexity.

■ **EXAMPLE 11.8.1**   Consider a simple one-rule fuzzy controller, having the following rule:

RULE: if e_1 is slow and e_2 is fast, then u is medium

The graphical representation of the rule involving the membership functions of the three members, slow, fast and medium, is given in Figure 11.13(a). Determine the fuzzy control u.

■ **SOLUTION**   To determine the fuzzy control u we distinguish the following two steps:

Step 1

Consider the particular time instant k. For this time instant, let the fuzzy variable e_1 have the value 25 and the fuzzy variable e_2 the value 65, both on the scale 0 to 100. Through these points, two vertical lines are drawn, one for each column, intersecting the fuzzy sets e_1 and e_2 (Figure 11.13(b)). Each of these two **intersection points** (also called **triggering points**) has a particular μ denoted as $\mu_{e_i}^k$, $i = 1, 2$. This results in the following:

first column (fuzzy variable e_1): $\mu_{e_1}^k = 0.5$

second column (fuzzy variable e_2): $\mu_{e_2}^k = 0.2$

Next, determine the value of s_1^k, defined as follows:

$$s_1^k = \min\{\mu_{e_1}^k, \mu_{e_2}^k\} = \min\{0.5, 0.2\} = 0.2 \qquad (11.13)$$

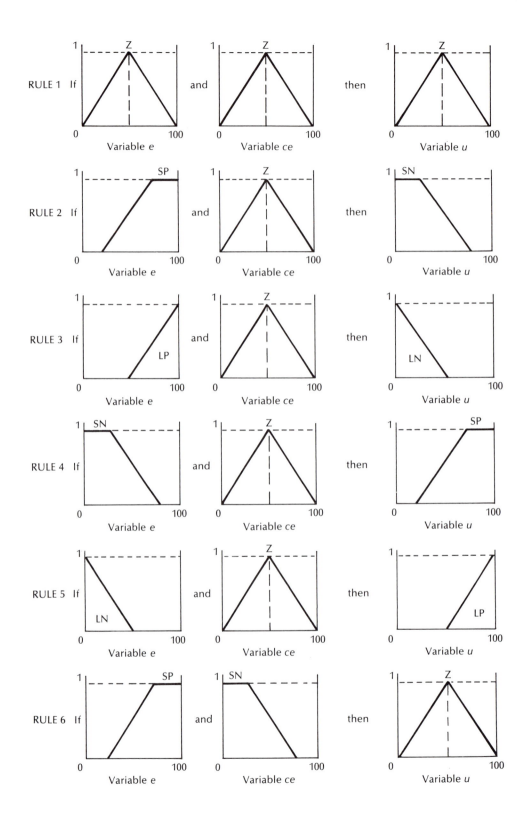

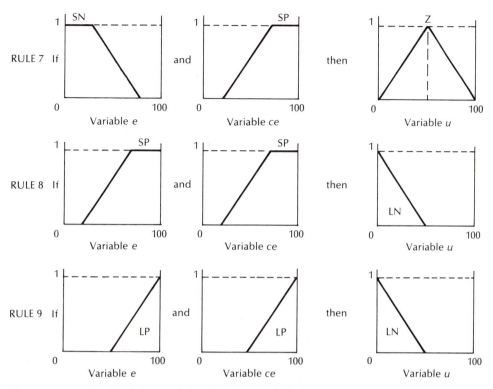

Figure 11.12 Graphical forms of the nine rules of Table 11.1.

This completes the first step, i.e. the determination of s_1^k. Note that s_1^k is related only to the 'if' part of the rule. This step is depicted in Figure 11.13(b). Clearly, when e_1, e_2, ..., e_n variables are involved in the 'if' part of the rule and there is a total of r rules, then

$$s_p^k = \min\{\mu_{e_1}^k, \mu_{e_2}^k, ..., \mu_{e_n}^k\} \quad \text{where} \quad p = 1, 2, ..., r \tag{11.14}$$

where p indicates the particular rule under consideration.

Step 2

The second step is the most important step in the inference engine, since it deduces the **result of the rule**, for the particular instant of time k. One way to deduce this result is to multiply the fuzzy variable u (third column) by s_1^k. The resulting curve is the fuzzy control sought and constitutes the **result of the rule**. This curve is the shaded area depicted in Figure 11.13(c).

To state this procedure more formally, let $\mu_1(u)$ and $\lambda_1^k(u)$ denote the membership functions of the given output fuzzy set u (first row, last column of Figure 11.13)

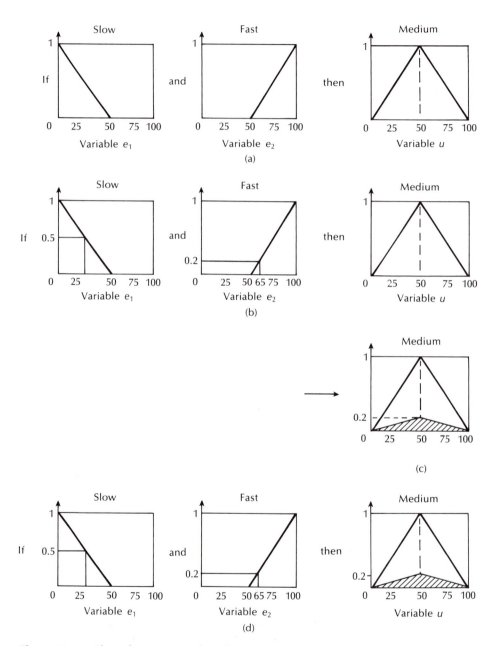

Figure 11.13 The inference procedure for Example 11.8.1: (a) graphical representation of the rule; (b) determination of the triggering points and of μ_{min}; (c) the result of the rule; (d) compact presentation of the three figures (a), (b) and (c).

and of the curve depicted in Figure 11.13(c), respectively. Then $\lambda_1^k(u)$ is given by the following expression:

$$\lambda_1^k(u) = [\min\{\mu_{e_1}^k, \mu_{e_2}^k\}] [\mu_1(u)] = [s_1^k] [\mu_1(u)] \tag{11.15}$$

For the general case, where n variables are involved in the 'if' part of the rule and there is a total of r rules, we have the following expression:

$$\lambda_p^k(u) = [\min\{\mu_{e_1}^k, \dots, \mu_{e_n}^k 1\}] [\mu_p(u)] = [s_p^k] [\mu_p(u)] \quad \text{where} \quad p = 1, 2, \dots, r \tag{11.16}$$

where p indicates the particular rule under consideration.

In practice, the above two steps are presented compactly, as shown in Figure 11.13(d). Clearly, these two steps are repeated for all desirable instants of time k, in order to construct u for the particular time interval of interest.

■ **EXAMPLE 11.8.2** Consider a fuzzy controller which is to apply a control strategy described by the following three if–then rules:

> RULE 1: if e_1 is negative and e_2 is negative, then u is negative
> RULE 2: if e_1 is zero and e_2 is zero, then u is zero
> RULE 3: if e_1 is positive and e_2 is positive, then u is positive

The graphical representation of the three rules involving the membership functions of the three fuzzy members negative, zero and positive is given in Figure 11.14. (To facilitate the presentation of the method, the members negative and positive are crisp. A fuzzy presentation is given in Figure 11.22 of Problem 2.) Determine the fuzzy control u.

■ **SOLUTION** Making use of the results of Example 11.8.1, we carry out the first two steps, as follows.

Step 1

Consider the particular time instant k. For this time instant, let the fuzzy variable e_1 have the value 25 and the fuzzy variable e_2 also have the value 25, all on the scale 0 to 100. Through these points, two vertical lines are drawn, one for each column. These vertical lines intersect the fuzzy curves at different **triggering points**, having a particular μ. This results in the following:

first column: in rule 1, $\mu_{e_1}^k = 1$; in rule 2, $\mu_{e_1}^k = 0.5$; and in rule 3, $\mu_{e_1}^k = 0$

second column: in rule 1, $\mu_{e_2}^k = 1$; in rule 2, $\mu_{e_2}^k = 0.5$; and in rule 3, $\mu_{e_2}^k = 0$

Next, determine s_p^k, $p = 1, 2, 3$, using definition (11.14) to yield

For rule 1: $s_1^k = \min\{\mu_{e_1}^k, \mu_{e_2}^k\} = \min\{1, 1\} = 1$

For rule 2: $s_2^k = \min\{\mu_{e_1}^k, \mu_{e_2}^k\} = \min\{0.5, 0.5\} = 0.5$

For rule 3: $s_3^k = \min\{\mu_{e_1}^k, \mu_{e_2}^k\} = \min\{0, 0\} = 0$

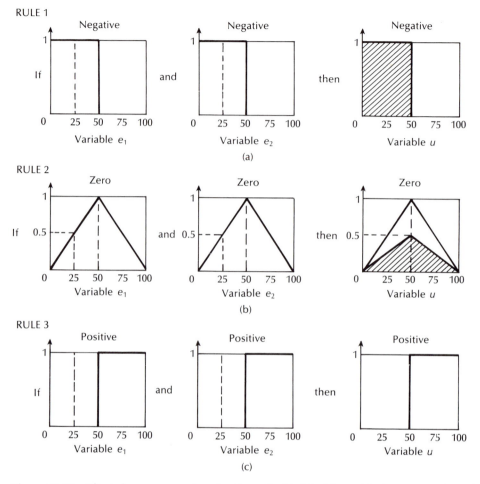

Figure 11.14 The inference procedure for Example 11.8.2: (a) graphical representation of rule 1; (b) graphical representation of rule 2; (c) graphical representation of rule 3.

It is evident that only rules 1 and 2 have a non-zero contribution on u, while rule 3 plays no part in the final value of u. Furthermore, rule 1 is seen to be dominant, while rule 2 plays a secondary role.

Step 2

Now multiply s_p^k of each rule by the corresponding curve of the third column, using definition (11.16). The result of this product is a curve (shaded area) for each of the three variables shown in the third column.

Since the present example involves more than one rule, as compared with Example 11.8.1, the following extra step is needed.

Step 3

The above procedure is the implementation of each one of the rules 1, 2 and 3 via their corresponding fuzzy curves of each of the elements appearing in each rule. The next step is to take the **result of each rule**, which is the shaded area in the third column, and 'unite' them as shown in Figure 11.15a. A popular method of constructing the shaded area of Figure 11.15a is as follows. We take the **union** of the three shaded areas of the third column of Figure 11.14. Then, the actual control u is the set (envelope) of these 'united' (superimposed) areas, shown in Figure 11.15(a).

Strictly speaking, the term 'union' refers to two or more sets and it is defined as the maximum of the corresponding values of the membership functions. More specifically, if we let $\lambda^k(u)$ be the membership function of the fuzzy control u depicted in Figure 11.15(a), then $\lambda^k(u)$ is evaluated as follows:

$$\lambda^k(u) = \max\{\lambda_1^k(u), \lambda_2^k(u), \lambda_3^k(u)\} \tag{11.17}$$

where $\lambda_1^k(u)$, $\lambda_2^k(u)$ and $\lambda_3^k(u)$ are the membership functions of the **result of each rule**. For the general case of r rules, (11.17) becomes

$$\lambda^k(u) = \max\{\lambda_1^k(u), ..., \lambda_r^k(u)\} \tag{11.18}$$

Now consider another instant of time k and let both fuzzy variables in the first and second column of Figure 11.14 have the value 75 on the scale from 0 to 100. Then, following the same procedure, one may similarly construct the corresponding third column of Figure 11.14. The resulting u for this second case is the shaded area in Figure 11.15(b).

■ **EXAMPLE 11.8.3** Consider a fuzzy controller which is to apply a control strategy described by the following two if–then rules:

RULE 1: if e_1 is positive and e_2 is zero, then u is negative
RULE 2: if e_1 is negative and e_2 is zero, then u is zero

The graphical representation of the two rules involving the membership functions of the three fuzzy members, positive, zero and negative, is given in Figure 11.16. Determine the fuzzy control u.

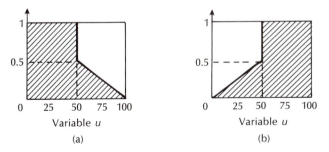

(a)

(b)

Figure 11.15 Fuzzy control signal graphical construction for Example 11.8.2. Actual control signal u when triggering at (a) 25 and (b) 75.

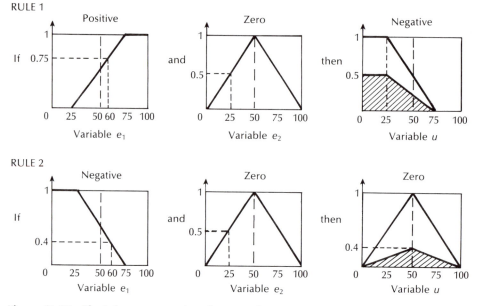

Figure 11.16   The inference procedure for Example 11.8.3.

■■ **SOLUTION**   Using the results of Examples 11.8.1 and 11.8.2 we have the following.

Step 1

Consider the particular time instant k. For this time instant, let the fuzzy variable e_1 have the value of 60 and the fuzzy variable e_2 have the value of 25. Through these points, two vertical lines are drawn, one for each column. These vertical lines intersect the fuzzy curves at different triggering points, having a particular μ. This results in the following:

first column: in rule 1, $\mu_{e_1}^k = 0.75$; and in rule 2, $\mu_{e_1}^k = 0.5$

second column: in rule 1, $\mu_{e_2}^k = 0.4$; and in rule 2, $\mu_{e_2}^k = 0.5$

Next, determine s_p^k, $p = 1, 2$, using definition (11.14) to yield

For rule 1: $s_1^k = \min\{\mu_{e_1}^k, \mu_{e_2}^k\} = \min\{0.75, 0.4\} = 0.4$

For rule 2: $s_2^k = \min\{\mu_{e_1}^k, \mu_{e_2}^k\} = \min\{0.5, 0.5\} = 0.5$

Step 2

Multiply s_p^k of each rule by the corresponding curve of the third column, according to definition (11.16). The resulting curves are the fuzzy curves of the output u (third column, shaded areas).

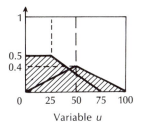

Figure 11.17 Fuzzy control signal graphical construction for Example 11.8.3.

Step 3

Using definition (11.18), the envelope of the actual fuzzy control u is constructed by superimposing the two shaded areas of the third column to yield the curve shown in Figure 11.17.

■ **REMARK 11.8.1** We may now make the following remark, regarding the overall philosophy of an FLC. To estimate the fuzzy control signal at each instant of time k, the FLC works as follows. Each rule contributes an 'area' (i.e. the shaded areas in the last column of Figure 11.13, 11.14 or 11.16). This area describes the output u of the controller as a fuzzy set. All these areas are subsequently superimposed in the manner explained above (see Figure 11.15 or 11.17), to give the fuzzy set of the output u. The envelope of this total area is the **final conclusion** of the inference engine for the instant of time k, deduced using the rule base. One can conclude, therefore, that the end product of the inference engine, given in Figure 11.15 or 11.17, is a rule base result, where **all** rules are **simultaneously** taken into consideration.

The very last action in an FLC is, by using Figure 11.15 or 11.17, to determine the crisp values for the control signal, which will serve as an input to the process. This is defuzzification, which is explained below.

11.9 **Defuzzification**

There are several methods for defuzzification. A rather simple method is the centre of area method which is defined as follows:

$$u = \frac{\sum_i \mu(x_i) x_i}{\sum_i \mu(x_i)} \tag{11.19}$$

or

$$u = \frac{\int \mu(x) x \, dx}{\int \mu(x) \, dx} \tag{11.20}$$

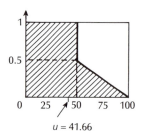

Figure 11.18 Defuzzified value of control signal.

where u is the crisp function sought, x_i is the member of the set and $\mu(x_i)$ is the associated membership function. Clearly, expressions (11.19) and (11.20) correspond to the discrete- and continuous-time cases, respectively.

■ **EXAMPLE 11.9.1** Consider the shaded area in Figure 11.15(a). Calculate the centre of area of this shaded area using (11.19).

■ **SOLUTION** We have:

$$\sum_i \mu(x_i)x_i = \text{the shaded area in Figure 11.15(a)}$$

$$= (50) + (0.5)(0.5)(50) = 50 + 12.5 = 62.5$$

$$\sum_i \mu(x_i) = 1 + 0.5 = 1.5$$

Hence $u = 62.5/1.5 = 41.66$. Therefore, the defuzzification procedure yields the crisp value of u, which is depicted in Figure 11.18.

There are other types of defuzzification, such as the mean of maxima, first of maxima, last of maxima, etc. For more information on these techniques see [8], [19].

11.10 Performance assessment

Up to now, no systematic procedures for the design of an FLC have been proposed (such as root locus, Nyquist plots, pole placement, stability tests, etc.). The basic difficulty in developing such procedures is the fact that the rule base has no mathematical description. As a consequence, it is not obvious how the rules and gains affect the overall performance of the closed-loop system.

The problem of stability of a closed-loop system incorporating an FLC essentially remains an unsolved question, even though an increasing number of research results have appeared recently in the literature. For linear time-variant systems with a known transfer function or state-space model, if the describing function approach is applied, together with a Nyquist plot, one may reach some safe conclusions regarding the stability margins of the systems.

To evaluate the performance of FLCs a theoretical approach has been proposed, which yields a partial evaluation of the performance. This approach refers to the **integrity** of the rule base and aims to secure the accuracy of the rule base. One way of investigating integrity is to plot the input and output signal of the FLC. Comparison of these two waveforms provides some idea of the integrity of the rule base. Clearly, the objective of this investigation is to study the behaviour of the control system and, if it is not satisfactory, to suggest improvements.

11.11 Application example: kiln control

The kilning process in the manufacture of cement has attracted much attention from the control viewpoint and indeed was one of the first applications of fuzzy control in the process industry [3], [18], [24]. The kilning process is one of the most complex industrial processes to control and has, till the advent of fuzzy control, defied automatic control. Human operators, however, can successfully control this process using rules which are the result of years of experience. These rules now form the basis for fuzzy control and there are many successful applications world-wide.

Briefly, in the kilning process a blend of finely ground raw materials is fed into the upper end of a long, inclining, rotating cylinder and slowly flows to the lower end, while undergoing chemical transformation due to the high temperatures produced by a flame at the lower end. The resultant product, clinker, constitutes the major component of cement. A measure of the burning zone temperature at the lower end of the rotary kiln can be obtained indirectly by measuring the torque of the motor rotating the kiln, whereas a measure of the quality of the end product is its free lime content (FCAO). These two quantities (or process output measurements) are essential in specifying the fuel feed to the kiln (i.e. the control strategy).

The block diagram of Figure 11.19 is a simplified controller for the kilning process. There are two inputs e_1 and e_2 and one output u, defined as follows [3], [24]:

e_1 = change in kiln torque drive (DELTQUE or ΔTQUE)

e_2 = free lime content (FCAO)

u = output fuel rate (DELFUEL or ΔFUEL)

where Δ stands for change. The corresponding ranges of e_1, e_2 and u are $(-3, 0, 3)$,

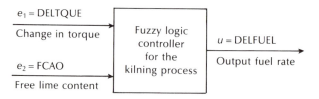

Figure 11.19 Simplified controller for the kilning process.

(0.3, 0.9, 1.5) and (−0.2, 0, 0.2), respectively, where the middle number indicates the centre of the fuzzy membership function.

The rule base is composed of nine if–then rules, as shown in Table 11.2. The graphical representation of this rule base is shown in Figure 11.20, where each row represents a rule. The first column represents the membership function for the first input e_1 = change in kiln torque drive (DELTQUE or ΔTQUE), the second column the membership function of the second input e_2 = free lime content (FCAO) and the third column the membership function of the output u = output fuel rate (DELFUEL or ΔFUEL). Clearly, in this example we have three sets with their corresponding members as follows:

first set: ΔTQUE (ZERO(ZE), NEGATIVE(NE), POSITIVE(PO))
second set: FCAO (LOW(LO), OK(OK), HIGH(HI))
third set: ΔFUEL (LARGE POSITIVE(LP), MEDIUM POSITIVE(MP), SMALL POSITIVE(SP), NO CHANGE(NC), SMALL NEGATIVE(SN), MEDIUM NEGATIVE(MN), LARGE NEGATIVE(LN)).

To determine the fuzzy controller output at some particular instant k, assume that DELTQUE and FCAO are $e_1 = -1.2\%/h$ and $e_2 = 0.54\%/h$, respectively. Thus for the first controller input variable e_1 (corresponding to the first column) a vertical line centred at $-1.2\%/h$ is drawn to intercept the fuzzy sets for the change in kiln torque drive for every rule. Likewise a vertical line, centred at $0.54\%/h$, is drawn to intercept the fuzzy sets for the second input e_2, free lime, for every rule.

Table 11.2 The nine if–then rules for the kiln process FLC.

RULE 1: if DELTQUE is zero and FCAO is low, then DELFUEL is medium negative
RULE 2: if DELTQUE is zero and FCAO is OK, then DELFUEL is zero
RULE 3: if DELTQUE is zero and FCAO is high, then DELFUEL is medium positive
RULE 4: if DELTQUE is negative and FCAO is low, then DELFUEL is small positive
RULE 5: if DELTQUE is negative and FCAO is OK, then DELFUEL is medium positive
RULE 6: if DELTQUE is negative and FCAO is high, then DELFUEL is large positive
RULE 7: if DELTQUE is positive and FCAO is low, then DELFUEL is large negative
RULE 8: if DELTQUE is positive and FCAO is OK, then DELFUEL is medium negative
RULE 9: if DELTQUE is positive and FCAO is high, then DELFUEL is small negative

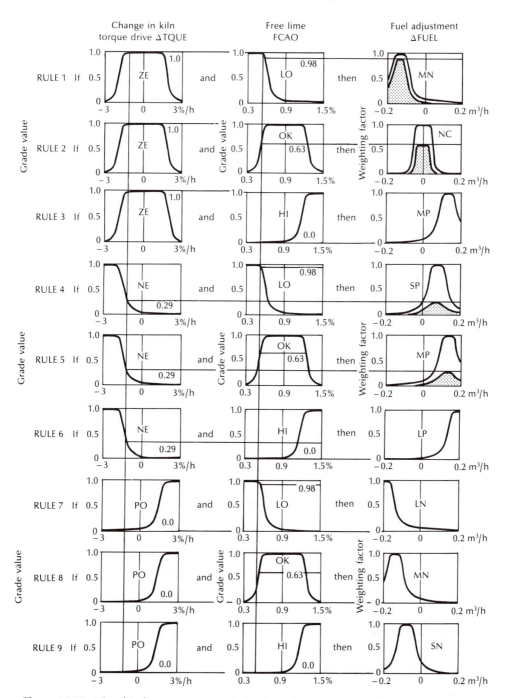

Figure 11.20 Graphical representation of the fuzzy logic interpretation of the nine control rules (Table 11.2) for the kilning process (reproduced by kind permission of FLS Automation, Denmark).

To obtain the final fuzzy output u of the controller, for this particular instant of time k, we follow the procedure presented in the examples of Section 11.8, as follows.

Step 1

Using the relationship (11.14) for the case of a two-input controller with nine rules at the kth time instant, we have that for each rule

$$s_p^k = \min\{\mu_{e_1}^k, \mu_{e_2}^k\} = \min\{\mu_{\text{DELTQUE}}^k, \mu_{\text{FCAO}}^k\} \tag{11.21}$$

As a result, the set of minima at this time instant, which is a measure of the strength or contribution of each rule on the final decision, is

$$\{s_1^k, s_2^k, s_3^k, s_4^k, s_5^k, s_6^k, s_7^k, s_8^k, s_9^k\} = \{0.98, 0.63, 0, 0.29, 0.29, 0, 0, 0, 0\}$$

It is evident here that only rules 1, 2, 4 and 5 have a non-zero contribution, the remainder playing no part in the final decision. Furthermore, rule 1 is seen to be dominant while rule 2 has a significant contribution. In contrast, rules 4 and 5 have only a small contribution.

Step 2

To determine the contribution of each rule to u, i.e. to determine $\lambda_p^k(u)$, $p = 1, 2, ..., 9$, we apply relation (11.16), i.e. the relation

$$\lambda_p^k(u) = [s_p^k][\mu_p(u)] \tag{11.22}$$

As a result, the shaded curves in the third column of Figure 11.20 are produced.

Step 3

To determine the final fuzzy control u, **simultaneously** taking into account all nine rules, we make use of (11.16) to yield

$$\lambda^k(u) = \max\{\lambda_1^k(u), ..., \lambda_9^k(u)\} \tag{11.23}$$

The fuzzy set $\lambda^k(u)$ is given in Figure 11.21.

Finally, we defuzzify $\lambda^k(u)$ by obtaining the centroid of this resultant output fuzzy set $\lambda^k(u)$. The final crisp output to the fuel actuator at this sampling instant is the centre of area (COA) of the envelope of the resultant output fuzzy set $\lambda^k(u)$ and is calculated to be (see Figure 11.21)

$$\Delta\text{FUEL}(k) = -0.048 \text{ m}^3/\text{h}$$

It is clear that this procedure must be repeated at every sampling instant k. The sequence of these control decisions is then the desired rule-based **control strategy**. For more details see [3], [24].

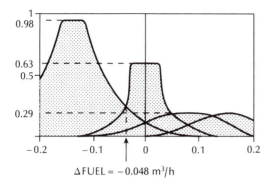

$\Delta FUEL = -0.048 \text{ m}^3/\text{h}$

Figure 11.21 The graphical representation of the control output *u*.

11.12 Problems

1. In Example 11.8.2, construct the curves of the third column for the case $k = k_2$, where the fuzzy variables have the value 75.

2. Solve Example 11.8.2 when the graphical representation of the membership functions of the three fuzzy members negative, positive and zero are as in Figure 11.22.

3. Solve Example 11.8.3 when the graphical representation of the membership functions of the three fuzzy members negative, positive and zero are as in Figure 11.23.

4. A fuzzy controller is to apply a control strategy described by the following three if–then rules:

RULE 1: if the temperature is low and the pressure is zero, then the speed is low
RULE 2: if the temperature is medium and the pressure is low, then the speed is medium
RULE 3: if the temperature is high and the pressure is high, then the speed is high

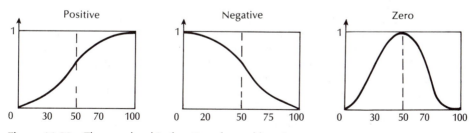

Figure 11.22 The membership functions for problem 2.

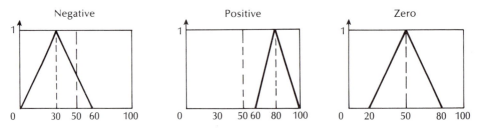

Figure 11.23 The membership functions for problem 3.

The ranges of the variables are: temperature from 0 to 100; pressure from 0 to 10; and speed from 0 to 1000 units.

(a) Describe temperature, pressure and speed by a graphical representation as fuzzy sets.
(b) Determine the three rules using the above fuzzy sets.
(c) For the instant of time k, the values of temperature and pressure are 30 and 5 units, respectively. Determine the fuzzy output (speed) set.
(d) Determine the crisp value by defuzzifying the above fuzzy output set.

11.13 References

Books

[1] P.J. Antsaklis and K.M. Passino (eds), *An Introduction to Intelligent and Autonomous Control*, Kluwer, Boston, Massachusetts, 1992.
[2] D. Driankov, H. Hellendoorn and M. Reinfrank, *An Introduction to Fuzzy Control*, Springer Verlag, Berlin, 1993.
[3] M.M. Gupta and Sanchez (eds), *Fuzzy Information and Decision Processes*, North-Holland, Amsterdam, 1982. Special chapter by L.P. Homblad and J.J. Ostergaard, 'Control of a Cement Kiln by Fuzzy Logic'.
[4] M.M. Gupta, G.N. Saridis and B.R. Gaines (eds), *Fuzzy Automata and Decision Processes*, North-Holland, Amsterdam, 1977. Special chapter by J.J. Ostergaard, 'Fuzzy Logic Control of a Heat Exchanger Process'.
[5] M.M. Gupta, Ragade and R.R. Yager (eds), *Advances in Fuzzy Set Theory Applications*, North-Holland, New York, 1979. Special chapter by M. Mizumoto, S. Fukami and K. Tanaka, 'Some Methods of Fuzzy Reasoning'.
[6] C.J. Harris, C.G. Moore and M. Brown, *Intelligent Control: Aspects of Fuzzy Logic and Neural Nets*, World Scientific, London, 1990.
[7] A. Kaufman, *Introduction to the Theory of Fuzzy Sets*, Academic Press, New York, 1975.
[8] B. Kosko, *Neural Networks and Fuzzy Systems*, Prentice Hall, Englewood Cliffs, New Jersey, 1992.

[9] E.H. Mamdani and B.R. Gaines (eds), *Fuzzy Reasoning and its Applications*, Academic Press, London, 1981. Special chapter by P.M. Larsen, 'Industrial Application of Fuzzy Logic Control'.

[10] W. Pedrycz, *Fuzzy Control and Fuzzy Systems*, John Wiley, New York, 1993.

[11] G.N. Saridis, *Self-organizing Control of Stochastic Systems*, Marcel Dekker, New York, 1977.

[12] M. Sugeno (ed.), *Industrial Applications of Fuzzy Control*, North-Holland, Amsterdam, 1985.

[13] R.R. Yager, S. Ovchinnikov, R.M. Tong and H.T. Nguyen, *Fuzzy Sets and Applications: Selected Papers by L.A. Zadeh*, John Wiley, New York, 1987.

[14] H.J. Zimmermann, *Fuzzy Set Theory and its Applications*, Kluwer, Boston, Massachusetts, 1993.

Papers

[15] S. Assilian and E.H. Mamdani, 'An Experiment in Linguistic Synthesis with a Fuzzy Logic Controller', *International Journal of Man–Machine Studies*, Vol. 7, pp. 1–13, 1974.

[16] S. Fukami, M. Mizumoto and K. Tanaka, 'Some Consideration of Fuzzy Conditional Inference', *Fuzzy Sets and Systems*, Vol. 4, pp. 243–273, 1980.

[17] W.J.M. Kickert and H.R. van Nauta Lemke, 'Application of a Fuzzy Controller in a Warm Water Plant', *Automatica*, Vol. 12, pp. 301–308, 1976.

[18] R.E. King, 'Expert Supervision and Control of a Large-scale Plant', *Journal of Intelligent Systems and Robotics*, Vol. 5, pp. 167–176, 1992.

[19] C.C. Lee, 'Fuzzy Logic in Control Systems: Fuzzy Logic Controller', *IEEE Transactions on Systems, Man and Cybernetics*, Vol. SMC-20, pp. 404–435, 1990.

[20] E.H. Mamdani, 'Application of Fuzzy Algorithms for the Control of a Simple Dynamic Plant', *Proceedings of the IEE*, Vol. 121, pp. 1585–1588, 1974.

[21] E.H. Mamdani, 'Application of Fuzzy Logic to Approximate Reasoning', *IEEE Transactions on Computers*, Vol. C-26, pp. 1182–1191, 1977.

[22] N.J. Mandic, E.M. Scharf and E.H. Mamdani, 'Practical Application of a Heuristic Fuzzy Rule-based Controller to the Dynamic Control of a Robot Arm', *IEE Proceedings D*, Vol. 132, pp. 190–203, 1985.

[23] S. Murakami, F. Takemoto, H. Fulimura and E. Ide, 'Weldline Tracking Control of Arc Welding Robot Using Fuzzy Logic Controller', *Fuzzy Sets and Systems*, Vol. 32, pp. 221–237, 1989.

[24] J.J. Ostergaard, 'FUZZY II: The New Generation of High Level Kiln Control', *Zement Kalk Gips (Cement-Lime-Gypsum)*, Vol. 43, pp. 539–541, 1990.

[25] T.J. Procyk and E.H. Mamdani, 'A Linguistic Self-organizing Process Controller', *Automatica*, Vol. 15, pp. 15–30, 1979.

[26] M. Sugeno, T. Murofushi, T. Mori, T. Tatematsu and J. Tanaka, 'Fuzzy Algorithmic Control of a Model Car by Oral Instructions', *Fuzzy Sets and Systems*, Vol. 32, pp. 207–219, 1989.

[27] K.L. Tang and R.J. Mulholland, 'Comparing Fuzzy Logic with Classical Controller Design', *IEEE Transactions on Systems, Man and Cybernetics*, Vol. SMC-17, pp. 1085–1087, 1987.

[28] S.G. Tzafestas, 'Fuzzy Systems and Fuzzy Expert Control: An Overview', *The Knowledge Engineering Review*, Vol. 9, No. 3, pp. 229–268, 1994.

[29] B.A.M. Wakileh and K.F. Gill, 'Use of Fuzzy Logic in Robotics', *Computers in Industry*, Vol. 10, pp. 35–46, 1988.

[30] T. Yamakawa and T. Miki, 'The Current Mode Fuzzy Logic Integrated Circuits Fabricated by the Standard CMOS Process', *IEEE Transactions on Computers*, Vol. C-35, pp. 161–167, 1986.

[31] S. Yasunobu, S. Miyamoto and H. Ihara, 'Fuzzy Control for Automatic Train Operation System', *Proceedings IFAC/IFIP/IFORS International Congress on Control in Transportation Systems, Baden-Baden*, 1983.

[32] L.A. Zadeh, 'Fuzzy Sets', *Information and Control*, Vol. 8, pp. 338–353, 1965.

[33] L. A. Zadeh, 'Outline of a New Approach to the Analysis of Complex Systems and Decision Processes', *IEEE Transactions on Systems, Man and Cybernetics*, Vol. SMC-3, pp. 43–80, 1973.

[34] L.A. Zadeh, 'The Concept of a Linguistic Variable and its Application to Approximate Reasoning', *Information Sciences*, Vol. 8, pp. 43–80, 1975.

[35] L.A. Zadeh, 'Making Computers Think Like People', *IEEE Spectrum*, pp. 26–32, 1984.

Appendix A

Matrix theory

A.1 Matrix definitions and operations

A.1.1 Matrix definitions

Consider the system of linear algebraic equations

$$
\begin{aligned}
a_{11}x_1 + a_{12}x_2 + \cdots + a_{1n}x_n &= b_1 \\
a_{21}x_1 + a_{22}x_2 + \cdots + a_{2n}x_n &= b_2 \\
\vdots \qquad \vdots \qquad \vdots \qquad \quad \vdots \\
a_{n1}x_1 + a_{n2}x_2 + \cdots + a_{nn}x_n &= b_n
\end{aligned}
\tag{A.1}
$$

This system can be written compactly as

$$\mathbf{Ax = b} \tag{A.2}$$

where

$$
\mathbf{A} =
\begin{bmatrix}
a_{11} & a_{12} & \cdots & a_{1n} \\
a_{21} & a_{22} & \cdots & a_{2n} \\
\vdots & \vdots & & \vdots \\
a_{n1} & a_{n2} & \cdots & a_{nn}
\end{bmatrix},
\quad
\mathbf{x} =
\begin{bmatrix}
x_1 \\
x_2 \\
\vdots \\
x_n
\end{bmatrix}
\quad \text{and} \quad
\mathbf{b} =
\begin{bmatrix}
b_1 \\
b_2 \\
\vdots \\
b_n
\end{bmatrix}
$$

The rectangular array, which has n columns and n rows, designated by the upper case letter $\mathbf{A}$, is called a **matrix**. Matrices which have only one column, designated by lower case letters (i.e. $\mathbf{x}$ and $\mathbf{b}$), are called **vectors** and have $n \times 1$ dimensions. The matrix $\mathbf{A}$ can also be written as $\mathbf{A} = [a_{ij}]$, $i, j = 1, 2, .., n$. The parameters a_{ij} are called the **elements** of matrix $\mathbf{A}$.

One of the basic reasons which led to the use of matrices is that they provide a concise description of multivariable systems. For example, relation (A.2) is a concise description of the algebraic equations (A.1). Furthermore, investigation of the solution of (A.1) is simplified when using (A.2). We say that equation (A.2) has a unique solution if the determinant (see Section A.2) of $\mathbf{A}$ is non-zero.

We present below certain useful types of matrices.

1. The column vector

This matrix is composed of a single column, i.e. it has the form

$$\mathbf{a} = \begin{bmatrix} a_1 \\ a_2 \\ \vdots \\ a_n \end{bmatrix} \qquad\qquad (A.3)$$

2. The row vector

This matrix is composed of a single row, i.e. it has the form

$$\mathbf{a}^T = [a_1 \ a_2 \ \cdots \ a_n] \qquad\qquad (A.4)$$

3. The square and non-square matrices

A square matrix is one which has an equal number of rows and columns, whereas a non-square matrix has an unequal number of rows and columns.

4. The diagonal matrix

The diagonal matrix is a square matrix whose non-zero elements lie on the main diagonal and has the form

$$\mathbf{A} = \begin{bmatrix} a_{11} & 0 & \cdots & 0 \\ 0 & a_{22} & \cdots & 0 \\ \vdots & \vdots & & \vdots \\ 0 & 0 & \cdots & a_{nn} \end{bmatrix} = \text{diag}\{a_{11}, \quad a_{22}, \quad \ldots, \quad a_{nn}\} \qquad (A.5)$$

5. The identity matrix

A diagonal matrix which has ones along the main diagonal and zeros elsewhere is called an $(n \times n)$ identity matrix. This matrix is denoted by $\mathbf{I}$ and has the form

$$\mathbf{I} = \begin{bmatrix} 1 & 0 & \cdots & 0 \\ 0 & 1 & \cdots & 0 \\ \vdots & \vdots & & \vdots \\ 0 & 0 & \cdots & 1 \end{bmatrix} \qquad\qquad (A.6)$$

6. The zero matrix

This is a matrix whose elements are all zero.

7. The singular and non-singular matrices

If the determinant of a square matrix is zero, the matrix is called **singular**, whereas if the determinant is non-zero, the matrix is called **non-singular**.

8. The transpose of a matrix

The matrix $\mathbf{A}^T$ is the transpose of $\mathbf{A}$ if the rows of the first matrix are the columns of the second. Hence if $\mathbf{A} = [a_{ij}]$, then $\mathbf{A}^T = [a_{ji}]$. Therefore, if $\mathbf{A}$ has dimensions $n \times m$, then $\mathbf{A}^T$ has dimensions $m \times n$. The superscript T denotes transposition.

9. The symmetric matrix

The symmetric matrix is a square matrix for which $\mathbf{A} = \mathbf{A}^T$ or $a_{ij} = a_{ji} \; \forall \; i, j$.

10. The triangular matrix

The triangular matrix is a square matrix which has one of the following two forms:

$$\mathbf{A} = \begin{bmatrix} a_{11} & a_{12} & \cdots & a_{1n} \\ 0 & a_{22} & \cdots & a_{2n} \\ \vdots & \vdots & & \vdots \\ 0 & 0 & \cdots & a_{nn} \end{bmatrix} \quad \text{or} \quad \mathbf{B} = \begin{bmatrix} a_{11} & 0 & \cdots & 0 \\ a_{21} & a_{22} & \cdots & 0 \\ \vdots & \vdots & & \vdots \\ a_{n1} & a_{n2} & \cdots & a_{nn} \end{bmatrix} \tag{A.7}$$

Matrix $\mathbf{A}$ is called **upper triangular**, whereas matrix $\mathbf{B}$ is called **lower triangular**.

11. The conjugate matrix

The matrix $\bar{\mathbf{A}}$ is called the conjugate matrix of $\mathbf{A}$ and its elements are the conjugate elements of $\mathbf{A}$. That is, if $\mathbf{A} = [a_{ij}]$, then $\bar{\mathbf{A}} = [\bar{a}_{ij}]$.

12. The Hermitian matrix

If $\mathbf{A} = \bar{\mathbf{A}}^T$, then the matrix $\mathbf{A}$ is called Hermitian.

13. The orthogonal matrix

A matrix $\mathbf{A}$ is called orthogonal if it is a square matrix with real elements and the following relation holds:

$$\mathbf{A}^T \mathbf{A} = \mathbf{A} \mathbf{A}^T = \mathbf{I} \tag{A.8}$$

Other useful definitions regarding matrices are the following:

(a) The trace

The trace of an $n \times n$ square matrix $\mathbf{A} = [a_{ij}]$ is denoted as tr $\mathbf{A}$ or trace $\mathbf{A}$ and is defined as the sum of all the elements of the main diagonal, i.e.

$$\text{tr } \mathbf{A} = \sum_{i=1}^{n} a_{ii} \tag{A.9}$$

(b) The rank

The rank of a matrix is equal to the dimension of its largest non-singular (square) submatrix.

A.1.2 Matrix operations

1. Matrix addition

Consider the matrices $\mathbf{A} = [a_{ij}]$ and $\mathbf{B} = [b_{ij}]$ both of whose dimensions are $n \times m$. Then, their addition $\mathbf{A} + \mathbf{B}$ is the $n \times m$ matrix $\mathbf{C} = [c_{ij}]$, whose elements c_{ij} are $c_{ij} = a_{ij} + b_{ij}$.

2. Matrix multiplication

Consider the matrices $\mathbf{A} = [a_{ij}]$ and $\mathbf{B} = [b_{ij}]$ of dimensions $n \times m$ and $m \times p$, respectively. Then, their product $\mathbf{AB}$ is the $n \times p$ matrix $\mathbf{C} = [c_{ij}]$, whose elements c_{ij} are given by

$$c_{ij} = \sum_{k=1}^{m} a_{ik}b_{kj} = \mathbf{a}_i^{\mathrm{T}} \mathbf{b}_j$$

where $\mathbf{a}_i^{\mathrm{T}}$ is the ith row of matrix $\mathbf{A}$ and $\mathbf{b}_j$ is the jth column of matrix $\mathbf{B}$. Hence, every c_{ij} element is determined by multiplying the ith row of matrix $\mathbf{A}$ by the jth column of matrix $\mathbf{B}$.

3. Multiplying a matrix by a constant

Consider the matrix $\mathbf{A} = [a_{ij}]$ and the constant k. Then, every element of the matrix $\mathbf{C} = k\mathbf{A}$ is simply $c_{ij} = ka_{ij}$.

4. Transpose of a matrix product

Consider the product $\mathbf{A}_1 \mathbf{A}_2 \dots \mathbf{A}_m$. Then

$$(\mathbf{A}_1 \mathbf{A}_2 \dots \mathbf{A}_m)^{\mathrm{T}} = \mathbf{A}_m^{\mathrm{T}} \mathbf{A}_{m-1}^{\mathrm{T}} \dots \mathbf{A}_1^{\mathrm{T}} \tag{A.10}$$

5. Derivatives of a matrix

Consider the matrix $\mathbf{A} = [a_{ij}]$ whose elements a_{ij} are functions of time t, and the function f which is a scalar function of time. Then, the following relations hold:

$$\frac{d\mathbf{A}}{dt} = \left[\frac{da_{ij}}{dt}\right] \tag{A.11}$$

$$\frac{d(f\mathbf{A})}{dt} = \left[\frac{df}{dt}\right]\mathbf{A} + f\left[\frac{d\mathbf{A}}{dt}\right] \tag{A.12}$$

$$\frac{d(\mathbf{A} + \mathbf{B})}{dt} = \frac{d\mathbf{A}}{dt} + \frac{d\mathbf{B}}{dt} \tag{A.13}$$

$$\frac{d(\mathbf{A}\mathbf{B})}{dt} = \left[\frac{d\mathbf{A}}{dt}\right]\mathbf{B} + \mathbf{A}\left[\frac{d\mathbf{B}}{dt}\right] \tag{A.14}$$

$$\frac{d(\mathbf{A}^{-1})}{dt} = -\mathbf{A}^{-1}\left[\frac{d\mathbf{A}}{dt}\right]\mathbf{A}^{-1} \tag{A.15}$$

6. Derivatives of a matrix with respect to a vector

The following relations hold:

$$\frac{\partial f}{\partial \mathbf{x}} = \begin{bmatrix} \dfrac{\partial f}{\partial x_1} \\ \dfrac{\partial f}{\partial x_2} \\ \vdots \\ \dfrac{\partial f}{\partial x_n} \end{bmatrix} \quad \text{and} \quad \frac{\partial \mathbf{y}^{\mathrm{T}}}{\partial \mathbf{x}} = \begin{bmatrix} \dfrac{\partial y_1}{\partial x_1} & \dfrac{\partial y_2}{\partial x_1} & \cdots & \dfrac{\partial y_m}{\partial x_1} \\ \dfrac{\partial y_1}{\partial x_2} & \dfrac{\partial y_2}{\partial x_2} & \cdots & \dfrac{\partial y_m}{\partial x_2} \\ \vdots & \vdots & & \vdots \\ \dfrac{\partial y_1}{\partial x_n} & \dfrac{\partial y_2}{\partial x_n} & \cdots & \dfrac{\partial y_m}{\partial x_n} \end{bmatrix} \tag{A.16}$$

$$\frac{\partial[\mathbf{Q}(t)\mathbf{y}(t)]}{\partial \mathbf{y}} = \mathbf{Q}^{\mathrm{T}}(t) \tag{A.17}$$

$$\frac{1}{2}\frac{\partial[\mathbf{y}^{\mathrm{T}}(t)\mathbf{Q}(t)\mathbf{y}(t)]}{\partial \mathbf{y}} = \mathbf{Q}(t)\mathbf{y}(t) \tag{A.18}$$

$$\frac{1}{2}\frac{\partial[\mathbf{y}^{\mathrm{T}}(t)\mathbf{Q}(t)\mathbf{y}(t)]}{\partial \mathbf{x}} = \left[\frac{\partial \mathbf{y}^{\mathrm{T}}(t)}{\partial \mathbf{x}}\right]\mathbf{Q}(t)\mathbf{y}(t) \tag{A.19}$$

where $\mathbf{y}^{\mathrm{T}} = (y_1, y_2, \ldots, y_m)$ and $\mathbf{Q}(t)$ is a symmetric matrix with dimensions $m \times m$.

7. Matrix integration

Here

$$\int \mathbf{A} \, dt = \left[\int a_{ij} \, dt \right] \tag{A.20}$$

A.2 Determinant of a matrix

1. Calculation of the determinant of a matrix

The determinant of an $n \times n$ matrix $\mathbf{A}$ is denoted by $|\mathbf{A}|$ or det $\mathbf{A}$ and is a scalar quantity. A popular method of calculating the determinant of a matrix is the Laplace expansion in which the determinant of an $n \times n$ matrix $\mathbf{A} = [a_{ij}]$ is the sum of the elements of a row or a column, where each element is multiplied by the determinant of an appropriate matrix, i.e.

$$|\mathbf{A}| = \sum_{i=1}^{n} a_{ij} c_{ij} \quad \text{for} \quad j = 1 \text{ or } 2 \text{ or } \dots \text{ or } n \text{ (column expansion)} \tag{A.21}$$

or

$$|\mathbf{A}| = \sum_{j=1}^{n} a_{ij} c_{ij} \quad \text{for} \quad i = 1 \text{ or } 2 \text{ or } \dots \text{ or } n \text{ (row expansion)} \tag{A.22}$$

where c_{ij} is defined as follows:

$$c_{ij} = (-1)^{i+j} |\mathbf{M}_{ij}|$$

where $\mathbf{M}_{ij}$ is the $(n-1) \times (n-1)$ square matrix formed from $\mathbf{A}$ by deleting the ith row and the jth column from the original matrix.

2. The determinant of a matrix product

Consider the square matrices $\mathbf{A}_1, \mathbf{A}_2, \mathbf{A}_3, \dots, \mathbf{A}_m$ and their product $\mathbf{B} = \mathbf{A}_1 \mathbf{A}_2 \dots \mathbf{A}_m$. Then

$$|\mathbf{B}| = |\mathbf{A}_1| \, |\mathbf{A}_2| \, \dots \, |\mathbf{A}_m| \tag{A.23}$$

A.3 The inverse of a matrix

1. Calculation of the inverse of a matrix

The inverse of an $n \times n$ square matrix $\mathbf{A}$ is denoted by $\mathbf{A}^{-1}$ and has the following property:

$$\mathbf{A}^{-1}\mathbf{A} = \mathbf{A}\mathbf{A}^{-1} = \mathbf{I} \tag{A.24}$$

The inverse matrix $\mathbf{A}^{-1}$ is determined as follows:

$$\mathbf{A}^{-1} = \frac{\text{adj } \mathbf{A}}{|\mathbf{A}|} \tag{A.25}$$

where the matrix adj $\mathbf{A}$, called the adjoint matrix of $\mathbf{A}$, is the transpose of the matrix whose (i, j) element is given by c_{ij} defined in Section A.2.

2. The inverse of a matrix product

Consider the $n \times n$ square matrices $\mathbf{A}_1$, $\mathbf{A}_2$, ..., $\mathbf{A}_m$ and their product $\mathbf{B} = \mathbf{A}_1 \mathbf{A}_2 \ldots \mathbf{A}_m$. Then

$$\mathbf{B}^{-1} = \mathbf{A}_m^{-1} \mathbf{A}_{m-1}^{-1} \ldots \mathbf{A}_1^{-1}$$

A.4 Matrix eigenvalues and eigenvectors

The eigenvalues of a matrix are of significant importance in the study of control systems since their behaviour is strongly influenced by their eigenvalues. The issue of the eigenvalues of an $n \times n$ square matrix $\mathbf{A} = [a_{ij}]$ stems from the following problem. Consider the n-dimensional vectors $\mathbf{u}^T = (u_1, u_2, \ldots, u_n)$ and $\mathbf{y}^T = (y_1, y_2, \ldots, y_n)$ and the relation

$$\mathbf{y} = \mathbf{A}\mathbf{u} \tag{A.26}$$

The foregoing relation is a transformation of the vector $\mathbf{u}$ onto the vector $\mathbf{y}$ through the matrix $\mathbf{A}$. One may ask the following question: are there non-zero vectors $\mathbf{u}$ which maintain their direction after such a transformation? If there exists such a vector $\mathbf{u}$, then the vector $\mathbf{y}$ is proportional to the vector $\mathbf{u}$; that is, the following holds:

$$\mathbf{y} = \mathbf{A}\mathbf{u} = \lambda\mathbf{u} \tag{A.27}$$

where λ is a constant. From relation (A.27) we have that $\mathbf{A}\mathbf{u} = \lambda\mathbf{u}$ or $\mathbf{A}\mathbf{u} - \lambda\mathbf{u} = \mathbf{0}$, and if we set $\lambda\mathbf{u} = \lambda\mathbf{I}\mathbf{u}$, then we have

$$(\lambda\mathbf{I} - \mathbf{A})\mathbf{u} = 0 \tag{A.28}$$

The system of equations (A.28) has a non-zero solution if the determinant of the matrix $\lambda\mathbf{I} - \mathbf{A}$ is equal to zero, i.e. if

$$|\lambda\mathbf{I} - \mathbf{A}| = \mathbf{0} \tag{A.29}$$

Equation (A.29) is a polynomial equation of degree n and has the general form

$$|\lambda\mathbf{I} - \mathbf{A}| = p(\lambda) = \lambda^n + a_1\lambda^{n-1} + a_2\lambda^{n-2} + \cdots + a_{n-1}\lambda + a_n = \prod_{i=1}^{n}(\lambda - \lambda_i) = 0 \tag{A.30}$$

The roots λ_1, λ_2, ..., λ_n of equation (A.30) are called the **eigenvalues** of the matrix $\mathbf{A}$. These eigenvalues, if set in (A.28), produce non-zero solutions to the problem of determining the vectors $\mathbf{u}$ which maintain their direction after the transformation $\mathbf{A}\mathbf{u}$.

The polynomial $p(\lambda)$ is called the **characteristic polynomial** of the matrix $\mathbf{A}$ and equation (A.30) is called the **characteristic equation** of the matrix $\mathbf{A}$.

A vector $\mathbf{u}_i$ is an eigenvector of matrix $\mathbf{A}$ and corresponds to the eigenvalue λ_i if $\mathbf{u}_i \neq \mathbf{0}$ and relation (A.27) is satisfied, i.e.

$$\mathbf{A}\mathbf{u}_i = \lambda_i \mathbf{u}_i \qquad\qquad\qquad (A.31)$$

In determining the characteristic polynomial $p(\lambda)$ of $\mathbf{A}$, considerable computational effort may be required to compute the coefficients $a_1, a_2, \ldots, a_n$ of $p(\lambda)$ from the elements of matrix $\mathbf{A}$, particularly as n becomes large. Several numerical methods have been proposed, one of the most popular of which is Bocher's recursive relation

$$a_k = -\frac{1}{k}\,(a_{k-1}T_1 + a_{k-2}T_2 + \cdots + a_1 T_{k-1} + T_k) \quad \text{for} \quad k = 1, 2, \ldots, n \qquad (A.32)$$

where $T_k = \operatorname{tr}\mathbf{A}^k$, $k = 1, 2, \ldots, n$. The description of this method is simple but it has a serious numerical drawback in that it requires computation up to the nth power of the matrix $\mathbf{A}$.

Certain interesting properties of matrix $\mathbf{A}$, of its eigenvalues and of the coefficients of $p(\lambda)$, which may be determined from the relations (A.30) to (A.32), are the following:

1. If the matrix $\mathbf{A}$ is singular, then it has at least one zero eigenvalue and vice versa.
2. $\operatorname{tr}\mathbf{A} = \lambda_1 + \lambda_2 + \cdots + \lambda_n$ $\qquad\qquad\qquad\qquad\qquad (A.33)$
3. $|\mathbf{A}| = \lambda_1 \lambda_2 \ldots \lambda_n$ $\qquad\qquad\qquad\qquad\qquad\qquad\quad (A.34)$
4. $a_n = (-1)^n |\mathbf{A}| = (-1)^n \lambda_1 \lambda_2 \ldots \lambda_n = p(0)$ $\qquad\quad (A.35)$

■ **EXAMPLE A.4.1** Find the characteristic polynomial, the eigenvalues and the eigenvectors of the matrix

$$\mathbf{A} = \begin{bmatrix} -1 & 1 \\ 0 & -2 \end{bmatrix}$$

■ **SOLUTION** The characteristic polynomial $p(\lambda)$ of the matrix $\mathbf{A}$, according to definition (A.30), is

$$p(\lambda) = |\lambda\mathbf{I} - \mathbf{A}| = \det\begin{bmatrix} \lambda + 1 & -1 \\ 0 & \lambda + 2 \end{bmatrix} = (\lambda + 1)(\lambda + 2) = \lambda^2 + 3\lambda + 2$$

The coefficients of $p(\lambda)$ may be determined from the recursive equation (A.32). In this case, we first compute T_1 and T_2. We have $T_1 = \operatorname{tr}\mathbf{A} = -3$ and $T_2 = \operatorname{tr}\mathbf{A}^2 = 5$. Hence

$$a_1 = -T_1 = 3 \quad \text{and} \quad a_2 = -\tfrac{1}{2}(a_1 T_1 + T_2) = -\tfrac{1}{2}(-9 + 5) = 2$$

Therefore, $p(\lambda) = \lambda^2 + a_1\lambda + a_2 = \lambda^2 + 3\lambda + 2$. We observe that, working in two different ways, we arrive at the same characteristic polynomial, as expected. The eigenvalues of matrix $\mathbf{A}$ are the roots of the characteristic polynomial $p(\lambda) = \lambda^2 + 3\lambda + 2 = (\lambda + 1)(\lambda + 2)$ and hence we immediately have that the eigenvalues of matrix $\mathbf{A}$ are $\lambda_1 = -1$ and $\lambda_2 = -2$. The eigenvectors can be determined from relation (A.31). Let the vector $\mathbf{u}_1$, which corresponds to the eigenvalue $\lambda_1 = -1$, have the form $\mathbf{u}_1 = (\omega_1, \omega_2)^\mathsf{T}$. Then, relation (A.31) becomes

$$\begin{bmatrix} -1 & 1 \\ 0 & -2 \end{bmatrix}\begin{bmatrix} \omega_1 \\ \omega_2 \end{bmatrix} = -\begin{bmatrix} \omega_1 \\ \omega_2 \end{bmatrix} \quad \text{or} \quad \begin{bmatrix} 0 & 1 \\ 0 & -1 \end{bmatrix}\begin{bmatrix} \omega_1 \\ \omega_2 \end{bmatrix} = \begin{bmatrix} 0 \\ 0 \end{bmatrix}$$

Thus, the eigenvector $\mathbf{u}_1$ is $\mathbf{u}_1 = (\omega_1, \omega_2)^\mathsf{T} = (d, 0)^\mathsf{T} \ \forall\, d \in \mathbb{R}$, where $\mathbb{R}$ is the space of real numbers. In the same way, we calculate the eigenvector $\mathbf{u}_2 = (v_1, v_2)^\mathsf{T}$, which corresponds to the eigenvalue $\lambda_2 = -2$. From (A.31) we have

$$\begin{bmatrix} -1 & 1 \\ 0 & -2 \end{bmatrix}\begin{bmatrix} v_1 \\ v_2 \end{bmatrix} = -2\begin{bmatrix} v_1 \\ v_2 \end{bmatrix} \quad \text{or} \quad \begin{bmatrix} 1 & 1 \\ 0 & 0 \end{bmatrix}\begin{bmatrix} v_1 \\ v_2 \end{bmatrix} = \begin{bmatrix} 0 \\ 0 \end{bmatrix}$$

Hence, the eigenvector $\mathbf{u}_2$ is $\mathbf{u}_2 = (v_1, v_2)^\mathsf{T} = (-k, k)^\mathsf{T} \ \forall\, k \in \mathbb{R}$. Setting $d = k = 1$, we have the following two eigenvectors:

$$\mathbf{u}_1 = \begin{bmatrix} 1 \\ 0 \end{bmatrix} \quad \text{and} \quad \mathbf{u}_2 = \begin{bmatrix} -1 \\ 1 \end{bmatrix}$$

A.5 Similarity transformations

Consider the $n \times n$ square matrices $\mathbf{A}$ and $\mathbf{B}$. The matrix $\mathbf{B}$ is **similar** to the matrix $\mathbf{A}$ if there exists an $n \times n$ non-singular matrix $\mathbf{T}$ such that

$$\mathbf{B} = \mathbf{T}^{-1}\mathbf{A}\mathbf{T} \tag{A.36}$$

It is obvious that if relation (A.36) is true, then it is also true that

$$\mathbf{A} = \mathbf{T}\mathbf{B}\mathbf{T}^{-1}$$

Hence, if $\mathbf{B}$ is similar to $\mathbf{A}$, then $\mathbf{A}$ must be similar to $\mathbf{B}$.

Relation (A.36) is called the **similarity transformation**. This transformation usually aims to simplify the matrix $\mathbf{A}$. Such a simplification facilitates the solution of certain problems. For example, consider the linear system of first-order differential equations

$$\dot{\mathbf{x}} = \mathbf{A}\mathbf{x} \quad \text{with} \quad \mathbf{x}(0) = \mathbf{x}_0 \tag{A.37}$$

where $\mathbf{x}^\mathsf{T} = (x_1, x_2, \ldots, x_n)$. The vector $\mathbf{x}(0)$ is the initial vector of the system. By setting $\mathbf{x} = \mathbf{T}\mathbf{z}$, where $\mathbf{z}^\mathsf{T} = (z_1, z_2, \ldots, z_n)$ is a new vector, then (A.37) becomes

$$\dot{\mathbf{z}} = \mathbf{T}^{-1}\mathbf{A}\mathbf{T}\mathbf{z} \quad \text{with} \quad \mathbf{z}(0) = \mathbf{T}^{-1}\mathbf{x}(0) \tag{A.38}$$

If the similarity transformation matrix $\mathbf{T}$ is chosen so that $\mathbf{T}^{-1}\mathbf{AT} = \mathbf{\Lambda} =$ diag$(\lambda_1, \lambda_2, \ldots, \lambda_n)$, where $\lambda_1, \lambda_2, \ldots, \lambda_n$ are the eigenvalues of matrix $\mathbf{A}$, then (A.38) becomes

$$\dot{\mathbf{z}} = \mathbf{\Lambda z} \quad \text{or} \quad \dot{z}_i = \lambda_i z_i \quad \text{for} \quad i = 1, 2, \ldots, n \tag{A.39}$$

Comparing relations (A.37) and (A.39) we can see that the similarity transformation has simplified (decoupled) the vector differential system of equations (A.37) into n first-order scalar differential equations whose solution is given by the simple relation

$$z_i(t) = z_i(0)e^{\lambda_i t}$$

In using the similarity transformation (A.36), one has to determine the transformation matrix $\mathbf{T}$. In the case where the similarity transformation aims to diagonalize the matrix $\mathbf{A}$, then the determination of $\mathbf{T}$ is directly linked to the problem of eigenvalues and eigenvectors of $\mathbf{A}$. In this case, the following holds true:

1. If the eigenvalues $\lambda_1, \lambda_2, \ldots, \lambda_n$ of the matrix $\mathbf{A}$ are distinct, then the eigenvectors $\mathbf{u}_1, \mathbf{u}_2, \ldots, \mathbf{u}_n$ are linearly independent. In this case, the similarity transformation matrix $\mathbf{T}$, which diagonalizes $\mathbf{A}$, is denoted by $\mathbf{M}$ and has the form

$$\mathbf{M} = [\mathbf{u}_1 \mid \mathbf{u}_2 \mid \cdots \mid \mathbf{u}_n] \tag{A.40}$$

   The matrix $\mathbf{M}$ is called the **eigenvector matrix**.

2. If the eigenvalues $\lambda_1, \lambda_2, \ldots, \lambda_n$ of the matrix $\mathbf{A}$ are not distinct, then the eigenvectors $\mathbf{u}_1, \mathbf{u}_2, \ldots, \mathbf{u}_n$ are not always linearly independent. Since the diagonalization of $\mathbf{A}$ has as a necessary and sufficient condition the existence of n linearly independent eigenvectors of the matrix $\mathbf{A}$, it follows that in the case where the eigenvalues of $\mathbf{A}$ are not distinct, the matrix $\mathbf{A}$ can be diagonalized only if it has n linearly independent eigenvectors. When $\mathbf{A}$ does not have n linearly independent eigenvectors, the matrix cannot be transformed into a diagonal form. However, it can be transformed to an 'almost diagonal' form, known as the **Jordan canonical form** and denoted by $\mathbf{J}$. The general form of the matrix $\mathbf{J}$ is the following:

$$\mathbf{J} = \begin{bmatrix} \mathbf{J}_{11}(\lambda_1) & & & & & & \\ & \mathbf{J}_{21}(\lambda_1) & & & & \mathbf{0} & \\ & & \ddots & & & & \\ & & & \mathbf{J}_{k1}(\lambda_1) & & & \\ & & & & \mathbf{J}_{12}(\lambda_2) & & \\ & \mathbf{0} & & & & \ddots & \\ & & & & & & \mathbf{J}_{mp}(\lambda_p) \end{bmatrix} \tag{A.41}$$

where

$$\mathbf{J}_{ji}(\lambda_i) = \begin{bmatrix} \lambda_i & 1 & 0 & \cdots & 0 & 0 \\ 0 & \lambda_i & 1 & \cdots & 0 & 0 \\ \vdots & \vdots & \vdots & & \vdots & \vdots \\ 0 & 0 & 0 & \cdots & \lambda_i & 1 \\ 0 & 0 & 0 & \cdots & 0 & \lambda_i \end{bmatrix}$$

where $\mathbf{J}_{ji}(\lambda_i)$ is called the **Jordan submatrix**.

Certain useful properties of the similarity transformation are that the characteristic polynomial (and therefore the eigenvalues), the trace and the determinant of a matrix are invariant under the similarity transformation. However, it should be noted that in general the reverse does not hold true. If, for example, two matrices have the same characteristic polynomial, this does not necessarily mean that they are similar.

Moreover, if $\mathbf{B} = \mathbf{T}^{-1}\mathbf{A}\mathbf{T}$ and $\mathbf{u}$ is an eigenvector of $\mathbf{A}$, then the vector $\mathbf{T}^{-1}\mathbf{u}$ is an eigenvector of $\mathbf{B}$.

■ **EXAMPLE A.5.1** Diagonalize the matrix $\mathbf{A}$ of Example A.4.1.

■ **SOLUTION** From Example A.4.1 we have the two eigenvectors $\mathbf{u}_1 = (1, 0)^T$ and $\mathbf{u}_2 = (-1, 1)^T$. According to the relation (A.40), the similarity matrix $\mathbf{M}$ is

$$\mathbf{M} = [\mathbf{u}_1 \mid \mathbf{u}_2] = \begin{bmatrix} 1 & -1 \\ 0 & 1 \end{bmatrix} \quad \text{and} \quad \mathbf{M}^{-1} = \begin{bmatrix} 1 & 1 \\ 0 & 1 \end{bmatrix}$$

Since $\mathbf{u}_1$ and $\mathbf{u}_2$ are linearly independent, the matrix $\mathbf{M}$ diagonalizes the matrix $\mathbf{A}$. Indeed

$$\mathbf{\Lambda} = \mathbf{M}^{-1}\mathbf{A}\mathbf{M} = \begin{bmatrix} 1 & 1 \\ 0 & 1 \end{bmatrix}\begin{bmatrix} -1 & 1 \\ 0 & -2 \end{bmatrix}\begin{bmatrix} 1 & -1 \\ 0 & 1 \end{bmatrix} = \begin{bmatrix} -1 & 0 \\ 0 & -2 \end{bmatrix}$$

A.6 The Cayley–Hamilton theorem

The Cayley–Hamilton theorem relates a matrix to its characteristic polynomial, as follows:

■ **THEOREM A.6.1** Consider an $n \times n$ matrix $\mathbf{A}$ with characteristic polynomial $p(\lambda) = |\lambda\mathbf{I} - \mathbf{A}| = \lambda^n + a_1\lambda^{n-1} + \cdots + a_{n-1}\lambda + a_n$. Then, the matrix $\mathbf{A}$ satisfies its characteristic polynomial, i.e.

$$p(\mathbf{A}) = \mathbf{A}^n + a_1\mathbf{A}^{n-1} + \cdots + a_{n-1}\mathbf{A} + a_n\mathbf{I} = \mathbf{0} \tag{A.42}$$

The Cayley–Hamilton theorem has several interesting and useful applications. Some of them are given below.

1. Calculation of the inverse of a matrix

From relation (A.42) we have

$$a_n\mathbf{I} = -[\mathbf{A}^n + a_1\mathbf{A}^{n-1} + \cdots + a_{n-1}\mathbf{A}]$$

Let the matrix $\mathbf{A}$ be non-singular. Then, multiplying both sides of the foregoing relation by $\mathbf{A}^{-1}$ we obtain

$$a_n\mathbf{A}^{-1} = -[\mathbf{A}^{n-1} + a_1\mathbf{A}^{n-2} + \cdots + a_{n-1}\mathbf{I}]$$

or

$$\mathbf{A}^{-1} = -\frac{1}{a_n}[\mathbf{A}^{n-1} + a_1\mathbf{A}^{n-2} + \cdots + a_{n-1}\mathbf{I}] \tag{A.43}$$

where $a_n = (-1)^n|\mathbf{A}| \neq 0$. Relation (A.43) expresses the matrix $\mathbf{A}^{-1}$ as a matrix polynomial of $\mathbf{A}$ of degree $n-1$ and presents an alternative way of determining $\mathbf{A}^{-1}$, different from that of Section A.3, which essentially requires the calculation of the matrices $\mathbf{A}^k$, $k = 2, 3, \ldots, n-1$.

■ **EXAMPLE A.6.1** Consider the matrix

$$\mathbf{A} = \begin{bmatrix} 0 & 1 \\ -6 & -5 \end{bmatrix}$$

Show that the matrix $\mathbf{A}$ satisfies its characteristic equation and calculate $\mathbf{A}^{-1}$.

■ **SOLUTION** We have

$$p(\lambda) = |\lambda\mathbf{I} - \mathbf{A}| = \lambda^2 + 5\lambda + 6$$

whence

$$p(\mathbf{A}) = \mathbf{A}^2 + 5\mathbf{A} + 6\mathbf{I}$$

$$= \begin{bmatrix} -6 & -5 \\ 30 & 19 \end{bmatrix} + 5\begin{bmatrix} 0 & 1 \\ -6 & -5 \end{bmatrix} + 6\begin{bmatrix} 1 & 0 \\ 0 & 1 \end{bmatrix} = \begin{bmatrix} 0 & 0 \\ 0 & 0 \end{bmatrix}$$

Therefore $p(\mathbf{A}) = \mathbf{0}$. For the calculation of $\mathbf{A}^{-1}$, using relation (A.43), we have

$$\mathbf{A}^{-1} = -\frac{1}{6}(\mathbf{A} + 5\mathbf{I}) = -\frac{1}{6}\begin{bmatrix} 5 & 1 \\ -6 & 0 \end{bmatrix}$$

---

2. Calculation of $\mathbf{A}^k$

From relation (A.42) we have

$$\mathbf{A}^n = -[a_1\mathbf{A}^{n-1} + \cdots + a_{n-1}\mathbf{A} + a_n\mathbf{I}] \tag{A.44}$$

Multiplying both sides of the above equation by the matrix $\mathbf{A}$ we have

$$\mathbf{A}^{n+1} = -[a_1\mathbf{A}^n + \cdots + a_{n-1}\mathbf{A}^2 + a_n\mathbf{A}] \tag{A.45}$$

Substituting (A.44) in (A.45) we have

$$\mathbf{A}^{n+1} = -[-a_1(a_1\mathbf{A}^{n-1} + \cdots + a_{n-1}\mathbf{A} + a_n\mathbf{I}) + \cdots + a_{n-1}\mathbf{A}^2 + a_n\mathbf{A}] \tag{A.46}$$

Relation (A.46) expresses the matrix $\mathbf{A}^{n+1}$ as a linear combination of the matrices $\mathbf{I}, \mathbf{A}, \ldots, \mathbf{A}^{n-1}$. The general case is

$$\mathbf{A}^k = \sum_{i=0}^{n-1}(\beta_k)_i\mathbf{A}^i \quad \text{for} \quad k \geqslant n \tag{A.47}$$

where $(\beta_k)_i$ are constants depending on the coefficients $a_1, a_2, \ldots, a_n$ of the characteristic polynomial $p(\lambda)$.

■ **EXAMPLE A.6.2**   Calculate the matrix $\mathbf{A}^5$, where

$$\mathbf{A} = \begin{bmatrix} 0 & 1 \\ 2 & 1 \end{bmatrix}$$

■ **SOLUTION**   From the Cayley–Hamilton theorem we have $p(\mathbf{A}) = \mathbf{A}^2 - \mathbf{A} - 2\mathbf{I} = \mathbf{0}$ or $\mathbf{A}^2 = \mathbf{A} + 2\mathbf{I}$. From relation (A.46) we obtain $\mathbf{A}^3 = \mathbf{A}^2 + 2\mathbf{A} = (\mathbf{A} + 2\mathbf{I}) + 2\mathbf{A} = 3\mathbf{A} + 2\mathbf{I}$, $\mathbf{A}^4 = 3\mathbf{A}^2 + 2\mathbf{A} = 3(\mathbf{A} + 2\mathbf{I}) + 2\mathbf{A} = 5\mathbf{A} + 6\mathbf{I}$ and $\mathbf{A}^5 = 5\mathbf{A}^2 + 6\mathbf{A} = 5(\mathbf{A} + 2\mathbf{I}) + 6\mathbf{A} = 11\mathbf{A} + 10\mathbf{I}$. Therefore

$$\mathbf{A}^5 = 11\mathbf{A} + 10\mathbf{I} = 11\begin{bmatrix} 0 & 1 \\ 2 & 1 \end{bmatrix} + 10\begin{bmatrix} 1 & 0 \\ 0 & 1 \end{bmatrix} = \begin{bmatrix} 10 & 11 \\ 22 & 21 \end{bmatrix}$$

A.7 Quadratic forms and Sylvester theorems

Consider a second-order polynomial of the form

$$g(\mathbf{x}) = \sum_{j=1}^{n}\sum_{i=1}^{n} a_{ij}x_ix_j \tag{A.48}$$

where $\mathbf{x}^T = (x_1, x_2, \ldots, x_n)$ is a vector of real variables. The polynomial $g(\mathbf{x})$ is called the quadratic form of n variables. Some of the properties and definitions of $g(\mathbf{x})$ are the following:

1. The polynomial $g(\mathbf{x})$ may be written as

$$g(\mathbf{x}) = \mathbf{x}^T\mathbf{A}\mathbf{x}, \quad \mathbf{A} = [a_{ij}] \quad \text{for} \quad i, j = 1, 2, \ldots, n \tag{A.49}$$

where $\mathbf{A}$ is an $n \times n$ real symmetric matrix (i.e. $\mathbf{A} = \mathbf{A}^T$).

2. The polynomial $g(\mathbf{x})$ may also be written as an inner product as follows:

$$g(\mathbf{x}) = \langle \mathbf{x},\, \mathbf{Ax} \rangle \qquad\qquad (A.50)$$

where $\langle \mathbf{a},\, \mathbf{b} \rangle = \mathbf{a}^\mathrm{T}\mathbf{b}$, and $\mathbf{a}$ and $\mathbf{b}$ are vectors of the same dimension.
3. The matrix $\mathbf{A}$ is called the matrix of the quadratic form $g(\mathbf{x})$. The rank of the matrix $\mathbf{A}$ is the order of $g(\mathbf{x})$.
4. The polynomial $g(\mathbf{x})$ is positive (negative) if its value is positive (negative) or zero for every set of real values of $\mathbf{x}$.
5. The polynomial $g(\mathbf{x})$ is positive definite (negative definite) if its value is positive (negative) for all $\mathbf{x}$ and equal to zero only when $\mathbf{x} = \mathbf{0}$.
6. The polynomial $g(\mathbf{x})$ is positive semi-definite (negative semi-definite) if its value is positive (negative) or zero.
7. The polynomial $g(\mathbf{x})$ is undefined if its value can take positive as well as negative values.

A central problem referring to quadratic forms pertains to the determination of whether or not a quadratic form is positive, negative definite or semi-definite. Certain theorems due to Sylvester are presented below.

■ **THEOREM A.7.1**   Let rank $\mathbf{A} = m$. Then, the polynomial $g(\mathbf{x}) = \mathbf{x}^\mathrm{T}\mathbf{Ax}$ may be written in the following form:

$$p(\mathbf{y}) = y_1^2 + \cdots + y_\mu^2 - y_{\mu+1}^2 - \cdots - y_{\mu+\nu}^2 + O y_{\mu+\nu+1}^2 + \cdots + O y_n^2 \qquad (A.51)$$

where $\mathbf{x} = \mathbf{Ty}$. The integers μ and ν correspond to the number of positive and negative eigenvalues of $\mathbf{A}$ and are such that $\mu + \nu = m$.

■ **THEOREM A.7.2**   From (A.51) we obtain $p(\mathbf{y})$ and hence the polynomial $g(\mathbf{x})$ is:

1. Positive definite if $\mu = n$ and $\nu = 0$.
2. Positive semi-definite if $\mu < n$ and $\nu = 0$.
3. Negative definite if $\mu = 0$ and $\nu = n$.
4. Negative semi-definite if $\mu = 0$ and $\nu < n$.
5. Undefined if $\nu\mu \neq 0$.

■ **THEOREM A.7.3**   The necessary and sufficient conditions for $g(\mathbf{x})$ to be positive definite are

$$\Delta_i > 0 \quad \text{for} \quad i = 1, 2, \ldots, n \qquad\qquad (A.52)$$

where

$$\Delta_1 = a_{11}, \Delta_2 = \begin{vmatrix} a_{11} & a_{21} \\ a_{21} & a_{22} \end{vmatrix}, \ldots, \Delta_n = \begin{vmatrix} a_{11} & \cdots & a_{n1} \\ \vdots & & \vdots \\ a_{n1} & \cdots & a_{nn} \end{vmatrix}$$

■ **THEOREM A.7.4** The necessary and sufficient conditions for $g(\mathbf{x})$ to be negative definite are

$$\Delta_i \begin{cases} > 0 & \text{when } i \text{ is even} \\ < 0 & \text{when } i \text{ is odd} \end{cases} \tag{A.53}$$

for $i = 1, 2, \ldots, n$.

■ **THEOREM A.7.5** The necessary and sufficient conditions for $g(\mathbf{x})$ to be positive semi-definite are

$$\Delta_i \geqslant 0 \quad \text{for} \quad i = 1, 2, \ldots, n-1 \quad \text{and} \quad \Delta_n = 0 \tag{A.54}$$

■ **THEOREM A.7.6** The necessary and sufficient conditions for $g(\mathbf{x})$ to be negative semi-definite are

$$\Delta_1 \leqslant 0, \Delta_2 \geqslant 0, \Delta_3 \leqslant 0, \ldots \quad \text{and} \quad \Delta_n = 0 \tag{A.55}$$

We note that Theorem A.7.2 has the drawback, over Theorems A.7.3 to A.7.6, that it requires the calculation of the integers μ and ν, the eigenvalues and the rank m of matrix $\mathbf{A}$. For this reason, Theorems A.7.3 to A.7.6 are used more often.

■ **EXAMPLE A.7.1** Find if the quadratic forms

$$g_1(\mathbf{x}) = 4x_1^2 + x_2^2 + 2x_3^2 + 2x_1x_2 + 2x_1x_3 + 2x_2x_3$$

$$g_2(\mathbf{x}) = -2x_1^2 - 4x_2^2 - x_3^2 + 2x_1x_2 + 2x_1x_3$$

$$g_3(\mathbf{x}) = -4x_1^2 - x_2^2 - 12x_3^2 + 2x_1x_2 + 6x_2x_3$$

are positive or negative definite.

■ **SOLUTION** We have

$$g_1(\mathbf{x}) = \mathbf{x}^T \mathbf{A} \mathbf{x} \quad \text{where} \quad \mathbf{x} = \begin{bmatrix} x_1 \\ x_2 \\ x_3 \end{bmatrix} \quad \text{and} \quad \mathbf{A} = \begin{bmatrix} 4 & 1 & 1 \\ 1 & 1 & 1 \\ 1 & 1 & 2 \end{bmatrix}$$

Thus

$$\Delta_1 = 4 > 0, \quad \Delta_2 = \begin{vmatrix} 4 & 1 \\ 1 & 1 \end{vmatrix} = 3 > 0 \quad \text{and} \quad \Delta_3 = |\mathbf{A}| = 3 > 0$$

Hence, according to Theorem A.7.3, $g_1(\mathbf{x})$ is positive definite. Also

$$g_2(\mathbf{x}) = \mathbf{x}^T\mathbf{A}\mathbf{x} \quad \text{where} \quad \mathbf{x} = \begin{bmatrix} x_1 \\ x_2 \\ x_3 \end{bmatrix} \quad \text{and} \quad \mathbf{A} = \begin{bmatrix} -2 & 1 & 1 \\ 1 & -4 & 0 \\ 1 & 0 & -1 \end{bmatrix}$$

Thus

$$\Delta_1 = -2 < 0 \quad \Delta_2 = \begin{vmatrix} -2 & 1 \\ 1 & -4 \end{vmatrix} = 7 > 0 \quad \Delta_3 = |\mathbf{A}| = -3 < 0$$

Hence, according to Theorem A.7.4, $g_2(\mathbf{x})$ is negative definite. Finally

$$g_3(\mathbf{x}) = \mathbf{x}^T\mathbf{A}\mathbf{x} \quad \text{where} \quad \mathbf{x} = \begin{bmatrix} x_1 \\ x_2 \\ x_3 \end{bmatrix} \quad \text{and} \quad \mathbf{A} = \begin{bmatrix} -4 & 1 & 0 \\ 1 & -1 & 3 \\ 0 & 3 & -12 \end{bmatrix}$$

Thus

$$\Delta_1 = -4 < 0, \quad \Delta_2 = \begin{vmatrix} -4 & 1 \\ 1 & -1 \end{vmatrix} = 3 > 0 \quad \text{and} \quad \Delta_3 = |\mathbf{A}| = 0$$

Hence, according to Theorem A.7.6, $g_3(\mathbf{x})$ is negative semi-definite.

Appendix B

Z-transform tables

B.1 Properties and theorems of the Z-transform

	Property or theorem	$f(kT)$	$F(z)$
1	Definition of Z-transform	$f(kT)$	$\displaystyle\sum_{k=0}^{\infty} f(kT)z^{-k}$
2	Definition of the inverse Z-transform	$\displaystyle\frac{1}{2\pi j}\oint F(z)z^{k-1}\,dz$	$F(z)$
3	Linearity	$c_1 f_1(kT) + c_2 f_2(kT)$	$c_1 F_1(z) + c_2 F_2(z)$
4	Shift to the left (advance)	$f(kT + \sigma T)$	$z^{\sigma}\left(F(z) - \displaystyle\sum_{k=0}^{\sigma-1} f(kT)z^{-k}\right)$
5	Shift to the right (delay)	$f(kT - \sigma T)$	$z^{-\sigma}F(z)$
6	Change in z-scale	$a^{\mp kT}f(kT)$	$F(a^{\pm T}z)$
7	Change in kT-scale	$f(mkT)$	$F(z^{-m})$
8	Multiplying by k	$kf(kT)$	$-z\,\dfrac{d}{dz}F(z)$
9	Summation	$\displaystyle\sum_{k=0}^{m} f(kT)$	$\dfrac{z}{z-1}F(z)$
10	Convolution	$f(kT) * h(kT)$	$F(z)H(z)$
11	Periodic function	$f(kT) = f(kT + pT)$	$\dfrac{z^{p}}{z^{p}-1}F_1(z)$
12	Initial value theorem	$f(0)$	$\displaystyle\lim_{z\to\infty} F(z)$
13	Final value theorem	$\displaystyle\lim_{k\to\infty} f(kT)$	$\displaystyle\lim_{z\to1}(1 - z^{-1})F(z)$

B.2   Z-transform pairs

	$f(kT)$	$F(s) = \int_0^\infty f(t)e^{-st}\,dt$	$F(z) = \sum_{k=0}^\infty f(kT)z^{-k}$
1	$\delta(kT - \sigma T)$	$e^{-\sigma Ts}$	$z^{-\sigma}$
2	$\delta(kT)$	1	1
3	$\beta(kT - \sigma T)$	$\dfrac{e^{-\sigma Ts}}{s}$	$\dfrac{z^{-\sigma+1}}{z-1}$
4	$\beta(kT)$	$\dfrac{1}{s}$	$\dfrac{z}{z-1}$
5	$kT - \sigma T$	$\dfrac{e^{-\sigma Ts}}{s^2}$	$\dfrac{Tz^{-\sigma+1}}{(z-1)^2}$
6	kT	$\dfrac{1}{s^2}$	$\dfrac{Tz}{(z-1)^2}$
7	$\dfrac{1}{2!}k^2T^2$	$\dfrac{1}{s^3}$	$\dfrac{T^2z(z+1)}{2(z-1)^3}$
8	$\dfrac{1}{3!}k^3T^3$	$\dfrac{1}{s^4}$	$\dfrac{T^3z(z^2+4z+1)}{6(z-1)^4}$
9	$\dfrac{1}{k!}k^mT^m$	$\dfrac{1}{s^{m+1}}$	$\lim_{a\to 0}\dfrac{(-1)^m}{m!}\dfrac{\partial^m}{\partial a^m}\left(\dfrac{z}{z-e^{-aT}}\right)$
10	a^{kT}	$\dfrac{1}{s - T\ln a}$	$\dfrac{z}{z-a^T}$
11	e^{-akT}	$\dfrac{1}{s+a}$	$\dfrac{z}{z-e^{-aT}}$
12	kTe^{-akT}	$\dfrac{1}{(s+a)^2}$	$\dfrac{Tze^{-aT}}{(z-e^{-aT})^2}$
13	$\dfrac{k^2T^2}{2}e^{-akT}$	$\dfrac{1}{(s+a)^3}$	$\dfrac{T^2e^{-aT}z}{2(z-e^{-aT})^2} + \dfrac{T^2e^{-2aT}z}{(z-e^{-aT})^3}$
14	$\dfrac{k^mT^m}{m!}e^{-akT}$	$\dfrac{1}{(s+a)^{m+1}}$	$\dfrac{(-1)^m}{m!}\dfrac{\partial^m}{\partial a^m}\left(\dfrac{z}{z-e^{-aT}}\right)$
15	$1 - e^{-akT}$	$\dfrac{a}{s(s+a)}$	$\dfrac{(1-e^{-aT})}{(z-1)(z-e^{-aT})}$

(continued)

Continued

$f(kT)$	$F(s) = \int_0^\infty f(t)e^{-st}\,dt$	$F(z) = \sum\limits_{k=0}^{\infty} f(kT)z^{-k}$
16	$kT - \dfrac{1 - e^{-akT}}{a}$	$\dfrac{a}{s^2(s+a)}$	$\dfrac{Tz}{(z-1)^2} - \dfrac{(1 - e^{-aT})z}{a(z-1)(z - e^{-aT})}$
17	$\sin \omega_0 kT$	$\dfrac{\omega_0}{s^2 + \omega_0^2}$	$\dfrac{z \sin \omega_0 T}{z^2 - 2z \cos \omega_0 T + 1}$
18	$\cos \omega_0 kT$	$\dfrac{s}{s^2 + \omega_0^2}$	$\dfrac{z(z - \cos \omega_0 T)}{z^2 - 2z \cos \omega_0 T + 1}$
19	$\sinh \omega_0 kT$	$\dfrac{\omega_0}{s^2 - \omega_0^2}$	$\dfrac{z \sinh \omega_0 T}{z^2 - 2z \cosh \omega_0 T + 1}$
20	$\cosh \omega_0 kT$	$\dfrac{s}{s^2 - \omega_0^2}$	$\dfrac{z(z - \cosh \omega_0 T)}{z^2 - 2z \cosh \omega_0 T + 1}$
21	$\cosh \omega_0 kT - 1$	$\dfrac{\omega_0^2}{s(s^2 - \omega_0^2)}$	$\dfrac{z(z - \cosh \omega_0 T)}{z^2 - 2z \cosh \omega_0 T + 1} - \dfrac{z}{z-1}$
22	$1 - \cos \omega_0 kT$	$\dfrac{\omega_0^2}{s(s^2 + \omega_0^2)}$	$\dfrac{z}{z-1} - \dfrac{z(z - \cos \omega_0 T)}{z^2 - 2z \cos \omega_0 T + 1}$
23	$e^{-akT} - e^{-bkT}$	$\dfrac{b - a}{(s+a)(s+b)}$	$\dfrac{z}{z - e^{-aT}} - \dfrac{z}{z - e^{-bT}}$
24	$(c - a)e^{-akT} + (b - c)e^{-bkT}$	$\dfrac{(b-a)(s+c)}{(s+a)(s+b)}$	$\dfrac{(c-a)z}{z - e^{-aT}} - \dfrac{(b-c)z}{z - e^{-bt}}$
25	$1 - (1 + akT)e^{-akT}$	$\dfrac{a^2}{s(s+a)^2}$	$\dfrac{z}{z-1} - \dfrac{z}{z - e^{-aT}} - \dfrac{aTz\,e^{-aT}}{(z - e^{-aT})^2}$
26	$b - b\,e^{-bkT} + a(a - b)kT\,e^{-akT}$	$\dfrac{a^2(s+b)}{s(s+a)^2}$	$\dfrac{bz}{z-1} - \dfrac{bz}{z - e^{-aT}} + \dfrac{a(a-b)T\,e^{-aT}z}{(z - e^{-aT})^2}$
27	$e^{-bkT} - e^{-akT} + (a - b)kT\,e^{-akT}$	$\dfrac{(a-b)^2}{(s+b)(s+a)^2}$	$\dfrac{z}{z - e^{-bT}} - \dfrac{z}{z - e^{-aT}} + \dfrac{(a-b)T\,e^{-aT}z}{(z - e^{-aT})^2}$
28	$e^{-akT} \sin \omega_0 kT$	$\dfrac{\omega_0}{(s+a)^2 + \omega_0^2}$	$\dfrac{z\,e^{-aT} \sin \omega_0 T}{z^2 - 2z\,e^{-aT} \cos \omega_0 T + e^{-2aT}}$
29	$e^{-akT} \cos \omega_0 kT$	$\dfrac{s+a}{(s+a)^2 + \omega_0^2}$	$\dfrac{z^2 - z\,e^{-aT} \cos \omega_0 T}{z^2 - 2z\,e^{-aT} \cos \omega_0 T + e^{-2aT}}$

(continued)

Continued

$f(kT)$	$F(s) = \int_0^\infty f(t)e^{-st}\,dt$	$F(z) = \sum\limits_{k=0}^{\infty} f(kT)z^{-k}$
30 $\quad e^{-bkT} - e^{-akT}\sec\theta\cos(\omega_0 kT - \theta)$ where $\theta = \tan^{-1}\left(-\dfrac{b-a}{\omega_0}\right)$	$\dfrac{(a-b)^2 + \omega_0^2}{(s+b)[(s+a)^2 + \omega_0^2]}$	$\dfrac{z}{z - e^{-bT}} - \dfrac{z^2 - z\,e^{-aT}\sec\theta\cos(\omega_0 T - \theta)}{z^2 - 2z\,e^{-aT}\cos\omega_0 T + e^{-2aT}}$
31 $\quad 1 - e^{-akT}\sec\theta\cos(\omega_0 kT + \theta)$ where $\theta = \tan^{-1}\left(-\dfrac{a}{\omega_0}\right)$	$\dfrac{a^2 + \omega_0^2}{s[(s+a)^2 + \omega_0^2]}$	$\dfrac{z}{z-1} - \dfrac{z^2 - z\,e^{-aT}\sec\theta\cos(\omega_0 T + \theta)}{z^2 - 2z\,e^{-aT}\cos\omega_0 T + e^{-2aT}}$
32 $\quad b - b\,e^{-akT}\sec\theta\cos(\omega_0 kT + \theta)$ where $\theta = \tan^{-1}\left(\dfrac{a^2 + \omega_0^2 - ab}{b\omega_0}\right)$	$\dfrac{(a^2 + \omega_0^2)(s+b)}{s[(s+a)^2 + \omega_0^2]}$	$\dfrac{bz}{z-1} - \dfrac{b[z^2 - z\,e^{-aT}\sec\theta\cos(\omega_0 T + \theta)]}{z^2 - 2z\,e^{-aT}\cos\omega_0 T + e^{-2aT}}$

Subject index